NATURAL PRODUCTS IN PLANT PEST MANAGEMENT

NATURAL PRODUCTS IN PLANT PEST MANAGEMENT

Edited by

Nawal K. Dubey

Centre for Advanced Studies in Botany
Banaras Hindu University, Varanasi, India

www.cabi.org

CABI is a trading name of CAB International

CABI Head Office
Nosworthy Way
Wallingford
Oxfordshire OX10 8DE
UK

Tel: +44 (0)1491 832111
Fax: +44 (0)1491 833508
E-mail: cabi@cabi.org
Website: www.cabi.org

CABI North American Office
875 Massachusetts Avenue
7th Floor
Cambridge, MA 02139
USA

Tel: +1 617 395 4056
Fax: +1 617 354 6875
E-mail: cabi-nao@cabi.org

A catalogue record for this book is available from the British Library, London, UK.

Library of Congress Cataloging-in-Publication Data

Natural products in plant pest management / edited by N.K. Dubey.
    p. cm.
   Includes bibliographical references and index.
   ISBN 978-1-84593-671-6 (alk. paper)
  1. Natural pesticides. 2. Agricultural pests–Control. 3. Plant products.
I. Dubey, N. K. II. Title.

   SB951.145.N37N394 2011
   632'.96–dc22

2010020068

ISBN-13: 978 1 84593 671 6

Commissioning editor: Sarah Mellor
Production editor: Fiona Chippendale

Typeset by AMA Dataset, Preston, UK.
Printed and bound in the UK by CPI Antony Rowe, Chippenham.

Contents

Contributors

V.K. Baranwal, *Advanced Centre of Plant Virology, I.A.R.I, New Delhi, India. E-mail: vbaranwal2001@yahoo.com*

Francisco Carretero, *Plant Production Department. University of Almeria, La Cañada de San Urbano. 04120 Almería, Spain*

N.K. Dubey, *Department of Botany, Banaras Hindu University, Varanasi-221005, India. E-mail: nkdubey2@rediffmail.com*

Fernando Diánez, *Plant Production Department. University of Almeria, La Cañada de San Urbano. 04120 Almería, Spain*

Sanath Hettiarachi, *Department of Botany, University of Ruhuna, Matara, Sri Lanka. Email: sanath@bot.ruh.ac.lk*

R.N. Kharwar, *Mycopathology and Microbial Technology Lab, Department of Botany, Banaras Hindu University, Varanasi-221005, India. E-mail: rnkharwar@yahoo.com*

Moshe Kostyukovsky, *Agricultural Research Organization, the Volcani Center, Israel, Bet Dagan, P.O. Box 6, 50250, Israel. E-mail: inspect@volcani.agri.gov.il*

Ashok Kumar, *Department of Botany, Banaras Hindu University, Varanasi-221005, India*

Sonia Marín, *Food Technology Department, Lleida University, XaRTA-UTPV, Lleida, Spain. E-mail: smarin@tecal.udl.cat*

D.B. Olufolaji, *Department of Crop, Soil and Pest Management, The Federal University of Technology, P.M.B. 704, Akure. Nigeria. E-mail: tundeolufolaji@yahoo.co.uk*

Roman Pavela, *Crop Research Institute, Drnovska 507, Prague 6 – Ruzyne, Czech Republic. E-mail: pavela@vurv.cz*

María Porras, *Department of Crop Protection, IFAPA Centro Las Torres – Tomejil, Aptdo. 41200-Alcalá del Río, Sevilla, Spain. E-mail: mariaa.porras@juntadeandalucia.es*

Bhanu Prakash, *Department of Botany, Banaras Hindu University, Varanasi-221005, India*

J.C. Pretorius, *Department of Soil, Crop and Climate Sciences, University of the Free State, P.O. Box 339, Bloemfontein 9300, South Africa. E-mail: pretorjc.sci@ufs.ac.za*

Antonio J. Ramos, *Food Technology Department, Lleida University, XaRTA-UTPV, Lleida, Spain*

K.A. Raveesha, *Herbal Drug Technology Laboratory, Department of Studies in Botany, University of Mysore, Manasagangotri, Mysore-570 006, India. E-mail: karaveesha@gmail.com*

Vicente Sanchis, *Food Technology Department, Lleida University, XaRTA-UTPV, Lleida, Spain*

Mila Santos, *Plant Production Department. University of Almeria, La Cañada de San Urbano, 04120 Almería, Spain. E-mail: msantos@ual.es*

Eli Shaaya, *Agricultural Research Organization, the Volcani Center, Israel, Bet Dagan, P.O. Box 6, 50250, Israel*

L.A. Shcherbakova, *Russian Research Institute of Phytopathology, VNIIF, B., Vyazyomy, Moscow reg., 143050, Russia. E-mail: larisa@vniif.rosmail.com*

Ravindra Shukla, *Department of Botany, Banaras Hindu University, Varanasi-221005, India*

Priyanka Singh, *Department of Botany, Banaras Hindu University, Varanasi-221005, India*

Gary Strobel, *Department of Plant Sciences, Montana State University, Bozeman, MT 59717, USA*

E. van der Watt, *Department of Soil, Crop and Climate Sciences, University of the Free State, P.O. Box 339, Bloemfontein 9300, South Africa*

H.N. Verma, *Jaipur National University, Jaipur, India. E-mail: vermulko@yahoo.co.uk*

Preface

The ever increasing global population needs substantial resources for food production. However, food production as well as its protection is imperative. The situation gets particularly critical in developing countries where the net food production rate is slowing down relative to the population rise. The world food situation is aggravated by the fact that, in spite of all the available means of plant protection, a major fraction of the yearly output of food commodities gets destroyed by various pests including bacteria, fungi, viruses, insects, rodents and nematodes. The production of mycotoxins by fungi has added new dimensions to the gravity of the problem. Losses at times are severe enough to lead to famine in large areas of the world that are densely populated and dependent on agriculture.

The use of synthetic pesticides has undoubtedly contributed to a green revolution in different countries through increased crop protection. However, recent years witnessed considerable pressure on consumers and farmers to reduce or even eliminate the deployment of synthetic pesticides in agriculture owing to environmental risks emerging from their indiscriminate use. Thus, there has been renewed interest in botanical pesticides as the alternative and eco-chemical option in pest management. It is also imperative for sustainable agriculture to reduce the incidence of pests and crop diseases to a degree that does not seriously damage the farmer's products and also to develop cost-effective strategies with minimal ecological side effects.

The use of locally available plants in the control of pests is an age-old technology in many parts of the world. Some plants, namely *Derris*, *Nicotiana* and *Ryania*, were used to combat agricultural pests during the prehistoric era. Used widely until the 1940s, such botanical pesticides have been partially replaced by synthetic pesticides that are easier to procure and longer lasting.

Higher plants, in this respect, harbour numerous compounds that may offer resistance to pathogens. There has been a renewed interest in botanical antimicrobials owing to various distinct advantages. Botanicals, being the natural derivatives, are biodegradable, and do not leave toxic residues or byproducts to contaminate the environment. The main thrust of recent research in the area has been to evolve alternative control strategies that eventually reduce dependency on synthetic fungicides. Recently, in different parts of the world, emphasis has been placed on exploitation of higher plant products as the novel chemotherapeutants for plants because of their non-toxicity, systemicity and biodegradability. To date, different plant products have been formulated as botanical pesticides for large-scale applications for the eco-friendly management of plant pests and as an alternative to synthetic pesticides in crop management. These products are cost-effective and have low toxicity to humans and livestock. Therefore, such products from higher plants may be exploited as the eco-chemical and biorational approach in integrated plant protection programmes. The interests and trends in the modern society for 'green consumerism' involving fewer synthetic ingredients in the agriculture industry may favour the approval of plant-based products as the safe alternative strategy in managing plant pests. Natural plant chemicals will certainly play a pivotal role in the near future for pest control in industrialized and developing countries as well. Because of their biodegradable nature, systemicity following application, ability to alter the behaviour of target pests and their favourable safety profile, the botanical pesticides are considered the vital tools in achieving the evergreen revolution and in the ecological strategy of controlling agricultural pests.

To date, natural plant products for pest management form the frontier areas of research on a global scale. The present book deals with the current state and future prospects of botanical pesticides in the eco-friendly management of plant pests. Looking at the growing interest of professionals and the general public alike, it seemed imperative to produce a book that encompasses as much information as possible, with an emphasis on the exploitation of plant products in the possible management of agricultural pests. The book may be useful to many, including plant pathologists, microbiologists, entomologists, plant scientists and natural product chemists. It is a compilation of invited chapters from eminent scientists and professors from educational institutions and research organizations of different countries. The book addresses different topics under the domain of natural compounds in agricultural pest management and is an attempt to include notable and diversified scientific works in the field carried out by leading scientists in the world.

The book harbours a total of 13 chapters contributed by eminent scientists working in the field so as to offer relevant and practical information on each topic. Different issues, including the global scenario on the application of botanical pesticides, plant products in the control of mycotoxins, the commercial application of botanical pesticides and their prospects in green consumerism, natural products as allelochemicals, their efficacy against viral diseases and storage pests, and bioactive products from fungal endophytes, have been covered in the book. I am very grateful to the contributors for their generous and timely responses in spite of their busy academic schedules.

I wish to extend my appreciation to all the contributors for their cooperation and encouragement. Without their enthusiasm and the timely submission of their excellent chapters, this work would not have been possible.

The book focuses on newer developments in agricultural pest management using natural products and provides up-to-date information from world experts in different fields. It is expected that the book will be a source of inspiration to many for future developments in the field. It is also hoped that the book will become useful for those engaged in such an extraordinary and attractive area. The book would serve as the key reference for recent developments in frontier research on natural products in the management of agricultural pests and also for the scientists working in this area.

I convey the strength of my feelings to my wife Dr Nirmala Kishore, daughter Dr Vatsala Kishore and son Navneet Kishore who have always been the excellent intellectual companions without whose constant loving support, patience, and unmatched help and sacrifices it would have been difficult to bring out the book on time. I also bow my head to my parents (Sri G.N. Dubey and Smt Shanti Devi) for their blessings. My sincere thanks are also due to my current research students, especially Ravindra Shukla, for help and cooperation.

Thanks are due to CABI Publishers for publishing the book with utmost interest; I am thankful to the staff members of CABI for sparing no pains to ensure a high standard of publication. Special thanks goes to Sarah Mellor, the Commissioning Editor, who initially motivated me to bring out this book during the ICPP 2008 held in Turin, Italy.

N.K. Dubey

1 Global Scenario on the Application of Natural Products in Integrated Pest Management Programmes

N.K. Dubey, Ravindra Shukla, Ashok Kumar, Priyanka Singh and Bhanu Prakash

Department of Botany, Banaras Hindu University, Varanasi, India

Abstract

In recent years there has been considerable pressure in agriculture to reduce chemical pesticides and to look for their better alternatives. The plant kingdom is recognized as the most efficient producer of different biologically active compounds, which provide them with resistance against different pests. Some higher plant products have been currently formulated as botanical pesticides and are used on a large scale as eco-friendly and biodegradable measures in managing agricultural pests. Botanicals used in agricultural pest management are safer to the user and the environment. The interest in the possible use of natural compounds to control agricultural pests has notably increased in response to consumer pressure to reduce or eliminate chemically synthesized additives in foods. There is a wide scope of use of plant-based pesticides in the integrated management of different agricultural pests. A consolidated and continuous search of natural products may yield safer alternative control measures comparable to azadirachtin and pyrethryoids, which are being used in different part of the world as ideal natural fungicides. The products from higher plants are safe and economical and would be in high demand in the global pesticide market because of their diverse mode of application.

1.1 Introduction

Agriculture plays an important role in the survival of humans and animals. It is the driving force for broad-based economic growth, particularly in developing countries. Tropical and subtropical regions have a greater potential for food production and can grow multiple crops annually. Agricultural crops suffer a colossal loss due to the ravages of insects and diseases thus causing a serious threat to our agricultural production. In some years, losses are much greater, producing catastrophic results for those who depend on the crop for food. Major disease outbreaks among food crops have led to famines and mass migrations throughout history. Loss of crops from plant diseases may result in hunger and starvation, especially in less developed countries where access to disease-control methods is limited and annual losses of 30–50 % are common for major crops. Owing to the congenial climatic conditions and particular environment, the agriculture in tropical and subtropical countries suffers severe losses due to pests (Varma and Dubey, 2001; Roy, 2003).

Even during storage, foods are severely destroyed by fungi, insects and other pests. The deterioration in the stored food commodities is mainly caused by three agents, fungi, insects and rodents, under different conditions of storage. Insect pests cause heavy losses to stored grains, especially in humid and warm areas of the world. The production of mycotoxins by several fungi has added a new dimension to the gravity of the problem. Fungi are significant destroyers of foodstuffs during storage, rendering them unfit for human consumption by retarding their nutritive value and sometimes by the production of mycotoxins. According to FAO estimates, 25% of the world food crops are affected by mycotoxins each year (Dubey *et al.*, 2008). Generally, tropical conditions such as high temperature and moisture, unseasonal rains during harvest, and flash floods lead to mycotoxins. Poor harvesting practices and improper storage during transport and marketing can also contribute to the proliferation of mycotoxins. Among the mycotoxins, aflatoxins raise the most concern posing a great threat to human and livestock health as well as international trade. Aflatoxins are the most dangerous and about 4.5 billion people in developing countries are exposed to aflatoxicoses (Williams *et al.*, 2004; Srivastava *et al.*, 2008). Aflatoxins are potent toxic, carcinogenic, mutagenic, immunosuppressive agents, produced as secondary metabolites by the fungi *Aspergillus flavus* and *A. parasiticus* on a variety of food products.

The entire effort of growing a crop will be lost in the absence of crop protection, resulting in financial loss to the grower. Therefore crop protection against various pests is a must in agriculture. There is a need to reduce if not eliminate these losses by protecting the crops from different pests through appropriate techniques. Currently, the role of crop protection in agriculture is of great importance and is a more challenging process than before. Hence, there is an urgent need to pay proper attention to control quantitative losses due to pest infestations of crops and their produce as well as qualitative losses due to mycotoxin contamination.

1.2 The Need to Search for Eco-friendly Control Measures

The most conventional and common method of pest and disease control is through the use of pesticides. Pesticides are the substances or mixture of substances used to prevent, destroy, repel, attract, sterilize or mitigate the pests. Generally pesticides are used in three sectors, viz. agriculture, public health and consumer use. These pesticides are largely synthetic compounds that kill or deter the destructive activity of the target organism.

Many farmers and crop growers use insecticides to kill infesting insects. The consumption of pesticide in some of the developed countries is almost 3000 g/hectare. Unfortunately, there are reports that these compounds possess inherent toxicities that endanger the health of the farm operators, consumers and the environment (Cutler and Cutler, 1999). Pesticides are generally persistent in nature. Upon entering the food chain they destroy the microbial diversity and cause ecological imbalance. Their indiscriminate use has resulted in the development of a resistance problem among insects to insecticides, pesticide residue hazards, upsetting the balance of nature and a resurgence of treated populations. Insect resistance to phosphine is a matter of serious concern (Rajendran, 2001; Benhalima *et al.*, 2004; Thie and Mills, 2005; Dubey *et al.*, 2008). The repeated use of certain chemical fungicides in packing houses has led to the appearance of fungicide-resistant populations of storage pathogens (Brent and Hollomon, 1998). The ability of some of these pests to develop resistance curbs the effectiveness of many commercial chemicals. Resistance has accelerated in many insect species and it was reported by the World Resources Institute that more than 500 insect and mite species are immune to one or more insecticides (WRI, 1994). Similarly about 150 plant pathogens such as fungi and bacteria are now shielded against fungicides (Shetty and Sabitha, 2009). Moreover, the control of such pests has become increasingly difficult because of reduced effectiveness of pesticides caused by the emergence of pesticidal resistance in arthropod pests.

Reliance on synthetic chemicals to control pests has also given rise to the destruction of beneficial non-target organisms (parasitoids and predators), thereby affecting the food chain and impacting on biological diversity. There have also been cases of pests becoming tolerant to insecticides, resulting in the use of double and triple application rates (Stoll, 2000).

Furthermore, the use of synthetic chemicals has also been restricted because of their carcinogenicity, teratogenicity, high and acute residual toxicity, ability to create hormonal imbalance, spermatotoxicity, long degradation period, environmental pollution and their adverse effects on food and side effects on humans (Omura, *et al.*, 1995; Unnikrishnan and Nath, 2002; Xavier, *et al.*, 2004; Konstantinou, *et al.*, 2006; Feng and Zheng, 2007). Pesticides can be remarkably persistent in biological systems. They have been reported to be accumulated in ecosystems. Osprey eggs in the Queen Charlotte Islands, polar bear fat in the high Arctic, and the blubber of whales in all the oceans of the world are contaminated with pesticide residues, even though all these creatures live far from point sources of pesticide application (CAPE, 2009). Water and wind, as well as the bodies of animals that serve as prey for others (including humans) higher on the food chain, are the universal vectors for pesticide dispersal. Highest on the food chain, human breast milk contamination is of great concern because of high levels of bio-accumulated pesticides. The World Health Organization (WHO) estimates that 200,000 people are killed worldwide, every year, as a direct result of pesticide poisoning (CAPE, 2009). Pest control strategies, therefore, need proper regulation in the interest of human health and environment. In recent years there has been considerable pressure on consumers to reduce or eliminate chemical fungicides in foods. There is increasing public concern over the level of pesticide residues in food. This concern has encouraged researchers to look for alternative solutions to synthetic pesticides (Sharma and Meshram, 2006).

Effort is therefore needed to find alternatives or formulations for currently used pesticides. There is, therefore, still a need for new methods of reducing or eliminating food-borne pathogens, possibly in combination with existing methods (Leistner, 1978). At the same time, Western society appears to be experiencing a trend of 'green consumerism' (Tuley de Silva, 1996; Smid and Gorris, 1999), desiring fewer synthetic food additives and products with a smaller impact on the environment.

Considering all these drawbacks, many components of the integrated pest management (IPM) concept were developed in the late 19th and early 20th centuries, and biopesticides and bioagents became important tools in pest management strategies that are practical, economical, eco-friendly and protective for both public health and the environment. Sustainable agriculture aims to reduce the incidence of pests and diseases to such a degree that they do not seriously damage the farmer's crops — but without upsetting the balance of nature. One of the aims of sustainable agriculture is to rediscover and develop methods whose cost and ecological side-effects are minimal.

1.3 Historical Use of Botanical Products in Pest Management

Plant diseases are known from times preceding the earliest writings. The Bible and other early writings mention diseases, such as rusts, mildews, blights and blast. Disease control measures are as old as modern civilizations and were first recorded long before the Renaissance in both the Western and the Oriental worlds. The use of locally available plants in the control of pests is an ancient technology in many parts of the world. Rotenone was originally employed in South America to paralyse fish, causing them to come towards the surface and be easily captured (Fig. 1.1a; Ware, 2002). It has been used in fruit gardens to combat such pests as Japanese beetles, mites and stinkbugs. Preparations of roots from the genera *Derris elliptica*, *Lonchocarpus nicou* and *Tephrosia* sp. containing rotenone were commercial insecticides in the 1930s. Rotenone is a flavonoid derivative that strongly inhibits mitochondrial function by inhibiting NADH-dependent coenzyme Q oxidoreductase (Singer and Ramsay, 1994). Rotenone persists 3–5 days on the foliage after application and is easily biodegradable. Nicotine obtained from several members of the genus *Nicotiana* has been used commercially as an insecticide (Fig. 1.1b). However, presently it is not accepted by most organic certification programmes because of its residue and the potential harm it can cause to humans and the agroecosystem (MMVR, 2003). Ryanodine, an alkaloid from the tropical shrub *Ryania speciosa*, has been used as a commercial insecticide against the European corn borer (Fig 1.1c). Ryania controls many insects, particularly those with chewing mouthparts (Coats, 1994; Regnault-Roger and Philogene, 2008).

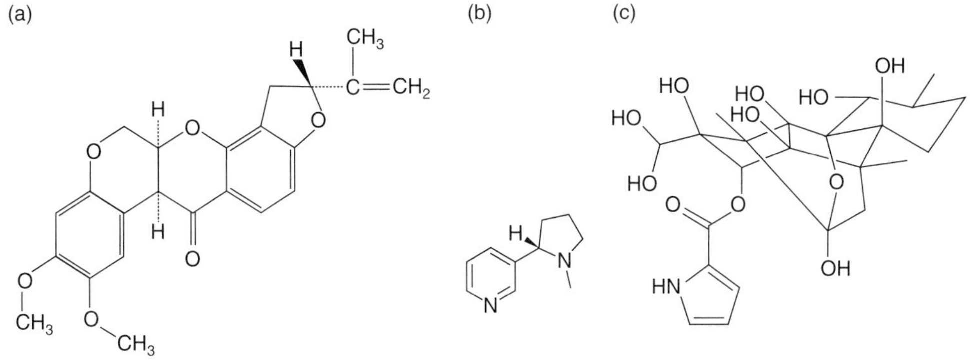

Fig. 1.1. (a) Rotenone (b) Nicotine (c) Ryanodine.

Used widely until the 1940s, such botanical pesticides were partly displaced by synthetic pesticides that at the time seemed easier to handle and longer lasting. With the knowledge of the adverse effects of synthetic pesticides, worldwide attention is currently being given to shifting to nonsynthetic safer pesticides. There is renewed interest in the application of botanical pesticides for crop protection, and scientists are now experimenting and working to protect crops from pest infestations using indigenous plant materials (Roy *et al.*, 2005).

1.4 Plant Products in Current Worldwide Use as Pesticides

The plant kingdom is recognized as the most efficient producer of chemical compounds, synthesizing many products that are used in defence against different pests (Prakash and Rao, 1986; 1997; Charleston *et al.*, 2004; Isman and Akhtar, 2007). Higher plants contain a wide spectrum of secondary metabolites such as phenolics, flavonoids, quinones, tannins, essential oils, alkaloids, saponins and sterols. Tens of thousands of secondary products of plants have been identified and there are estimates that hundreds of thousands of such compounds exist. These secondary compounds represent a large reservoir of chemical structures with biological activity. Therefore, higher plants can be exploited for the discovery of new bioactive products that could serve as lead compounds in pesticide development because of their novel modes-of-action (Philogene *et al.*, 2005). The rainforest plants are particularly thought to have developed a complete array of defence-providing chemicals. This resource is largely untapped for use as pesticides (Tripathi *et al.*, 2004).

Extracts prepared from different plants have been reported from time to time to have a variety of properties including insecticidal activity, repellence to pests, antifeedant effects, insect growth regulation, toxicity to nematodes, mites and other agricultural pests, and also antifungal, antiviral and antibacterial properties against pathogens (Prakash and Rao, 1986, 1997).

Natural pest controls using botanicals are safer to the user and the environment because they break down into harmless compounds within hours or days in the presence of sunlight. Botanical pesticides are biodegradable (Devlin and Zettel, 1999) and their use in crop protection is a practical sustainable alternative. Pesticidal plants have been in nature for millions of years without any ill or adverse effects on the ecosystem. Botanical pesticides are also very close chemically to those plants from which they are derived, so they are easily decomposed by a variety of microbes common in most soils. Their use maintains the biological diversity of predators (Grange and Ahmed, 1988), and reduces environmental contamination and human health hazards. Botanical pesticides tend to have broad-spectrum activity and are sometimes stimulatory to the host metabolism (Mishra and Dubey, 1994). Botanical insecticides can often be easily produced by farmers and small-scale industries. Recently, attention has been paid towards the exploitation of higher plant products as novel chemotherapeutics in plant protection. Such plant products have also been formulated for their large-scale application in crop protection, and are regarded as pro-poor and cost-effective (Dubey *et al.*, 2009).

Neem

Neem (*Azadirachta indica*) is regarded as the 'Wonder Tree', 'Botanical Marvel', 'Gift of Nature' and 'Village Pharmacy' in India. From prehistoric times, neem has been used primarily against household and storage pests, and to some extent against pests related to field crops in the Indian subcontinent. Neem oil and seeds are known to have inherent germicidal properties and have been in use for Ayurvedic (herbal) medicines in India for a long time. Burning neem leaves in the evening is a common practice in rural India to repel mosquitoes. Neem is widely grown in other Asian countries and tropical and subtropical areas of Africa, America and Australia. It grows well in poor, shallow, degraded and saline soil. Neem can be considered as the most important among all biopesticides for controlling pests. Neem pesticides do not leave any residue on the crop and therefore are preferred over chemical pesticides. In the past decade, neem has become a source of natural pesticide due to its non-toxicity, environmental safety and so on, thereby replacing synthetic pesticides. Neem derivatives have been applied against several species of storage pests and crop pests as leaves, oil, cake, extracts and as formulations in neem oil (Gahukar, 2000; Dhaliwal *et al.*, 2004). Neem pesticides are thus a potential alternative to chemical-based pesticides and their use can avoid the dumping of thousands of tonnes of agrochemicals on Earth every year.

Neem-based pesticides are sold under trade names such as Margosan-O, Azatin Rose Defense, Shield-All, Triact and Bio-neem. They have been shown to control gypsy moths, leaf miners, sweet potato whiteflies, western flower thrips, loopers, caterpillars and mealybugs as well as some of the plant diseases, including certain mildews and rusts.

Neem products also function as insect growth regulators (IGRs). The treated insects are usually prevented from moulting to develop into the next life stage and they die. The treatment may also deter egg laying. Generally, chewing insects are affected more than sucking insects. Insects that undergo complete metamorphosis are also generally affected more than those that do not undergo metamorphosis.

Neem seeds are a rich storehouse of over 100 tetranortriterpenoids and diverse non-isoprenoids (Devkumar and Sukhdev, 1993). The neem tree contains more than 100 different limonoids in its different tissues (Isman *et al.*, 1996). Many of these are biologically active against insects as antifeedants. The most touted biologically active constituent of neem has been highly oxygenated azadirachtin and some of its natural analogues and derivatives. Azadirachtin (molecular formula: $C_{35}H_{44}O_{16}$, chemical structure shown in Fig. 1.2), a highly oxidized triterpenoid, is the most widely publicized bioactive molecule in neem. It is systemic in nature, absorbed into the plant and carried throughout the tissues, being ingested by insects when they feed on the plant. This may make it effective against certain foliage-feeders that cannot be reached with spray applications. Azadirachtin is more effective when formulated in a neem oil medium together with the other natural products of neem. Hence, it is preferable to use neem oil enriched with azadirachtin as a stable feed stock for making pesticide formulations.

Neem as a harmless and safe pesticide fits into integrated pest management and organic farming. In toxicological studies carried out in the USA and Germany, different neem products were neither mutagenic nor carcinogenic, and they did not produce any skin irritations or organic alterations in mice and rats, even at high concentrations. Azadirachtin is considered non-toxic to mammals, having a low mammalian toxicity with an LD_{50} of >5000 mg/kg for rat (oral acute) (Raizada *et al.*, 2001), fish (Wan *et al.*, 1996) and pollinators (Naumann and Isman, 1996). Hence it is classified by the

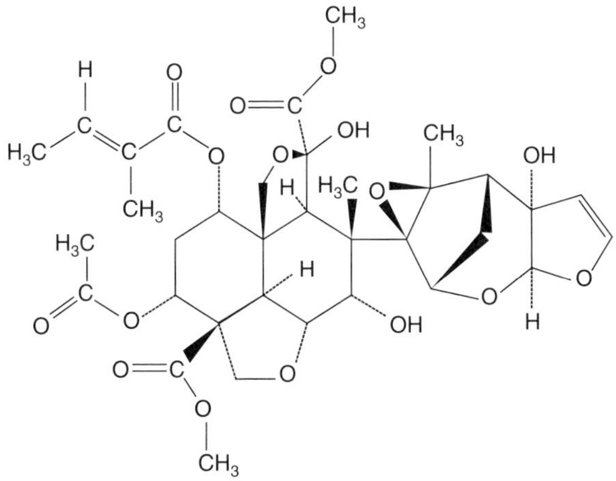

Fig. 1.2. Azadirachtin.

Environmemtal Protection Agency (EPA) in class IV. Regarding its environmental impact, neem is sensitive to light and the half-life of azadirachtin is one day (Kleeberg, 2006).

Pyrethrum

Pyrethrum is one of the oldest and safest insecticides, and is extracted from the dried flower buds of *Chrysanthemum* sp. The ground, dried flowers were used in the early 19th century to control body lice during the Napoleonic Wars. Even today, powders of the dried flowers of these plants are sold as insecticides. Pyrethrum is a mixture of four compounds: pyrethrins I and II and cinerins I and II (chemical structures shown in Fig. 1.3; Ware, 2002). Chrysanthemum plants, *Chrysanthemum cinerariaefolium*, are grown primarily in Kenya, Uganda, Tanzania and Ecuador. Pyrethrins affect the insect on contact, creating disturbances in the nervous system which eventually result in convulsions and death. Low doses, however, often cause temporary paralysis from which the insect may recover. For this reason, pyrethrums are mixed with a synergist such as piperonyl butoxide (PBO) derived from sassafras or *n*-octyl bicycloheptane dicarboximide to increase insect mortality and to extend their shelf life (Ware, 2002).

Pyrethrum products represent 80% of the total market of botanical insecticides and are favoured by organic growers because of their low mammalian toxicity and environmental non-persistence (Isman, 1994). Pyrethrum is non-toxic to most mammals, making it among the safest insecticides in use.

Pyrethroids versus pyrethrin

Pyrethroids are synthetic materials designed to imitate natural pyrethrum. They have been developed based on pyrethrins, but are much more toxic and long lasting (Singh and Srivasava, 1999). They are marketed under various trade names, for example Ambush or Decis. Pyrethroids can be useful insecticides, but some pyrethroids are extremely toxic to natural enemies. Pyrethroids are also toxic to honey bees and fish. Sunlight does not break them down and they stick to leaf surfaces for weeks, killing any insect that touches the leaves. This makes them less specific in action and more harmful to the environment than pyrethrin. In addition they irritate the human skin.

Sabadilla

Sabadilla, also known as cevadilla, is derived from the seeds of the sabadilla lily (*Schoenocaulon officinale*), a tropical lily that grows in Central and South America (Soloway, 1976). The active ingredient is an alkaloid known as veratrine which is commonly sold under the trade names 'Red Devil' or 'Natural Guard' (for the structure, see Fig. 1.4). This compound was first used in the 16th century, and grew in popularity during the Second World War, when other botanicals such as pyrethrum and rotenone were in short supply. The dust is made from the seeds and the active components are

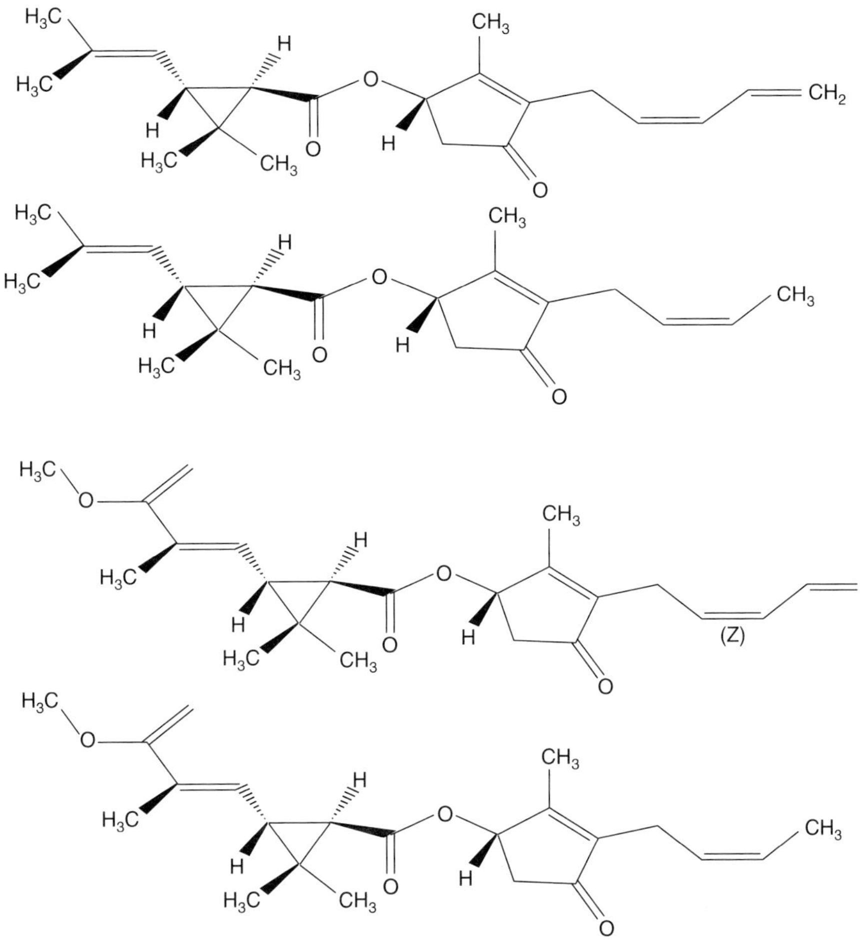

Fig. 1.3. Pyrethrum.

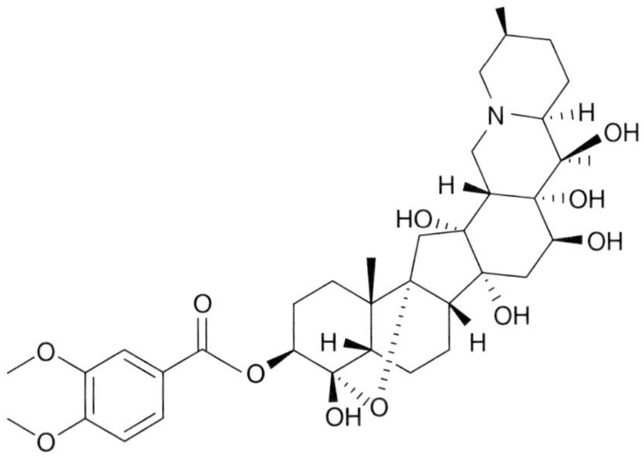

Fig. 1.4. Veratrine.

lacking in the other plant parts (roots, bulbs, stems and leaves). It is interesting that the toxic constituents actually become more powerful after storage.

Sabadilla is considered among the least toxic of botanical insecticides, with an oral LD_{50} of 4000–5000 mg/kg (in mice) (Dayan *et al.*, 2009). It slows down the shutting of Na^+ channels and disturbs membrane depolarization, causing paralysis before death (Bloomquist, 1996). No residue is left after the application of sabadilla because it breaks down rapidly in sunlight.

Carvone

Carvone is a monoterpene of the essential oil of *Carum carvi* (see Fig. 1.5 for the structure). It is a non-toxic botanical pesticide under the trade name TALENT. It inhibits the sprouting of potato tubers during storage and protects them from bacterial rotting without exhibiting mammalian toxicity. Thus, it enhances the shelf life of stored fruits and vegetables and inhibits microbial deterioration without altering the taste and odour of the fruits after treatment (Varma and Dubey, 1999). The LD_{50} value of carvone (in mice) is reported to be 1640 mg/kg (Isman, 2006).

Allyl isothiocyanate

Allyl isothiocyanate (Fig. 1.6) is an organosulfur compound that serves the plant as a defence against herbivores. Because it is harmful to the plant itself, it is stored in the harmless form of the glucosinolate, separate from the myrosinase enzyme present in plants. When an animal chews the plant, the allyl isothiocyanate is released due to action of myrosinase enzyme, thus repelling the animal.

Members of the plant family Brassicaceae are chemically linked by the almost universal presence of glucosinolates, a class of sulfur-containing glycosides, also called mustard oil glycosides or thioglucosides. These compounds are considered the first line of defence of crucifers against insects and other organisms (Renwick, 1996).

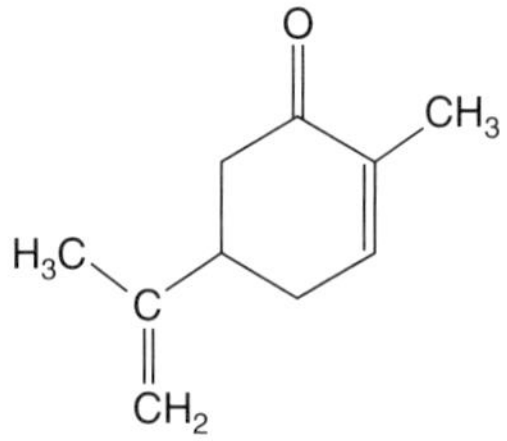

Fig. 1.5. Carvone.

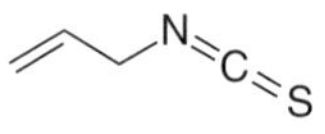

Fig. 1.6. Allyl isothiocyanate.

1.5 Other Plant Products

Other environmentally safe botanical pesticides are in use from plants, namely from *Annona squamosa* (seeds), *Pongamia pinnata* (seeds) and *Vitex negundo* (leaf) (Hiremath *et al.*, 1997). Some plant species are known to be highly resistant to nematodes. The best documented of these include marigolds (*Tagetes* spp.), rattlebox (*Crotalaria spectabilis*), chrysanthemums (*Chrysanthemum* spp.), and castor bean (*Ricinus communis*) (Duke, 1990). These plants may be recommended as intercropping plants to control the nematode population in soil through their exudates and leachates. In addition to protecting crops from infestation, many rainforest plants can be used as insect repellents. Bright orange berries of *Bixa orellana* are effective in deterring biting insects, in addition to being used as a body paint and dye (Butler, 2009). In some of the members of Asteraceae, the photodynamic compound alpha-terthienyl has been shown to account for the strong nematicidal activity of the roots (Fig. 1.7a). However, no plant-derived products are sold commercially for the control of nematodes. Strychnine formulations are used in commercial rodenticides (Fig. 1.7b). *Quassia amara* (Surinam Wood), belonging to the family Simaroubaceae, is a tree species naturally distributed in Suriname and several tropical countries. Traditionally, the bark and leaves are used in herbal remedies and as medicine because the major secondary metabolites of this tree, quassin and neo-quassin, exhibit pharmacological properties such as antimalarial, antifungal, anti-ulcerative, anti-edimogenic and anticancer activity. The male reproductive system, particularly spermatogenesis, sperm maturation and androgen biosynthesis, is highly sensitive to the metabolites of *Q. amara*, which would be useful for insect pest control but may also affect male reproduction in non-target organisms. Therefore, their pharmacological effects on mammals should be determined before recommendation to avoid any handling problems with such chemicals. However, quassin has been recently assessed in trials in Australia to control pests of Brassicaceae (Thacker, 2002).

Some plants have been reported to contain insect growth regulatory chemicals (IGRs), which disrupt insect maturation and emergence as adults. Juvabione (Fig. 1.8a), found in the wood of balsam fir, was discovered by accident when paper towels made from this source were used to line insect-rearing containers resulting in a suppression of insect development

(a)   (b)

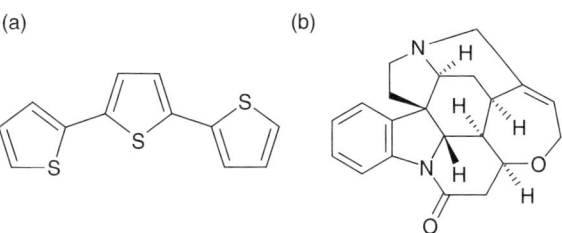

Fig. 1.7.   (a) α-terthienyl (b) Strychnine.

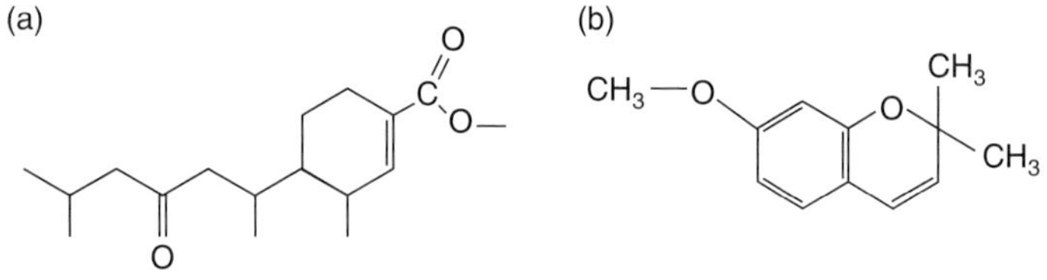

Fig. 1.8. (a) Juvabione (b) Precocenes.

(Varma and Dubey, 1999). Analogues of insect juvenile hormones such as juvocimenes in *Ocimum basilicum* have also been reported (Balandrin *et al.*, 1985). Precocenes isolated from essential oils of *Matricaria recutita* interfere with the normal function of insect glands that produce juvenile hormones resulting in the suppression of insect growth while moulting (Fig. 1.8b).

Many plant chemicals deter insects from feeding, thereby showing an antifeedant effect. Azadirachtin and limonoids such as limonin and nomilin from different plant species in Meliaceae and Rutaceae (e.g. from *Citrus* fruits) have long been used successfully for insect control, especially in India. Azadirachtin protects newly grown leaves of crop plants from feeding damage, thereby showing systemic antifeedant properties (Varma and Dubey, 1999).

1.6 Essential Oils

Since the middle ages, essential oils have been widely used for bactericidal, virucidal, fungicidal, antiparasite, insecticidal, medicinal and cosmetic applications, especially nowadays in the pharmaceutical, sanitary, cosmetic, and agricultural and food industries. In nature, essential oils play an important role in the protection of the plants as antibacterials, antivirals, antifungals, insecticides and also against herbivores by reducing their appetite for such plants. They also may attract some insects to favour the dispersion of pollens and seeds, or repel undesirable others. Some essential oils have been recognized as an important natural source of pesticides. Aromatic plants produce many compounds that are insect repellents or act to alter insect feeding behaviour, growth and development, ecdysis (moulting), and behaviour during mating and oviposition. Recently researchers have demonstrated such compounds showing larvicidal and antifeedant activity (Adebayo *et al.*, 1999; Larocque *et al.*, 1999; Gbolade, 2001), capacity to delay development, adult emergence and fertility (Marimuthu *et al.*, 1997), deterrent effects on oviposition (Naumann and Isman 1995; Oyedele *et al.*, 2000), and arrestant and repellent action (Landolt *et al.*, 1999). Plants with strong smells, such as French marigold and coriander, act as repellents and can protect the crops nearby.

Most insect repellents are volatile terpenoids such as terpenen-4-ol. Other terpenoids can act as attractants. In some cases, the same terpenoid can repel certain undesirable insects while attracting more beneficial insects. For instance, geraniol will repel houseflies while attracting honey bees (Duke, 1990).

Repellents and attractants modify the behavioural response of insects. This is the basis for the principle of behavioural insect control, whereby a given species is either attracted to a bait or pheromone, or repelled from a host plant by a repulsive agent (Fagoonee, 1981). Some of the plants contain chemicals which alter the behaviour and life cycle of insect pests without killing them. Such chemicals are termed as semio-chemicals by the Organisation for Economic Cooperation and Development (Jones, 1998). The most attractive aspect of using essential oils and/or their constituents as crop protectants is their favourable mammalian toxicity because many essential oils and their constituents are commonly used as culinary herbs and spices. Such products are generally exempted from toxicity data requirements by the Environmental Protection Agency, USA. Some American companies have recently taken advantage of this situation and have been able to bring essential-oil-based pesticides to market. Mycotech Corporation produces Cinnamite™, an aphidicide/miticide/fungicide for glasshouse and horticultural crops, and Valero™, a fungicide for use in grapes, berry crops, citrus and nuts. Both products are based on cinnamon oil, with cinnamaldehyde as the active ingredient. EcoSMART Technologies are aiming to become a world leader in essential-oil-based pesticides (Shaaya and Kostjukovsky, 1998). Several essential oil constituents are already in use as an alternative to conventional insecticides. For example, d-limonene is an active ingredient of commercially available flea shampoos, pulegone and citronellal are used as mosquito repellents, and 1,8-cineole is the structural base of the herbicide cinmethylin (Duke *et al.*, 2000). Many commercial products such as Buzz Away (containing oils of citronella, cedarwood, eucalyptus and lemongrass), Green Ban (containing oils of citronella, cajuput, lavender, safrole from sassafrass, peppermint and bergaptene from bergamot oil) and Sin-So-Soft® (containing various oils) are in use as insect repellents (Chou *et al.*, 1997).

A Push–Pull or stimulo–deterrent diversionary strategy has been developed in South Africa for minimizing damage due to maize stem borer insects (Cook *et al.*, 2006). This strategy involves the selection of plant species employed as trap crops to attract stem borer insects away from maize crops, or some plant species are used as intercrops to repel insects. The trap and repellent plants contain some semio-chemicals which attract or repel the insect. *Pennisetum purpureum* and *Sorghum vulgare* attract the stem borer insect, while *Milinis minutiflra*, *Desmodium uncinatum* and *D. intorium* are the repellent plants. The Push–Pull strategy is also employed in the control of *Heliothis* sp. in cotton fields (Pyke *et al.*, 1987). The Push–Pull strategy exploiting the chemical ecology of plants would prove an interesting, indigenous and readily available concept in the management of insect population in field crops. Plant flowers such as marigolds and certain types of vegetables can help to control pests in or around the main crop, which is sometimes called 'companion planting' (Kuepper and Dodson, 2001).

Some of the essential oils and their components show chemosterilant activity making the insect pests sterile. The compound β-asarone extracted from rhizomes of *Acorus calamus*, possesses antigonadial activity causing the complete inhibition of ovarian development of different insects (Varma and

Dubey, 1999) (Fig. 9.1). The products showing chemosterilant activity are highly required in integrated pest management programmes to limit the chances of physiological (resistant) race development by insects.

Some of the essential oils have been found useful against those species of pests that are resistant towards synthetic pesticides. These essential oils are a complex mixture of components including minor constituents, in contrast to synthetic pesticides based on single products, and they act synergistically within the plant as a defence strategy. Hence, it is likely that they are more durable towards pests evolving resistance (Feng and Isman, 1995). Due to their largely environmentally friendly nature, they can be efficiently used for pest management in urban areas, homes and other sensitive areas such as schools, restaurants and hospitals (Isman, 2006).

Octopamine (a biogenic amine found in insects) has a broad spectrum of biological roles in insects, acting as a neurotransmitter, neurohormone and circulating neurohormone–neuromodulator (Evans, 1980; Hollingworth *et al.*, 1984). Octopamine exerts its effects through interacting with at least two classes of receptors which, on the basis of pharmacological criteria, have been designated octopamine-1 and octopamine-2 (Evans, 1980). Interrupting the function of octopamine results in a total breakdown of the nervous system in insects. Therefore, the octopaminergic system of insects represents a bio-rational target for insect control. The lack of octopamine receptors in verte-brates probably accounts for the profound selectivity of certain essential oils as insecticides. A number of essential oil compounds have been demon-strated to act on the octopaminergic system of insects (Enan *et al.*, 1998). Mode-of-action studies on monoterpenoids also indicate the inhibition of acetylcholinesterase enzyme activity as the major site of action in insects (Rajendran and Sriranjini, 2008).

Encapsulation is the suitable technology for the formulation of essential-oil-based pesticides. The method reduces the loss of the active agents and offers the possibility of a controlled release of oil vapours (Moretti *et al.*, 1998). Essential oils can also be incorporated with polymers into sheets. Attractant adhesive films with essential oils have been prepared to control insects in agriculture and horticulture (Klerk's Plastic Industries B.V., 1990).

Many of the commercial products that include essential oils are on the 'Generally Recognised as Safe' (GRAS) list fully approved by the Food and Drug Administration (FDA) and Environmental Protection Agency (EPA) in USA for food and beverage consumption (Burt, 2004).

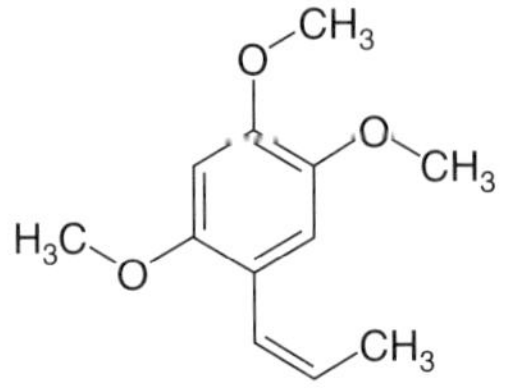

Fig. 1.9. β-Asarone.

1.7 Higher Plant Products as Inhibitors of Aflatoxin Secretions

The post-harvest colonization of various moulds on food commodities reduces their shelf life and market value, as well as rendering them unfit for human consumption because of the secretion of different types of myco-toxins, including aflatoxin, that cause undesirable effects on human health.

Hence, both qualitative as well as quantitative losses of food commodities have been reported due to fungal infestations. For the complete protection of stored food commodities from fungal biodeterioration, a fungitoxicant should be inhibitory to fungal growth as well as aflatoxin secretion. Reducing aflatoxin residue levels in food or feed can confer international trade advantages in developing countries and there may also be long-term benefits for the local population through health improvement (Dichter, 1987). Some natural products from plants such as allicin from garlic and onion extracts, clove oil, and black and white pepper have been reported to control fungal toxins (Ankri and Mirelman, 1999). Leaves of *Garcinia indica* (Selvi *et al.*, 2003), *Morinda lucida* and *Azadirachta indica* (Bankole, 1997) have been found effective in controlling aflatoxin production in food commodities. Some of the natural products, such as cinnamon and clove oil (Sinha *et al.*, 1993), phenols (Singh, 1983), some spices (Hasan and Mahmoud, 1993) and many essential oils (Razzaghi-Abyaneh *et al.*, 2008) have been reported as effective inhibitors of fungal growth and aflatoxin production. The extracts of several wild and medicinal plants have also been tested against aflatoxin-producing fungi (Bilgrami *et al.*, 1980). Essential oils from *Cymbopogon citratus, Monodora myristica*, *Ocimum gratissimum, Thymus vulgaris* and *Zingiber officinale* have been reported for their inhibitory effect on food spoilage and mycotoxin-producing fungi. Recently, the essential oils of *Cinnamomum camphora* (Singh *et al.*, 2008a), *Thymus vulgaris* (Kumar *et al.*, 2008) and *Pelargonium graveolens* (Singh *et al.*, 2008b) have been reported to suppress aflatoxin B_1 secretion by different toxigenic strains of *A. flavus*. However, there is little in the literature on the ability to monitor aflatoxin secretions by toxigenic fungal strains on food commodities.

1.8 Conclusion

Sustainable agriculture aims to reduce the incidence of pests and diseases to such a degree that they do not seriously damage the farmer's crop without upsetting the balance of nature. One of the aims of sustainable agriculture is to rediscover and develop strategies of which the cost and ecological side-effects are minimal. The secondary compounds of plants are a vast repository of compounds with a wide range of biological activities. Unlike compounds synthesized in the laboratory, secondary compounds from plants are virtually guaranteed to have biological activity and that activity is likely to function in protecting the producing plant from a pathogen, herbivore, or competitor. Among the variety of nature's ecosystem services, natural pest control is an important aspect. Hence, it is pertinent to explore the pesticidal activity of plant products.

Natural pest controls using botanicals are safer to the user and the environment because they break down into harmless compounds within hours or days in the presence of sunlight. They are also very close chemically to those plants from which they are derived, so they are easily decomposed by a variety of microbes common in most soils. Because of greater consumer awareness and negative concerns towards synthetic chemicals, crop protection using botanical pesticides is becoming more popular. There is a wide scope for the use of plant-based pesticides in the integrated management of different agricultural pests.

References

Adebayo, T.A., Gbolade, A.A. and Olaifa, J.I. (1999) Comparative study of toxicity of essential oils to larvae of three mosquito species. *Nigerian Journal of Natural Products and Medicine* 3, 74–76.

Ankri, S. and Mirelman, D. (1999) Antimicrobial properties of allicin from garlic. *Microbes and Infection* 1, 125–129.

Balandrin, M.F., Klocke, J.A., Wurtele, E.S. and Bollinger, W.H. (1985) Natural plant chemicals: sources of industrial and medicinal materials. *Science* 228, 1154–1160.

Bankole, S.A. (1997) Effect of essential oils from two Nigerian medicinal plants (*Azadirachta indica* and *Morinda lucida*) on growth and aflatoxin B_1 production in maize grain by a toxigenic *Aspergillus flavus*. *Letters in Applied Microbiology* 24, 190–192.

Benhalima, H., Chaudhry, M.Q., Mills, K.A. and Price, N.R. (2004) Phosphine resistance in stored-product insects collected from various grain storage facilities in Morocco. *Journal of Stored Products Research* 40, 241–249.

Bilgrami, K.S., Prasad, T., Misra, R.S. and Sinha, K.K. (1980) *Survey and Study of Mycotoxin Producing Fungi Associated with the Grains in Standing Maize Crops. 98, Final technical Report.* ICAR project, Bhagalpur University, India.

Bloomquist, J.R. (1996) Pesticides: Chemistries and characteristics. *Radcliffe National IPM textbook.* <http://ipmworl.umn.edu/chapters/bloomq.html>.

Brent, K.J. and Hollomon, D.W. (1998) *Fungicide resistance: the assessment of risk.* FRAC, Global Crop Protection Federation, Brussels, Monograph No. 2, pp. 1–48.

Burt, S. (2004) Essential oils: their antibacterial properties and potential applications in foods – a review. *International Journal of Food Microbiology* 94, 223–253.

Butler, R.A. (2009) Saving the rainforest with Medicinal Plants. <http://rainforests.mongabay.com/1007.htm>.

CAPE (2009) *Position Statement on Synthetic Pesticides.* <http://www.cape.ca/toxics/pesticidesps.html>.

Charleston, D.S., Dicke, M., Vet, L.E.M. and Kfir, R. (2004) Integration of biological control and botanical pesticides: evaluation in a tritrophic context. In: *Proceedings of the Fourth International Workshop on the Management of Diamondback moth and other Crucifer Pests.* Melbourne, Australia. 26–29 November 2001, pp. 207–216.

Chou, J.T., Rossignol, P.A. and Ayres, J.W. (1997) Evaluation of commercial insect repellents on human skin against *Aedes aegypti* (Diptera: Culicidae). *Journal of Medical Entomology* 34, 624–630.

Coats, J.R. (1994) Risks from natural versus synthetic insecticides. *Annual Review of Entomology* 39, 489–515.

Cook, S.M., Khan, Z.R. and Pickett, J.A. (2006) The Use of Push-Pull Strategies in Integrated Pest Management. *Annual Review of Entomology* 52, 375–400.

Cutler, H.G. and Cutler, S.J. (1999) *Biological Active Natural Products: Agrochemicals.* CRS Press, Boca Raton, USA.

Dayan, F.E., Cantrell, C.L. and Duke, S.O. (2009) Natural products in crop protection.

Bioorganic and Medicinal Chemistry 17, 4022–4034.

Delvin J.F. and Zettel, T. (1999) *Ecoagriculture: Initiatives in Eastern and Southern Africa.* Weaver Press, Harare, Zimbabwe.

Devkumar, C. and Sukhdev. (1993) Active properties in neem. In: Radhwa, N.S. and Parmar, B.S. (eds) *Neem Research and Development.* Society of Pesticide Science, New Delhi, India, pp. 63–96.

Dhaliwal, G.S., Arora, R. and Koul, O. (2004) Neem research in Asian continent: Present status and future outlook, In: Koul, O. and Wahab, S. (eds) *Neem: Today and in the New Millennium*, Springer, the Netherlands, pp. 65–96.

Dichter, G.R. (1987) *Cost-effectiveness analysis of aflatoxin control programmes.* A paper presented on the Joint FAO/WHO/UNEP, Second International Conference on Mycotoxins. Bangkok, Thailand.

Dubey, N.K., Kumar, A., Singh, P. and Shukla, R. (2009) Exploitation of Natural Compounds in Eco-Friendly Management of Plant Pests. In: Gisi, U., Chet, I. and Gullino, M.L. (eds) *Recent Developments in Management of Plant Diseases.* Springer, the Netherlands, pp. 181–198.

Dubey, N.K., Srivastava, B. and Kumar, A. (2008) Current Status of Plant Products as botanical pesticides in storage pest management. *Journal of Biopesticides* 1, 182–186.

Duke, S.O. (1990) Natural pesticides from plants. In: Janick J. and Simon J.E. (eds) *Advances in new crops.* Timber Press, Portland, USA, pp. 511–517.

Enan, E., Beigler, M. and Kende, A. (1998) Insecticidal action of terpenes and phenols to cockroaches: effect on octopamine receptors, *Proceedings of the International Symposium on Plant Protection*, Gent, Belgium, May 5–10.

Evans, P.D. (1980). Biogenic amines in the insect nervous system. *Advances in Insect Physiology* 15, 317–473.

Fagoonee, I. (1981) Behavioural response of *Crocidolomia binotalis* to neem. In: Schmutterer, H., Ascher., K.R.S. and Rembold, H. (eds) *Natural pesticides from the neem tree (Azadirachta indica* A. Juss)

Proceedings of the First International Neem Conference. Rottach-Egern, Germany 16–18 June 1980. GTZ, Germany.

Feng, R. and Isman, M.B. (1995) Selection for resistance to azadirachtin in the green peach aphid. *Myzus Persicae. Experientia* 51, 831–833.

Feng, W. and Zheng, X. (2007) Essential oil to control *Alternaria alternata in vitro* and *in vivo. Food Control* 18, 1126–1130.

Gahukar, R.T. (2000) Use of neem products/pesticides in cotton pest management. *International Journal of Pest Management* 46, 149–160.

Gbolade, A.A. (2001) Plant-derived insecticides in the control of malaria vector. *Journal of Tropical Medicinal Plants* 2, 91–97.

Grange, N. and Ahmed, S. (1988) *Handbook of Plants with Pest Control Properties.* John Wiley and Sons, New York.

Hasan, H.A. and Mahmoud, A.L.E. (1993) Inhibitory effect of spice oils on lipase and mycotoxin production. *Zentralbl Mikrobiol* 148, 543–548.

Hiremath, I.G., Ahn, Y.J. and Kim, S.L. (1997) Insecticidal activity of Indian plant extracts against *Nilaparvata lugens* (Homoptera: Delphacidae). *Applied Entomology and Zoology* 32, 159–166.

Hollingworth, R.M., Johnstone, E.M. and Wright, N. (1984) In: Magee, P.S., Kohn, G.K. and Menn, J.J. (eds) *Pesticide Synthesis through Rational Approaches, ACS Symposium Series No. 255*, American Chemical Society, Washington DC, pp. 103–125.

Isman M.B. (1994) Botanical insecticides. *Pesticide Outlook* 5, 26–31.

Isman, M.B. (2006) The role of botanical insecticides, deterrents and repellents in modern agriculture and an increasingly regulated world. *Annual Review of Entomology* 51, 45–66.

Isman, M.B. and Akhtar, Y. (2007) Plant natural products as a source for developing environmentally acceptable insecticides. In: Shaaya, I., Nauen, R., and Horowitz, A.R. (eds) *Insecticides Design Using Advanced Technologies.* Springer, Berlin, Heidelberg, pp. 235–248.

Isman, M.B., Matsuura, H., MacKinnon, S., Durst, T., Towers, G.H.N. and Arnason, J.T. (1996) Phytochemistry of the Meliaceae. So many terpenoids, so few insecticides. In: Romeo, J.T., Saunders, J.A. and Barbosa, P (eds) *Recent Advances in Phytochemistry volume 30: Phytochemical Diversity and Redundancy in Ecological Interactions*, Plenum Press, New York. pp. 155–178.

Jones, O.T. (1998) The commercial exploitation of semiochemicals and others plants derived pest managements chemicals. *Pesticide Science* 54, 293–296.

Kleeberg, H. and Ruch, B. (2006) Standardization of neem-extracts. *Proceedings of International Neem Conference*, Kunming, China, 11–15 November 2006, pp. 1–11.

Klerk's Plastic Industrie B.V., Netherland (1990) *Insect control agent for agriculture and horticulture*. Patent NL 90-1461 900626.

Konstantinou, I.K., Hela, D.G. and Albanis, T.A. (2006) The status of pesticide pollution in surface waters (rivers and lakes) of Greece. Part I. Review on occurrence and levels. *Environmental Pollution* 141, 555–557.

Kuepper, G. and Dodson, M. (2001) *Companion Planting: Basic Concepts and Resources*. Agronomy Systems Guide, National Sustainable Agriculture Information Service (ATTRA), Fayetteville, USA.

Kumar, A., Shukla, R., Singh, P., Prasad, C.S. and Dubey, N.K. (2008) Assessment of *Thymus vulgaris* L. essential oil as a safe botanical preservative against post harvest fungal infestation of food commodities. *Innovative Food Science and Emerging Technology* 9, 575–580.

Landolt, P.J., Hofstetter, R.W. and Biddick, L.L. (1999) Plant essential oils as arrestants and repellents for neonate larvae of the codling moth (Lepidoptera: Totricidae). *Environmental Entomology* 28, 954–960.

Larocque, N., Vincent, C., Belanger, A., Bourassa, J.P. (1999) Effects of tansy essential oil from *Tanacetum vulgare* on biology of oblique-banded leafroller, *Choristoneura rosaceana*. *Journal of Chemical Ecology* 25, 1319–1330.

Leistner, L. (1978) Microbiology of ready to serve foods. *Fleischwirtschaft* 58, 2008–2111.

Marimuthu, S., Gurusubramanian, G. and Krishna, S.S. (1997) Effect of exposure of eggs to vapours from essential oils on egg mortality, development and adult emergence in *Earias vittella* (F.) (Lepidoptera: Noctuidae). *Biological Agriculture and Horticulture* 14, 303–307.

Mishra, A.K. and Dubey, N.K. (1994) Evaluation of some essential oils for their toxicity against fungi causing deterioration of stored food commodities. *Applied and Environmental Microbiology* 60, 1101–1105.

MMVR (2003) *Nicotine poisoning after ingestion of contaminated ground beef: Morbidity and Mortality Weekly Report*. Michigan, 52, 413–416.

Moretti, M.D.L., Peana, A.T., Franceschini, A. and Carta, C. (1998) *In vivo* activity of *Salvia officinalis* oil against *Botrytis cinerea*. *Journal of Essential Oil Research* 10, 157–160.

Naumann, K. and Isman, M.B. (1995) Evaluation of neem *Azadirachta indica* seed extracts and oils as oviposition deterrents to noctuid moths. *Entomologia Experimentalis et Applicata* 76, 115–120.

Naumann, K. and Isman, M.B. (1996) Toxicity of neem (*Azadirachta indica* A. Juss.) seed extracts to larval honeybees and estimation of dangers from field application, *American Bee Journal* 136, 518–520.

Omura, M., Hirata, M., Zhao, M., Tanaka, A. and Inoue, N. (1995) Comparative testicular toxicities of 2 isomers of dichloropropanol, 2,3-dichloro-1-propanol, and 1,3-dichloro-2-propanol, and their metabolites alpha-chlorohydrin and epichlorohydrin, and the potent testicular toxicant 1,2-dibromo-3-chloropropane, *Bulletin of Environmental Contamination and Toxicology* 55, 1–7.

Oyedele, A.O., Orafidiya, L.O., Lamikanra, A. and Olaifa, J.I. (2000) Volatility and mosquito repellency of *Hemizygia welwitschii* oil and its formulations. *Insect Science and its Application* 20, 123–128.

Philogene, B.J.R., Regnault-Roger, C., and Vincent, C. (2005) Botanicals: yesterday's and today's promises. In: Regnault-Roger, C., Philogene, B.J.R. and Vincent, C.

(eds) *Biopesticides of Plant Origin*, Lavoisier, Paris and Intercept, Andover, UK, pp. 1–15.

Prakash, A. and Rao, J. (1986) Evaluation of plant products as antifeedants against the rice storage insects. *Proceedings from the Symposium on Residues and Environmental Pollution*, pp. 201–205.

Prakash, A. and Rao, J. (1997) *Botanical pesticides in agriculture*. CRC Lewis Publishers, Boca Raton, USA, p. 481.

Pyke, B., Rice, M., Sabine, G. and Zaluki, M. (1987) The Push–pull strategy-behavioral control of *Heliothis*. *Australian Cotton Grower* May–July, pp. 7–9.

Raizada, R.B., Srivastava, M.K., Kaushal, R.A. and Singh, R.P. (2001) Azadirachtin, a neem biopesticide: subchronic toxicity assessment in rats. *Food and Chemical Toxicology* 39, 477–483.

Rajendran, S. and Sriranjini, V. (2008) Plant products as fumigants for stored-product insect control. *Journal of Stored Products Research* 44, 126–135.

Rajendran, S. (2001) Insect resistance to phosphine – challenges and strategies. *International Pest Control* 43, 118–123.

Razzaghi-Abyaneh, M., Shams-Ghahfarokhi, M., Yoshinari, T., Rezaee, M-B, Jaimand, K., Nagasawa, H., Sakuda, S. (2008) Inhibitory effects of *Satureja hortensis* L. essential oil on growth and aflatoxin production by *Aspergillus parasiticus*. *International Journal of Food Microbiology* 123, 228–233.

Regnault Roger, C. and Philogene, B.J.R. (2008) Past and current prospects for the use of botanicals and plant allelochemicals in Integrated Pest Management. *Pharmaceutical Biology* 46, 41–52.

Renwick, J.A.A. (1996) Diversity and dynamics of crucifer defences against adults and larvae of cabbage butterflies. In: Romeo, J.T., Saunders, J.A. and Barbosa, P. (eds.) *Recent Advances in Phytochemistry, Volume 30, Phytochemical diversity and redundancy in ecological interactions*. Plenum Press, New York.

Roy, A.K. (2003) Mycological problems of crude drug research-overview and challenges. *Indian Phytopathology* 56, 1–13.

Roy, B., Amin R., Uddin, M.N., Islam, A.T.M.S., Islam, M.J. and Halder, B.C. (2005) Leaf extracts of Shiyalmutra (*Blumea lacera* Dc.) as botanical pesticides against lesser grain borer and rice weevil. *Journal of Biological Sciences* 5, 201–204.

Selvi, A.T., Joseph S.G., Jayaprakasha, G.K. (2003) Inhibition of growth and aflatoxin production in *Aspergillus flavus* by *Garcinia indica* extract and its antioxidant activity. *Food Microbiology* 20, 455–460.

Shaaya, E. and Kostjukovsky, M. (1998). Efficacy of phyto-oils as contact insecticides and fumigants for the control of stored-product insects. In: Shaaya, I., Degheele, D. (eds) *Insecticides with Novel Modes of Action: Mechanisms and Application*. Springer, Berlin, Germany, pp. 171–187.

Sharma, K. and Meshram, N.M. (2006) Bioactivity of essential oils from *Acorus calamus* Linn. and *Syzygium aromaticum* Linn. against *Sitophilus oryzae* Linn. in stored wheat. *Biopesticide International* 2, 144–152.

Shetty, P.K. and Sabitha, M. (2009) Economic and Ecological Externalities of Pesticide Use in India. In: Peshin, R. and Dhawan A.K. (eds) *Integrated Pest Management: Innovation-Development Process*. Springer, the Netherlands, pp. 113–129.

Singer, T.P. and Ramsay, R.R. (1994) The reaction sites of rotenone and ubiquinone with mitochondrial NADH dehydrogenase. *Biochimica et Biophysica Acta* 1187, 198–202.

Singh, A. and Srivastava, V.K. (1999) Toxic effect of synthetic pyrethroid permethrin on the enzyme system of the freshwater fish *Channa striatus*. *Chemosphere* 39, 1951–1956.

Singh, P. (1983) Control of aflatoxin through natural plant extracts. In: Bilgrami, K.S., Prasad, T. and Sinha, K.K. (eds) *The Proceedings of Symposium on Mycotoxin in Food and Feed*. Allied Press, Bhagalpur, India, pp. 307–315.

Singh, P., Srivastava, B., Kumar, A. and Dubey, N.K. (2008a) Fungal contamination of raw materials of some herbal drugs and recommendation of *Cinnamomum camphora* oil as herbal fungitoxicant. *Microbial Ecology* 56, 555–560.

Singh, P., Srivastava, B., Kumar, A., Kumar, R., Dubey, N.K. and Gupta, R. (2008b) Assessment of *Pelargonium graveolens* oil as plant-based antimicrobial and aflatoxin suppressor in food preservation. *Journal of the Science of Food and Agriculture* 88, 2421–2425.

Sinha, K.K., Sinha, A.K. and Prasad, G. (1993) The effect of clove and cinnamon oils on growth of and aflatoxin production by *Aspergillus flavus. Letters in Applied Microbiology* 16, 114–117.

Smid, E.J. and Gorris, L.G.M. (1999) Natural antimicrobial for food preservation. In: Rahman, M.S. (ed.) *Handbook of Food Preservation*. Marcel Dekker, New York, pp. 285–308.

Soloway, S.B. (1976) Naturally occurring insecticides. *Environmental Health Prospectives* 14, 109–117.

Srivastava, B., Singh, P., Shukla, R. and Dubey, N.K. (2008). A novel combination of the essential oils of *Cinnamomum camphora* and *Alpinia galanga* in checking aflatoxin B1 production by a toxigenic strain of *Aspergillus flavus. World Journal of Microbiology and Biotechnology* 24, 693–697.

Stoll, G. (2000) *Natural Crop protection in the tropics: Letting information come to life.* 2nd edn. Margraf Verlag, Weikersheim, p. 376.

Tamil, S.A., Joseph, G.S. and Jayaprakash, G.K. (2003) Inhibition of growth and aflatoxin production in *Aspergillus flavus* by *Garcinia indica* extract and its antioxidant activity. *Food Microbiology* 20, 455–460.

Thacker, J.M.R. (2002) *An Introduction to Arthropod Pest Control.* Cambridge University Press, Cambridge, UK, p. 343.

Thie, A.I. and Mills, K.A. (2005) Resistance to phosphine in stored-grain insect pests in Brazil. *Brazilian Journal of Food Technology* 8, 143–147.

Tripathi, P., Dubey, N.K., Banerji, R. and Chansouria, J.P.N. (2004) Evaluation of some essential oils as botanical fungitoxicants in management of post-harvest rotting of *Citrus* fruits. *World Journal of Microbiology and Biotechnology* 20, 317–321.

Tuley de Silva, K. (1996) *A Manual on the Essential Oil Industry.* United Nations Industrial Development Organization, Vienna.

Unnikrishnan, V. and Nath, B.S. (2002) Hazardous chemicals in foods. *Indian Journal of Dairy and Biosciences* 11, 155–158.

Varma, J. and Dubey, N.K. (1999) Perspective of botanical and microbial products as pesticides of tomorrow. *Current Science* 76, 172–179.

Varma, J. and Dubey, N.K. (2001) Efficacy of essential oils of *Caesulia axillaris* and *Mentha arvensis* against some storage pests causing biodeterioration of food commodities. *International Journal of Food Microbiology* 68, 207–210.

Wan, M.T., Watts, R.G., Isman, M.B. and Strub, R. (1996) An evaluation of the acute toxicity to juvenile. Pacific northwest salmon of azadirachtin, neem extract and neem-based products. *Bulletin of Environmental Contamination and Toxicology* 56, 32–439.

Ware, G.W. (2002) *An Introduction to Insecticides*, 3rd edn. University of Arizona, Tucson, USA.

Williams, H.J., Phillips, T.D., Jolly, E.P., Stiles, K.J., Jolly, M.C. and Aggarwal, D. (2004) Human aflatoxicosis in developing countries: a review of toxicology, exposure, potential health consequences and interventions. *American Journal of Clinical Nutrition* 80, 1106–1122.

WRI (1994) World Resources, 1994/1995. Oxford University Press, UK.

Xavier, R., Rekha, K. and Bairy, K.L. (2004) Health perspective of pesticide exposure and dietary management. *Malaysian Journal of Nutrition* 10, 39–51.

2 Plant Products in the Control of Mycotoxins and Mycotoxigenic Fungi on Food Commodities

SONIA MARÍN, VICENTE SANCHIS AND ANTONIO J. RAMOS

Food Technology Department, Lleida University, Lleida, Spain

Abstract

Mycotoxins are naturally occurring secondary metabolites of several toxigenic fungi that contaminate the whole food chain, from agricultural products, through to human consumption. Restrictions imposed by the food industries and regulatory agencies on the use of some synthetic food additives have led to a renewed interest in searching for alternatives, such as natural antimicrobial compounds, particularly those derived from plants. This chapter summarizes recent work on the antifungal activity of plant products and their potential for use as food additives. During the past two decades many publications have dealt with the inhibition of mycotoxigenic species by natural plant products. Most of them showed the high efficacy of such products as antifungals. Their final application to food products is, however, still in its infancy. The reasons for this are: (i) different origin, varieties and extraction methods of plant products result in essential oils and oleoresins that are widely varied in composition, preventing a direct extrapolation of results, unless experiments are carried out using pure components of these essential oils and oleoresins; (ii) plant products should be applied in such a way and at a concentration that does not affect sensorial quality of food products. Most *in vitro* studies used high concentrations of plant extracts and direct contact as the screening technique, so the application of these extracts to foods has not always been successful; and (iii) safety issues should be addressed prior to the widespread application of such extracts.

2.1 Mycotoxins in Foods

Mycotoxins are naturally occurring secondary metabolites of several toxigenic fungi that contaminate the whole food chain, from agricultural products such as peanuts and other nuts, fruits and dried fruits, and ultimately are consumed by man. These toxins can be produced in the field during the growth of the fungus on the crop or later, as a result of substandard handling or storage. Animal-derived foods such as milk, cheese and meat can be other sources of mycotoxins, if animals have been given contaminated feed. The occurrence of mycotoxins may differ from year to year. The FAO estimates that mycotoxins contaminate 25% of agricultural crops worldwide. They also have a significant impact on economics, by causing losses in farm animals or giving rise to difficulties in their management, or by rendering commodities unacceptable for national or international trade because they do not conform to existing regulations (EU, Commission Regulation 1881/2006 and 1126/2007, 105/2010 and 165/2010).

The generic term mycotoxicosis covers a variety of toxicities that target exposed organs of animals and humans. In this regard, several mycotoxins are potent animal carcinogens and have been classified by the International Agency for Research in Cancer as potential human carcinogens. The toxic effects of mycotoxins on human health are acute with a rapid onset and an obvious toxic or chronic response as characterized by low-dose exposure to mycotoxins over a long-time period. So, the real impact of the mycotoxins in human health depends on the type and the amount of the mycotoxin ingested.

2.2 Natural Preservatives for Control of Mycotoxigenic Fungi and Mycotoxins in Foods

Synthetic preservatives are nowadays the more effective and widespread way of chemical control of harmful microorganisms and diseases caused by them. Both field fungicides or antifungal food additives are efficient ways to control fungal growth in foods and raw materials. Antifungal agents are chemicals that prevent or delay mould growth. However, the presence of chemical residues in foods and labelling of preservatives in food packages are major concerns nowadays; furthermore, increasing fungal resistance has also become a problem. The interest in the search for natural preservatives and their application in food products appear to be stimulated by modern consumer trends and concerns about the safety of current food preservatives.

Restrictions imposed by the food industries and regulatory agencies on the use of some synthetic food additives have led to renewed interest in searching for alternatives, such as natural antimicrobial compounds, particularly those derived from plants.

2.3 Plant Products as Alternatives for Control of Mycotoxigenic Fungi and Mycotoxins

Plants are continuously exposed to a wide range of pathogens and, even though they do not possess an immune system, they have evolved a variety of defence mechanisms against these pathogens. After the perception of the primary wound, a signal is transducted locally and systemically through the activation of different pathways which are integrated in the regulation of the stress response (Doares *et al.*, 1995). Several plant pathogenesis-related proteins such as β-glucanases, chitinases, chitin-binding proteins, polygalacturonase-inhibiting proteins and α-amylases have been shown to possess antifungal activity against toxigenic fungi. When attacked by insects and pathogens, plants produce, together with defence proteins, a wide variety of volatile and non-volatile secondary metabolites, such as phytoalexins, alkaloids, terpenes, aldehydes, etc.

Much research has been published on the antimicrobial activity of plant products, including flours, plant extracts, oleoresins and essential oils. A very interesting option is the use of essential oils as antimicrobial additives, because they are rich sources of biologically active compounds. Essential oils are mainly obtained by steam distillation from various plant sources.

The antimicrobial activity of plant products has been extensively studied and demonstrated against a number of microorganisms, mostly *in vitro* rather than in tests with foods, and usually using a direct-contact antimicrobial assay. In the direct-contact method, active compounds are brought into contact with the selected microorganisms, and their inhibition is monitored by means of direct inspection or by measuring a physical property that is directly related to microorganism growth, such as optical density, impedance or conductance. The solution of active compound may be added to a test tube or to agar medium, and then microorganisms are inoculated. The disc diffusion test consists of spiking a sterile disc with the active compound; spiked discs are then added after inoculation of the medium, after which the inhibition zones are measured, giving an indication of the antimicrobial strength. For volatile compounds such as essential oils, the antimicrobial effectiveness in the vapour phase is of particular interest. This variety of testing techniques makes it difficult to compare results obtained by different researchers, mainly because of the difficulty in knowing the concentrations of active compounds applied.

In vitro assays against mycotoxigenic fungi

Many plant extracts have been tested *in vitro* for antifungal activity against a wide range of fungi associated with deterioration of food commodities and herbal drugs.

Studies carried out by Karthikeyan *et al.* (2007) revealed that aqueous leaf extract of zimmu (an interspecific hybrid of *Allium cepa* L. and *Allium sativum* L.) exhibited strong antifungal activity against *Aspergillus flavus*, *Fusarium moniliforme* and *Alternaria alternata* and caused *in vitro* fungal growth inhibition of 73.3%, 71.1% and 74.4%, respectively. These moulds cause mouldy sorghum grains. Essential oil and methanol extract of *Satureja hortensis* had strong activity against *A. flavus* isolated from lemon fruit (Dikbas *et al.*, 2008).

Extracts of *Cynara cardunculus* were effective in the control of *A. flavus*, *A. niger*, *A. ochraceus*, *Fusarium tricinctum*, *Penicillium funiculosum*, *P. ochrochloron*, *Trichoderma viride* and *Alt. alternata* (Kukic *et al.*, 2008). Similar results were obtained with aqueous extract of *Adenocalymma sativum* in order to control *A. flavus, A. niger, A. terreus, F. oxysporum, F. roseum, P. italicum, Clado-sporium cladosporioides* and *Alt. alternata*. The minimum inhibitory concentra-tion of extracts against *A. flavus* and *A. niger* were superior to those of two commonly used synthetic fungicides (Shukla *et al.*, 2008).

The essential oil of *Amomum subulatum* exhibited a fungitoxic spectrum against a wide range of moulds such as *A. niger, A. flavus, A. terreus, A. fumigatus, Alt. alternata, Cladosporium herbarum, Curvularia lunata, F. oxyspo-rum, Helminthosporium oryzae* and *Trichoderma viride*. Their mycelial growth was significantly inhibited at 750 µg/ml (Singh *et al.*, 2008a).

An *in vitro* initial screening of a range of 37 essential oils (including cinnamon leaf, *Cynnamomum zeylanicum*; clove, *Syzygium aromaticum*; lemon-grass, *Cymbopogon citrates*; oregano, *Origanum vulgare*; and palma rose, *Cymbopogon martinii*) on inhibition of mycelial growth of toxigenic strains of *F. verticillioides, F. proliferatum* and *F. graminearum* under different tempera-tures (20–30°C) and water activities (a_w) (0.95–0.995) was made. The basic medium was a 3% maize meal extract agar. The agar medium was modified with glycerol in order to get the assayed water activity and the essential oils were incorporated at different concentrations (0, 500, 1000 µg/ml). Although water activity was determinant for the growth of the isolates, in general, the preservative effects of the oils were not linked to a_w. However, a trend to a higher inhibition by the oils when a_w was low was observed. Temperature had a minor importance in the inhibitory effect of the essential oils (Velluti *et al.*, 2004a).

Essential oils of both cinnamon and clove were shown to effectively inhibit growth of both *A. flavus* and *P. islandicum* by direct contact, and in the vapour phase. In the atmosphere generated by these two essential oils euge-nol was the major compound, while other essential oils with low levels of eugenol proved to be ineffective, suggesting that eugenol was the antimicro-bial agent (López *et al.*, 2005).

Guynot *et al.* (2003) proved that the volatile fractions of cinnamon leaf, clove, bay (*Laurus nobilis*), lemongrass and thyme (*Thymus vulgaris*) showed potential antifungal activity against *A. flavus* and *A. niger*.

In vitro assays for the inhibition of mycotoxin production

Species in *Aspergillus* section *Flavi*, mainly *A. flavus* and *A. parasiticus* are among the most commonly occurring spoilage fungi. Some strains of these species have the ability to produce aflatoxins. These fungi invade agricul-tural commodities such as corn, peanuts and cottonseed, and herbal drugs, the resulting contamination with aflatoxin often making these products unfit for consumption. Moreover, these mycotoxins are considered carcinogenic substances and their presence is a health concern, so their control is one of the aims in order to obtain safe products. Little attention has been paid, how-ever, to the efficacy of essential oils in inhibiting aflatoxin production. Recently, work has been carried out to study the effect of these products as anti-aflatoxigenic agents.

Bluma *et al.* (2008) carried out a study with several plant extracts. A total of 96 extracts from 41 Argentinian plant species were screened against four strains of *Aspergillus* section *Flavi*. Studies on the percentage of germination, germ-tube elongation rate, growth rate, and aflatoxin B$_1$ (AFB$_1$) accumulation were carried out. Clove, mountain thyme (*Hedeoma multiflora*) and poleo (penny royal; *Lippia turbinate* var. *integrifolia*) essential oils showed the most antifungal effect in all growth parameters analysed as well as AFB$_1$ accumulation.

Essential oils from *Pelargonium graveolens* (Singh *et al.,* 2008b) and *Artabotrys odoratissimus* (Srivastava *et al.,* 2009) exhibited fungitoxicity against toxigenic strains of *A. flavus* at 0.75 g/l and 750 µl/l, respectively. The oils of *P. graveolens* and *Aelagonium odoratissimus* showed excellent anti-aflatoxigenic efficacy as they completely inhibited AFB$_1$ production even at 0.50 g/l and 750 µl/l, respectively.

In addition, an *in vitro* study with aqueous extracts of neem (*Azadirachta indica*) leaves on *A. flavus* and *A. parasiticus* has shown that the extracts fail to inhibit the vegetative growth of these moulds, while aflatoxin biosynthesis was essentially blocked *in vitro* (*A. flavus* 100% and *A. parasiticus* more than 95%, using extract concentrations at 10% v/v). AFB$_1$ synthesis was also inhibited at low extract concentrations of *Allium sativum* in a semi-synthetic medium (Shukla *et al.,* 2008).

The essential oil of *Cinnamomum camphora* was effective against other toxigenic *A. flavus* isolates detected in medicinal plants. Srivastava *et al.* (2008) evaluated other essential oils. In this work, the growth of a toxigenic strain of *A. flavus* decreased progressively with increasing concentrations of essential oils from leaves of *C. camphora* and the rhizome of *Alpinia galanga* incorporated into SMKY medium. Both oils showed complete inhibition of growth of the toxigenic strain of *A. flavus* at 1000 mg/l. The oils significantly arrested AFB$_1$ production by *A. flavus*. The oil of *C. camphora* completely blocked AFB$_1$ production at 750 mg/l, whereas that of *Alpina galanga* showed complete inhibition at 500 mg/l only. The combination of *C. camphora* and *Alpina galanga* oils showed more efficacy than the individual oils, showing complete inhibition of AFB$_1$ production even at 250 mg/l. The major components of *C. camphora* oil, as determined using GC–MS, were fenchone (34.82%), camphene (23.77%), α-thujene (17.45%), L-limolene (7.54%) and *cis-p*-menthane (5.81%). In case of *Alpina galanga* oil, bicyclo[4.2.0]oct-1-ene, 7-exo-ethenyl (58.46%), *trans*-caryophyllene (7.05%), α-pinene (14.94%) with camphene (2.15%), germacrene (1.78%) and citronellyl acetate (1.41%) were recorded as major components. In this study, the oils showed anti-aflatoxigenic properties at concentrations lower than their fungitoxic concentration. Thus, the inhibition of fungal mycelia by these oils may be through a mode other than the aflatoxin inhibition. The difference in antifungal and aflatoxin inhibition efficacy of essential oils may be attributed to the oil composition. The components of the oils may be acting by different modes of action for antifungal activity and aflatoxin inhibition. The interesting finding of this study is the better efficacy of the oil combination of *C. camphora* and *Alpina galanga* in controlling the mycelial growth as well as aflatoxin production at a concentration lower than with the individual oils (Srivastava *et al.*, 2008).

The results obtained by Sandosskumar *et al.* (2007) in *in vitro* experiments confirmed the antifungal activity of the zimmu extract against toxigenic strains of *A. flavus*. In addition, when the aflatoxigenic strains were grown in medium containing zimmu extract the production of AFB_1 was completely inhibited, even at a concentration of 0.5%. In addition, when AFB_1 was incubated with this extract a complete degradation of the toxin was observed 5 days after incubation. It is possible that the reduction in AFB_1 content may be due to detoxification or catabolism of AFB_1 by root exudates of zimmu.

Molyneux *et al.* (2007) suggested the hypothesis that aflatoxin biosynthesis is stimulated by oxidative stress on the fungus and the compounds capable of relieving oxidative stress should therefore suppress or eliminate aflatoxin biosynthesis.

Srivastava *et al.* (2008) concluded that, in general, the inhibitory action of natural products on fungal cells involves cytoplasm granulation, cytoplasmic membrane rupture and inactivation and/or inhibition of synthesis of intracellular enzymes. These actions can occur in an isolated or in a concomitant manner and culminate with mycelium germination inhibition. Phenolic compounds in the essential oils have been mostly reported to be responsible for their biological properties; however, some non-phenolic constituents of oils are more effective. The aldehyde group is also believed to be responsible for antimicrobial activity. Among the alcohols, longer chain (C6–C10) molecules in the oils have been reported to be more effective. Such compounds present in the oils may be held responsible for such biological activities.

In addition, the therapeutic use of essential oils and their combinations comprising more than one fungitoxic ingredient may also provide a solution for the rapid development of fungal resistance which is currently noticed in cases of different prevalent antifungal therapeutics. Results obtained by other researchers (Sidhu *et al.*, 2009) confirmed the synergistic effect of plant extracts. So, the combination of botanicals can be used for control of fungal growth and aflatoxin production.

Another mycotoxigenic mould is *P. expansum*, which is mainly responsible for decay in apples and pears kept in cold storage rooms. In addition, it is regarded as the major producer of the mycotoxin patulin. When apples invaded by *P. expansum* are used in making apple products (e.g. fruit juices), these products will probably be contaminated with patulin. Its presence is a health concern.

Neem leaf extracts inhibited patulin production at concentrations higher than 12.5 mg/ml, reaching a 96% inhibition at 50 mg/ml of neem extract. Patulin concentrations were reduced by neem extracts in cultures whose growth was not inhibited. So, the inhibition of patulin production does not appear to be simply a function of mycelial weight reduction (Mossini *et al.*, 2004).

The aldehydes hexanal, *trans*-2-hexenal, citral, *trans*-cinnamaldehyde and *p*-anisaldehide, the phenols carvacrol and eugenol, and the ketones 2-nonanone and (–)-carvone were screened for their ability to control *P. expansum* conidia germination and mycelial growth (Neri *et al.*, 2006).

The *in vitro* spore germination and mycelial growth assay showed a consistent fungicidal activity by *trans*-2-hexenal, whereas (–)-carvone, *p*-anisaldehyde, eugenol and 2-nonanone exhibited a progressively lower inhibition. The aldehyde *trans*-2-hexenal was the best inhibitor of conidial germination with a minimum inhibitory concentration (MIC) of 24.6 µl/l, while carvacrol was the best inhibitor of mycelial growth with a MIC of 24.6 µl/l. Other *in vitro* experiments carried out with this mould by Venturini *et al.* (2002) determined that thymol and citral were the essential oil components that showed the greatest inhibitory effects.

2.4 Research in Antimicrobial Plant Products Applied to Foods

Biologicals, because of their natural origin, are biodegradable and they do not leave toxic residues or by-products to contaminate the environment. Spices and herbs have been used for thousands of centuries by many cultures to enhance the flavour and aroma of foods. However, an increasing number of researchers have demonstrated that commonly used herbs and spices, such as garlic, clove, cinnamon, thyme, oregano, allspice, bay leaves, mustard and rosemary, possess antimicrobial properties. The activity was mainly due to the presence of essential oils (Table 2.1). Essential oils and their constituents have been used extensively as flavour ingredients in a wide variety of foods, beverages and confectionery products. Many such products are classified as 'Generally Recognized As Safe'.

Cereals

Cereals are among food commodities that are mostly prone to bear mycotoxins. *Fusarium* mycotoxins (including fumonisins, zearalenone and trichothecenes, among others) and ochratoxin A are mainly found in cereals. Maize, in particular, is a suitable substrate for aflatoxin accumulation. Therefore, plant extracts, essential oils, or their purified components have been assayed for their efficacy controlling fungi when applied directly to cereals. Together with the effect of these compounds on mycotoxigenic fungi development and on mycotoxin production, the phytotoxicity of the oils to the seeds has been evaluated, mainly calculated as the percentage of germination of the seeds, their viability or seedling growth. Maize has been the most investigated cereal, followed by wheat, sorghum, rice and others.

Maize

Lemongrass as powder or as essential oil has been one of the most frequently assayed for its efficacy against mycotoxigenic fungi on maize. Thus, Adegoke and Odesola (1996) found that the application of the essential oil of lemongrass to maize could avoid the development of inoculated *A. flavus* and *P. chrysogenum*. Authors suggested that the presence of phytochemical components such as tannins, alkaloids and glycosides, as well as the terpenes found in the essential oil, could play an important role in this effect.

Table 2.1. Most common essential oils investigated in relation to mycotoxigenic moulds prevention in foodstuffs.

Essential oil	Plant species	Major compounds	References
Thyme	*Thymus vulgaris*	Thymol, carvacrol, linalool	Montes-Belmont and Carvajal, 1998; Soliman and Badeaa, 2002
Oregano	*Origanum vulgare*	Carvacrol, thymol	Marín *et al.*, 2003; Velluti *et al.*, 2003; López *et al.*, 2004; Marín *et al.*, 2004; Souza *et al.*, 2007
Cinnamon	*Cinnamomum zeylanicum*	Eugenol, cinnamaldehyde, caryophyllene	Montes-Belmont and Carvajal, 1998; Soliman and Badeaa, 2002; Marín *et al.*, 2003; Velluti *et al.*, 2003; Marín *et al.*, 2004
Mustard	*Brassica hirta, Brassica juncea, Brassica nigra, Brassica rapa*	Allyl isothiocyanate	Nielsen and Rios, 2000; Mari *et al.*, 2002; Dhingra *et al.*, 2009
Clove	*Syzygium aromaticum*	Eugenol, cariophyllene	Montes-Belmont and Carvajal, 1998; Nielsen and Rios, 2000; Awuah and Ellis, 2002; Marín *et al.*, 2003; Velluti *et al.*, 2003; Marín *et al.*, 2004; Matan *et al.*, 2006; Bluma and Etcheverry, 2008; Reddy *et al.*, 2009
Lemongrass	*Cymbopogon citratus*	Geranial, neral	Adegoke and Odesola, 1996; Dubey *et al.*, 2000; Marín *et al.*, 2003; Velluti *et al.*, 2003; Fandohan *et al.*, 2004; Velluti *et al.*, 2004b; Marín *et al.*, 2004; Somda *et al.*, 2007; Souza *et al.*, 2007
Basil	*Ocimum basilicum*	Thymol	Montes-Belmont and Carvajal, 1998; Soliman and Badeaa, 2002; Fandohan *et al.*, 2004; Atanda *et al.*, 2007
Neem	*Azadirachta indica*	Hexadecanoic acid, oleic acid	Montes-Belmont and Carvajal, 1998; Owolade *et al.*, 2000; Fandohan *et al.*, 2004; Somda *et al.*, 2007; Reddy *et al.*, 2009

Lemongrass essential oil has also been assayed to control the growth and fumonisin B_1 (FB_1) production by *F. proliferatum* (Velluti *et al.*, 2003; Souza *et al.*, 2007). For example, Marín *et al.* (2003) found that the essential oil (500 or 1000 µg/g) inhibited growth of this fungus on artificially inoculated irradiated grains at 0.995 a_w and 20 or 30°C, but results where not satisfactory at 0.95 a_w. Inhibition of FB_1 production was only effective when it was applied to grains at 0.995 a_w and 30°C. Results were similar when assayed on naturally contaminated maize inoculated with *F. verticillioides* and *F. proliferatum*, showing that antimycotoxigenic ability of lemongrass only took place at the higher water activities (Marín *et al.*, 2003). From this study it was suggested that competing mycobiota play an important role in FB_1 accumulation and that the efficacy of essential oils in cereals may be much lower than that observed in the *in vitro* experiments using synthetic media. The main components found in the essential oil used in these studies were geranial and neral, and in much lower quantities limonene, geranyl acetate, geraniol and methyl heptenone.

Fandohan *et al.* (2004) also assayed lemongrass essential oil against *F. verticillioides*, finding a total inhibition of *F. verticillioides* growth at a concentration of 8 µl/g over 21 days. At 4.8 µl/g fumonisin production was not affected under open storage conditions, but a marked reduction was observed in closed conditions. Unfortunately, at this dose the oil adversely affected kernel germination.

When assayed on irradiated maize against a mycotoxigenic strain of *F. graminearum*, lemongrass essential oil (500 or 1000 µg/g) had an inhibitory effect on growth rate of grains at 0.995 a_w and at 20 or 30°C, but at 0.95 a_w only the higher dose was effective. Deoxynivalenol (DON) production was inhibited at 0.995 a_w/30°C at both doses, but no significant effect was observed for zearalenone (ZEA) control (Velluti *et al.*, 2004b) (DON and ZEA both being toxins produced by *F. graminearum*, amongst other *Fusarium* species). Similarly, several *F. graminearum* strains assayed on non-sterilized maize with lemongrass essential oil showed a limited efficacy for DON control (total prevention at 0.995 a_w/30°C, but no effect at 20°C) and a limited effect for ZEA control (prevention at 0.95 a_w /30°C) (Marín *et al.*, 2004).

Besides lemongrass, a huge number of essential oils have been tested for fungal control or mycotoxin prevention on maize. Thus it has been confirmed that essential oils from seeds of *Azadirachta indica* (neem tree; 500 and 1000 µg/g) and leaves of *Morinda lucida* (500 µg/g), two Nigerian medicinal plants, completely inhibited *A. flavus* aflatoxin synthesis in inoculated maize grains. Similarly, essential oils of cinnamon, peppermint (*Mentha piperita*), basil (*Ocimum basilicum*), oregano, *Teloxys ambrosioides* (the flavoring herb *epazote*), clove and thyme caused a total inhibition of *A. flavus* development on maize kernels (Montes-Belmont and Carvajal, 1998). Only around a 50% reduction was observed with the oils of *Eucalyptus globulus* (eucalyptus) and *Piper nigrum* (black pepper). Some of their constituents, when used at 2% concentration, such as thymol or *o*-methoxycinnamaldehyde significantly reduced *A. flavus* maize grain contamination. No effect was observed with the essential oils of *Allium cepa* (onion) and *Allium sativum* (garlic). However, other authors have found that *Allium* extracts at 2.5% have a significant effect on the reduction of the incidence of *F. proliferatum* in maize grains (Souza *et al.*, 2007).

Essential oils of *Pimpinella anisum* (anise), *Pëumus boldus* (boldus), mountain thyme, clove and poleo were assayed against *Aspergillus* section *Flavi* (*A. flavus* and *A. parasiticus*) in sterile maize grain under different water activities (0.982–0.90 a_w range) (Bluma and Etcheverry, 2008). Five essential oils were shown to influence lag phase, growth rate and AFB_1 accumulation, in a way that depends on their concentration, substrate water activity and time of incubation. Only the highest concentration assayed (3000 µg/g) showed the ability to maintain antifungal activity during a 35-day incubation period.

Ethanolic, methanolic and aqueous extracts of leaves, roots, scape and flowers of *Agave asperrima* (maguey cenizo) and *Agave striate* (espadin) have been tested for their capacity to inhibit growth and aflatoxin production by *A. flavus* and *A. parasiticus* on maize. Leaves and roots showed no inhibitory effect, and methanolic extracts from flowers were the most effective. It was found that 50% of the minimal inhibitory concentration of extracts (approximately 20 mg/ml) produced aflatoxin reductions higher than 99% in maize in storage conditions (Sánchez *et al.*, 2005).

Natural maize phenolic acids such as *trans*-cinnamic acid (CA) and ferulic acid (FA), alone or in combination, have been tested to control *A. flavus* and *A. parasiticus* growth and aflatoxin production on maize (Nesci *et al.*, 2007). A combination of 25 mM CA + 30 mM FA was very effective in the control of fungal growth and completely inhibited AFB_1 production at all a_w assayed (0.99–0.93 a_w range). However, some treatments, as a CA–FA mixture at 10 + 10 mM, could lead to stimulation of an *A. parasiticus* population, and stimulation of AFB_1 could be possible in some treatments.

Addition of whole or ground dry basil leaves at 50–100 mg/g to maize has shown to be effective in the reduction of aflatoxin contamination in maize stored for 32 days, with reductions in the range of 75–94% (Atanda *et al.*, 2007). The authors suggested that aflatoxins could be controlled by co-storing whole dry sweet basil leaves with aflatoxin-susceptible foods in a very simple manner.

Aqueous extracts of leaves of some indigenous plants from Nigeria, such as *Ocimum gratissimum*, *Acalypha ciliata*, *Vernonia amyygdalina*, *Mangifera indica* and *Azadirachta indica*, had a significant inhibitory effect on *F. verticillioides* development on maize, with *Acalypha ciliata* being the most effective (Owolade *et al.*, 2000).

The effect of cinnamon, clove, oregano and palma rose oils (500–1000 µg/g) on growth and FB_1 accumulation by *F. verticillioides* and *F. proliferatum* in maize grain, at 0.995 and 0.950 a_w and at 20 and 30°C, have also been evaluated, resulting in different efficacies depending on the treatment conditions (Marín *et al.*, 2003; Velluti *et al.*, 2003). These essential oils have also been used to study their effects on ZEA and DON production by *F. graminearum* in non-sterilized maize grain in the same initial conditions. In this case the efficacy of essentials oils was found to be poor, clove essential oil being that with a broader applicability (Marín *et al.*, 2004).

Ocimum vulgare (30 µg/g) and *Aloysia triphylla* (lemon verbena; 45 µg/g) essential oils were evaluated on *F. verticillioides* FB_1 production on maize grain. The oregano essential oil decreased the production level of FB_1, probably because of its content of monoterpenes (such as thymol, menthol and cinnamaldehyde) that act as antioxidants and inhibitors of toxicogenesis and sclerotial development. On the other hand, *Aloysia triphylla* increased the production of FB_1, probably because of the presence of myrcenone, alpha thujone and isomers of myrcenone in the oil, compounds that showed oxidant properties that increase lipid peroxidation and, consequently, fumonisin production (López *et al.*, 2004).

Essential oils of *O. basilicum* and *O. gratissimum* have been demonstrated to reduce the incidence of *F. verticillioides* in maize, and totally inhibited fungal growth at concentrations of 6.4 and 4.8 µl/g, respectively, although at the 4.8 µl/g dose they did not affect fumonisin production. On the other hand, oil of neem seeds has been shown to accelerate the growth of *F. verticillioides* on maize grains (Fandohan *et al.*, 2004).

Wheat

To control *A. flavus* development on wheat, the essential oils of *C. citratus*, *O. gratissimum*, *Zingiber cassumunar* and *Caesulia axillaris* have been assayed. The oils of *Caesulia citratus* and *Caesulia axillaris* showed fungistatic activity, indicating their *in vivo* applicability as herbal fumigants (Dubey *et al.*, 2000).

In the same way, vapours of the essential oil of *Caesulia axillaris* (1300 µg/g, v/v) have been shown to control *A. flavus*, *A. niger*, *A. fumigatus*, *A. sulphureus*, *Rhizopus* spp., *Mucor* spp., *Curvularia* spp., *Penicillium oxalicum* and *Absidia* spp. during 12 months of storage. The essential oil of *Mentha arvensis* (600 µg/g, v/v) showed similar results. These oils were also effective in controlling the insect pests *Sitophilus oryzae* (Varma and Dubey, 2001). Similarly, the practical applicability of essential oil from leaves of wormseed (*Chenopodium ambrosioides*) as a fumigant for protecting stored wheat has been assayed. A concentration of 100 µg/ml applied as a fumigant was able to control *A. flavus*, *A. niger*, *A. parasiticus*, *A. terreus*, *A. candidus* and *P. citrinum* during 12 months of storage (Kumar *et al.*, 2007).

The essential oils of thyme and cinnamon (≤500 µg/g), *Calendula officinalis* (marigold) (≤2000 µg/g), *Mentha viridis* (spearmint), basil and *Achillea fragantissima* (quyssum) (≤3000 µg/g) completely inhibited *A. flavus*, *A. parasiticus*, *A. ochraceus* and *F. moniliforme* on wheat grains. Caraway oil (*Carum carvi*) was inhibitory at 2000 µg/g against *A. flavus* and *A. parasiticus* and at 3000 µg/g against *A. ochraceus* and *F. moniliforme*. Anise oil completely inhibited the four fungi at ≤500 µg/g. Worse results were obtained with essential oils of chamomile (*Matricaria chamomilla*) and hazanbul (*Achillea millefalium*) (Soliman and Badeaa, 2002).

Resveratrol, an extract of grape skin, has been demonstrated to be effective in controlling the ochratoxigenic fungi *P. verrucosum* and *A. westerdijkiae* on naturally contaminated wheat (Aldred *et al.*, 2008). Total populations of fungi were significantly reduced by the presence of this compound, often by about 1–3 log colony forming units (CFU), but CFUs of the mycotoxigenic inoculated fungi were reduced only about 1–2 logs. In experiments developed at different water activities (0.995–0.80 a_w range) and temperatures (15–25°C) during a 28-day period of storage, grain treated with resveratrol (200 µg/g) had significantly less ochratoxin A (OTA) than the untreated controls. OTA contamination was reduced by >60% in most of the treatment conditions.

Extracts of *Argyreia speciosa* and *Oenothera biennis* have also been assayed. Hexadecanyl *p*-hydroxycinnamate and scopoletin isolated from roots of *A. speciosa* have demonstrated effective antifungal activity against the mycotoxigenic species *A. alternata* on wheat grains (more than 80% inhibition at 500 μg/g, and 100% at 1000 μg/g). Gallic acid from *O. biennis* showed activity against *F. semitectum* at 1000 μg/g (96.9% inhibition) (Shukla *et al.*, 1999).

Sorghum

Essential oils of sweet basil, cassia (*Cinnamomum cassia*), coriander (*Coriandrum sativum*) and bay leaf (*Laurus nobilis*) were tested for their efficacy to control an aflatoxigenic strain of *A. parasiticus* on non-sterilized sorghum grains stored at 28°C for 7 days (Atanda *et al.*, 2007). Sweet basil seemed to be the best option, as 5% dosage gave an optimal sorghum protective effect. The use of this oil is adequate in terms of the possibility of practical application, and is within the acceptable sensory levels of 1–5% for food consumption. Good results on fungal growth control were also obtained with the essential oils of cassia, and with the combination of cassia + sweet basil essential oils. It was also suggested that aflatoxin production could be controlled in sorghum grains by co-storing whole dry sweet basil leaves with the grains; 100 mg/g of basil leaves have been shown to reduce the total aflatoxin contamination by 90.6%.

The effect of essential oils of lemongrass, eucalyptus and crude oil of neem has been tested against *F. moniliforme*, as well as against *Colletotrichum graminicola* and *Phoma sorgina*, on sorghum grains (Somda *et al.*, 2007). Of the three plant extracts evaluated, lemongrass essential oil was the most potent. Concentrations of 6 and 8% caused the greatest reduction of seed infection at levels comparable to synthetic fungicides such as Dithane M 15. More than 50% of *F. moniliforme* growth was reduced by lemongrass essential oil.

An emulsifiable concentrate (EC) of zimmu has been tested under field conditions to check its efficacy controlling *A. flavus*, *F. moniliforme* and *A. alternata* on sorghum grains (Karthikeyan *et al.*, 2007). A leaf extract of zimmu, formulated to 50 EC, contaning 50% (v/v) zimmu extract, using an organic solvent (cyclohexanone), an emulsifier (Tween-80) and a stabilizer (epichloro-hydrin), was prepared. Foliar application of zimmu formulation 50 EC at 3 ml/l (v/v) concentration 60, 75 and 90 days after sowing significantly reduced the incidence of grain mould (about 70% reduction) and increased the grain weight and grain hardness. The plant extract compared favourably with foliar applications of the fungicide mancozeb (2.5 g/l). A significant reduction in the AFB_1 content in sorghum was observed when plants were sprayed with a 50 EC formulation at 0.3% concentration.

Rice

Plant extracts of different parts of *Allium cepa*, *Allium sativum*, *Azadirachta indica*, *Eucalyptus terticolis*, *Ocimum sanctum*, *Annona squamosa*, *Pongamia glaberrima*, *Curcuma longa* and *Syzgium aromaticum* have been assayed for their activities against aflatoxigenic *A. flavus* growth and AFB_1 production on rice (Reddy *et al.*, 2009). Among the plant extracts tested, *Syzgium aromaticum* (5 g/kg) showed complete inhibition of *A. flavus* growth and AFB_1 production.

Curcuma longa, Azadirachta sativum and *O. sanctum* also effectively inhibited the *A. flavus* growth (65–78%) and AFB_1 production (72.2–85.7%) at 5 g/kg.

Fruits

Penicillium expansum (blue mould) is the causative agent of the mycotoxin patulin in apples and pears. Some researchers in this field have focused on the use of certain compounds of the essential oils of plants (Table 2.1), instead of applying the essential oil directly. For example, marked fungicidal action was obtained with carvacrol against *Mucor piriformis* at 125 µg/ml, while p-anisaldehyde stopped mycelial growth of *P. expansum* at 1000 µg/ml (Caccioni and Guizzardi, 1994). Allyl-isothiocyanate (AITC) can be employed successfully in modified atmosphere packaging or as a gaseous treatment before storage. Blue mould was controlled by exposing inoculated Conference and Kaiser pears for 24 h to an AITC-enriched atmosphere. At 5 mg/l, AITC showed a high fungicidal activity on pears inoculated with *P. expansum* at 10^4 conidia/ml, without causing phytotoxic effects. Disease incidence appeared to be correlated with inoculum concentration and AITC treatment concentration. Increasing the *P. expansum* inoculum concentration at a constant AITC concentration resulted in increasing disease incidence. In fruits inoculated with the higher *P. expansum* concentration (10^6 conidia/ml) and treated with 5 mg/l AITC, the infected wounds were reduced by only 20% with respect to the control. At such a high inoculum concentration the efficacy of the most active molecules is usually significantly reduced. The good results obtained with AITC treatments delayed up to 24–48 h after inoculation suggest that this compound has potential for postharvest disease control. The efficacy of AITC in controlling *P. expansum* in pears could allow it to be used to control blue mould infection during packing-house fruit handling. The ability of AITC to control thiabendazole-resistant strains of *P. expansum* is particularly useful in this context because the proportion of thiabendazole-resistant strains is usually high. The results of analysis on the skin and pulp of pears treated with AITC confirmed the extremely low concentration of AITC residue in fruit, which is unlikely to have any implications for human health. The use of AITC, produced from purified sinigrin or from *Brassica juncea* defatted meal, against *P. expansum* appears very promising as an economically viable alternative with a moderately low impact on the environment (Mari *et al.*, 2002). The potential use of volatile fungicides to control postharvest diseases requires a detailed examination of their biological activity and dispersion in fruit tissues, and the development of a formulation that inhibits growth of pathogens without producing phytotoxic effects on fruits.

Neri *et al.* (2006) showed that volatile eugenol at concentrations of 74 and 984 µl/l was necessary to inhibit mycelial growth and conidial germination, respectively, in *P. expansum*. Apples harvested at commercial maturity were subjected to water treatment, lecithin at 50 mg/ml, eugenol-ethoxylate (2 mg/ml) and pure eugenol (2 mg/ml) combined with

50 mg/ml lecithin. All treatments were tested at 18 and 50°C. Fruits were dipped in the treatment solutions. After 2 min of treatment, fruits were stored at 2°C at normal atmosphere. Significant disease incidence reductions were observed in fruits treated with eugenol mixed with lecithin at 50°C; this combination reduced the incidence of *P. expansum* by 60–90%. The lecithin–eugenol formulation did not induce immediate or delayed phytotoxicity at room temperature. Investigation of its vapour phase properties and its applicability in the storage room will surely facilitate the application of eugenol as a control agent for long periods and may avoid the problems of phytotoxicity induced by some liquid formulations (Amiri *et al.*, 2008).

Essential oils of *Caesulia axillaris* and *Mentha arvensis* were applied at 1500 and 1000 µl/l to the storage atmosphere of *P. italicum* inoculated oranges. The *Caesulia*-oil-treated oranges showed an increased storage life of 3 days, and the *Mentha* oil-treated oranges showed an increase of 7 days. No visual symptoms of possible injury caused by the oils were observed on the peel of the fruits (Varma and Dubey, 2001).

Aspergillus section *Nigri* (formerly *A. niger*) is an ubiquitous fungal contaminant of foodstuffs, such as fruits, vegetables, nuts and spices. Some species in this section have been recently shown to produce ochratoxin A.

Pepperfruit (*Dennetia tripetala*) extracts have been tested in tomato puree against common spoilage fungi, including *A. niger*. Extracts, as a single hurdle, failed to inhibit fungal growth when compared to counts before treatment. Moreover, the use of extracts alone would entail the use of concentrations that may affect the sensorial properties of the tomato; however, combining with heat treatment (80°C for 1 min) or NaCl addition (10 mg/g) resulted in effective treatments (Ejechi *et al.*, 1999). Cinnamon essential oil has been tested against a range of fungi isolated from tomato, including *A. niger*. Although *in vitro* experiments showed promising results in terms of colony development and fungal sporulation inhibition, when cinnamon oil was tested as a volatile in the atmosphere of stored tomatoes and peppers no major effects were observed (Tzortzakis, 2009). Tomatoes and strawberries exposed to an enriched oil vapour showed improved fruit-quality-related attributes, confirming the benefits observed after exposure to cinnamon oil vapour at different concentrations. These findings may have considerable commercial significance.

Extracts of *Zingiber officinale* and *Xylopia aetiopica* were added (1–3%) to *A. niger*, *A. flavus* or *Rhizopus stolonifer* inoculated orange and apple juices. Although growth was reduced, the extracts either alone or in combination did not impose enough stress to stop the growth of the fungi (Akpomeyade and Ejechi, 1999).

The advantage of essential oils is their bioactivity in the vapour phase and the limitation of aqueous sanitation for several commodities (e.g. strawberries and grapes) make the essential oils useful as possible fumigants for stored commodity protection. One limitation of the essential oils is the strong flavour they impart, thus restricting their applicability only to products with a compatible flavour.

Nuts

Salicylic acid, thymol, vanillyl acetate, vanillin and cinnamic acid completely inhibited the germination of fungi contaminating walnut kernels. All five compounds showed somewhat similar activity in inhibiting the growth of the potentially mycotoxigenic *A. niger, A. flavus* and *P. expansum* at the concentrations tested (up to 25 mM), with thymol showing the highest activity (i.e. complete inhibition of growth at 5 mM) (Kim *et al.*, 2006).

Powders from the leaves of *O. gratissimum* and cloves of *Syzgium aromaticum* were used as protectants at 3% in combination with various packaging methods to store 3.5 kg groundnut kernel samples (9.3% moisture) artificially inoculated with *A. parasiticus*. Selected treatments were repeated with naturally infected kernels. A high level of protection was obtained with *Syzygium* powder at 3% concentration using 12% moisture kernels (Awuah and Ellis, 2002).

Calori-Domingues and Fonseca (1995) found that treatment of unshelled peanuts with grapefruit seed extract was not efficient in controlling aflatoxin production during storage. Peanuts treated with grapefruit seed extract at 5,000 and 10,000 mg/kg had mean aflatoxin contamination in the range 2,757–56,334 µg/kg and 688–5,092 µg/kg, respectively, while the control had 3,362–108,333 µg/kg. Of all of the chemicals tested only propionic acid was effective in controlling aflatoxin production. Treatments were considered efficient when the aflatoxin content ($B_1 + G_1$) remained less than 30 µg/kg.

Shelled groundnut samples with moisture contents between 7.5 and 10.5% and inoculated with conidia of *A. glaucus* and *A. parasiticus* were stored for 15–90 days at 25°C, and fumigated with synthetic food grade essential oil of mustard (*Brassica rapa*) (100 µl/l space). Deterioration of the samples was assessed by estimating the percentage of kernels colonized by fungi, the number of CFUs/kernel, and the accumulation of ergosterol and free fatty acids. The values of these variables increased with the moisture content and storage period, independent of the fumigation treatment; however, the rate of increase was significantly lower in fumigated samples (Dhingra *et al.*, 2009).

2.5 Use of Plant Products in Active Packaging

Active packaging (AP) is an innovative food packaging concept that combines advances in food technology, food safety, and packaging and material sciences in an effort to better meet consumer demands for fresh-like, safe products. One specific application is to incorporate essential oil(s) into the packaging to prevent the growth of microorganisms.

The use of preservative food-packaging films offers several advantages compared to the direct addition of preservatives to food products, since the preservative agents are applied to the packaging material in such a way that only low levels of the preservatives come into contact with the food. Either of two approaches can be used to give packaging materials antimicrobial activities. In preservative-releasing or migrating approaches, preservatives are introduced into their bulk mass, or applied to their surfaces, which subsequently migrate into the food or the headspace surrounding the food. In non-migrating approaches, compounds are applied to the packaging surfaces that inhibit target microorganisms when they come into contact with them.

The most widely used materials in food and drink packaging are various kinds of papers and boards, which are usually wax-coated to improve their water-resistance and increase the shelf-life of the packaged products. Adding an active compound to the wax formulation before coating creates an AP material. The shelf-life of the packaging manufactured using cinnamaldehyde-fortified-cinnamon essential oil was evaluated against *A. flavus*, and it was found to retain its total activity over the whole 71-day test period. Finally, the efficacy of the coatings was tested in trials with two varieties of strawberries. Complete protection was obtained during 7 days storage at 4°C, during which no visible fungal contamination developed and there were no apparent visible or organoleptic changes in the strawberries (Rodriguez *et al.*, 2007).

López *et al.* (2007) demonstrated the potential utility of polypropylene and polyethylene/ethylene vinyl alcohol copolymer films with incorporated oregano or cinnamaldehyde-fortified-cinnamon essential oils at concentrations of 4% (w/w) as antifungal packaging materials, also against *A. flavus*. They maintain their antimicrobial properties for more than 2 months, and their use in contact with foodstuffs has been demonstrated not to be harmful to consumers' health. The main drawback could be the organoleptical alteration of the packaged food due to the chemicals released by the active package.

AP may be an alternative to modified atmosphere packaging (MAP) or could complement it. For example, the volatile gas phase of combinations of cinnamon oil and clove oil showed good potential to inhibit growth of spoilage fungi normally found in intermediate moisture foods when combined with a modified atmosphere comprising a high concentration of CO_2 (40%) and low concentration of O_2 (<0.05%). To prevent the growth of *A. flavus*, 4000 µl of mixed oils was required to be added to the active MAP system and higher ratios of cinnamon to clove oil were more effective (Matan *et al.*, 2006). *A. flavus* was found to be the most resistant microorganism; this is in accordance with other authors who found *A. flavus* to be the least inhibited by essential oils in different antimicrobial tests.

The effect of MAP active packaging using volatile essential oils and oleoresins from spices and herbs were tested against a range of fungi commonly found on bread (*P. commune, P. roqueforti, A. flavus* and *Endomyces fibuliger*). Mustard essential oil showed the strongest effect. Cinnamon, garlic and clove also had high activity, while oregano oleoresin only inhibited growth weakly. Vanilla (*Vanilla planifolia*) showed no inhibitory effect towards the tested microorganisms at the applied concentrations. *A. flavus* was more resistant than the other microorganisms. Mustard essential oil was investigated in greater detail. MIC for the active component, AITC, was determined for the same species and an additional three moulds and one yeast. The minimum inhibitory concentration (MIC) values ranged from 1.8 to 3.5 µg/ml gas phase. Results showed that whether AITC was fungistatic or fungicidal depended on its concentration, and the concentration of spores. When the gas phase contained at least 3.5 µg/ml, AITC was fungicidal to all tested fungi. Results of sensory evaluation showed that hot-dog bread was more sensitive to AITC than rye bread. The minimal recognizable concentration of AITC was 2.4 µg/ml gas phase for rye bread and between 1.8 and 3.5 µg/ml gas phase for hot-dog bread. These findings showed that the required shelf-life of rye bread could be achieved by AP with AITC. AP of hot-dog bread, may nevertheless require the additional effect of other preserving factors to avoid off-flavour formation (Nielsen and Rios, 2000).

Comparisons between the effectiveness of the volatile gas phases and the liquid phases of essential oils have shown that oil in the liquid phase is more effective in preventing spoilage than when added via the gas phase. Higher volumes are required if the essential oils contact only the contaminating microorganisms in the gas phase. However, advantages of using a volatile gas phase of essential oil for food products are that it may have a lesser influence of the final taste and aroma of the product and its release may be regulated more easily.

2.6 Conclusions

During the past decade many publications have dealt with the inhibition of mycotoxigenic species by natural plant products. Most of them showed a high efficacy of such products as antifungals. Their final application to food products is, however, still in its infancy. Several reasons are involved:

- Different origin, varieties and extraction methods of plant products result in essential oils and oleoresins that are widely varied in their com-position. This heterogeneity prevents a direct extrapolation of results, unless experiments are carried out using pure components of these essential oils and oleoresins.
- Plant products should be applied in such a way and at concentrations that do not affect grain viability or the sensorial quality of food products. Most *in vitro* studies used high concentrations of plant extracts and used direct contact as the screening technique, so the application of these extracts to foods was not always successful.
- Finally, safety issues should be fully addressed prior to the widespread application of such plant products.

References

Adegoke, G.O. and Odesola, B.A. (1996) Storage of maize and cowpea and inhibition of microbial agents of biodeterioration using the powder and essential oil of lemon grass (*Cymbopogon citratus*). *International Biodeterioration & Biodegradation* 37, 81–84.

Akpomeyade, D.E. and Ejechi, B.O. (1999) The hurdle effect of mild heat and two tropical spice extracts on the growth of three fungi in fruit juices. *Food Research International* 31, 339–341.

Aldred, D., Cairns-Fuller, V. and Magan, N. (2008). Environmental factors affect efficacy of some essential oils and resveratrol to control growth and ochratoxin A production by *Penicillium verrucosum* and *Aspergillus westerdijkiae* on wheat grain. *Journal of Stored Products Research* 44, 341–346.

Amiri, A., Dugas, R., Pichot, A.L. and Bompeix, G. (2008) *In vitro* and *in vivo* activity of eugenol oil (*Eugenia caryophylata*) against four important postharvest apple pathogens. *International Journal of Food Microbiology* 126, 13–19.

Atanda, O.O., Akpan, I. and Oluwafemi, F. (2007) The potential of some spice essential oils in the control of *A. parasiticus* CFR 223 and aflatoxin production. *Food Control* 18, 601–607.

Awuah, R.T. and Ellis, W.O. (2002) Effects of some groundnut packaging methods and protection with *Ocimum* and *Syzygium* powders on kernel infection by fungi. *Mycopahologia* 154, 29–36.

Bluma, R., Amaiden, M.R. and Etcheberry, M. (2008) Screening of Argentine plants extracts: Impact on growth parameters and aflatoxin B_1 accumulation by *Aspergillus* section *Flavi*. *International Journal of Food Microbiology* 122, 114 125.

Bluma, R.V. and Etcheverry, M.G. (2008) Application of essential oils in maize grain: impact on *Aspergillus* section *Flavi* growth parameters and aflatoxin accumulation. *Food Microbiology* 25, 324–334.

Caccioni, D. and Guizzardi, M. (1994) Inhibition of germination and growth of fruit and vegetable postharvest pathogenic fungi by essential oil components. *Journal of Essential Oil Research* 6, 173–179.

Calori-Domingues, M.A. and Fonseca, H. (1995) Laboratory evaluation of chemical control of aflatoxin production in unshelled peanuts (*Arachis hypogaea* L.). *Food Additives and Contaminants* 12, 347–350.

Dhingra, O.D., Jham, G.N., Rodrigues, F.A., Silva Jr., G.J. and Costa, M.L.N. (2009) Fumigation with essential oil of mustard retards fungal growth and accumulation of ergosterol and free fatty acid in stored shelled groundnuts. *Journal of Stored Products Research* 45, 24–31.

Dikbas, N., Kotan, R., Dadasoglu, F. and Sahin, F. (2008) Control of *Aspergillus flavus* with essential and methanol extract of *Satureja hortensis*. *International Journal of Food Microbiology* 124, 179–182.

Doares, S.H., Syrovets T., Weller E.W. and Ryan C.A. (1995) Oligalacturonides and chitosans activate plant defensive genes through the octadecanoid pathway. *Proceedings of the National Academy of Sciences USA* 92, 4095–4098.

Dubey, N.K., Tripathi, P. and Singh, H.B. (2000) Prospects of some essential oils as antifungal agents. *Journal of Medicinal and Aromatic Plant Sciences* 22, 350–354.

Ejechi, B.O., Nwafor, O.E. and Okoko, F.J. (1999) Growth inhibition of tomato-rot fungi by phenolic acids and essential oil extracts of pepperfruit (*Dennetia tripetala*). *Food Research International* 32, 395–399.

Fandohan, P., Gbenou, J.D., Gnonlonfin, B., Hell, K., Marasas, W.F.O. and Wingfield, M.J. (2004) Effect of essential oils on the growth of *Fusarium verticillioides* and fumonisin contamination in corn. *Journal of Agricultural and Food Chemistry* 52, 6824 6829.

Guynot, M.E., Ramos, A.J., Seto, L., Purroy, P., Sanchis, V. and Marin, S. (2003) Antifungal activity of volatile compounds generated by essential oils against fungi commonly causing deterioration of bakery products. *Journal of Applied Microbiology* 94, 893–899.

Karthikeyan, M., Sandosskumar, R., Radhajeyalakshmi, R., Mathiyazhagan, S., Khabbaz, S.E., Ganesamurthy, K., Selvi, B. and Velazhahan, R. (2007) Effect of formulated zimmu (*Allium cepa* L. x *Allium sativum* L.) extract in the management of grain mold of sorghum. *Journal of the Science of Food and Agriculture* 87, 2495–2501.

Kim, J.H., Mahoney, N., Chan, K.L., Molyneux, R.J. and Campbell, B.C. (2006) Controlling food-contaminating fungi by targeting their antioxidative stress-response system with natural phenolic compounds. *Applied Microbial and Cell Physiology* 70, 735–739.

Kukic, J., Popovic, V., Petrovic, S., Mucaji, P., Ciric, A., Stojkovic, D. and Sokovic, M. (2008) Antioxidant and antimicrobial activity of *Cynara cardunculus* extracts. *Food Chemistry* 107, 861–868.

Kumar, R., Mishra, A.K., Dubey, N.K. and Tripathi, Y.B. (2007) Evaluation of *Chenopodium ambrosioides* oil as a potential source of antifungal, antiaflatoxigenic and antioxidant activity. *International Journal of Food Microbiology* 115, 159–164.

López, A.G., Theumer, M.G., Zygadlo, J.A. and Rubinstein, H.R. (2004) Aromatic plants essential oils activity on *Fusarium verticillioides* Fumonisin B$_1$ production in corn grain. *Mycopathologia* 158, 343–349.

López, P., Sanchez, C., Batlle, R. and Nerin, C. (2005) Solid- and vapor-phase antimicrobial activities of six essential oils: Susceptibility of selected food borne bacterial and fungal strains. *Journal of Agricultural and Food Chemistry* 53, 6939–6946.

López, P., Sánchez, C., Batlle, R. and Nerín, C. (2007) Development of flexible antimicrobial films using essential oils as active agents. *Journal of Agricultural and Food Chemistry* 55, 8814–8824.

Mari, M., Leoni, O., Iori, R. and Cembali, T. (2002) Antifungal vapour-phase activity of allyl-isothiocyanate against *Penicillium expansum* on pears. *Plant Pathology* 51, 231–236.

Marín, S., Velluti, A., Muñoz, A., Ramos, A.J. and Sanchis, V. (2003) Control of fumonisin B$_1$ accumulation in naturally contaminated maize inoculated with *Fusarium verticillioides* and *Fusarium proliferatum*, by cinnamon, clove, lemongrass, oregano and palmarosa essential oils. *European Food Research Technology* 217, 332–337.

Marín, S., Velluti, A., Ramos, A.J. and Sanchis, V. (2004) Effect of essential oils on zearalenone and deoxynivalenol production by *Fusarium graminearum* in non-sterilized maize grain. *Food Microbiology* 21, 313–318.

Matan, N., Rimkeeree, H., Mawson, A.J., Chompreeda, P., Haruthaithanasan, V. and Parker, M. (2006) Antimicrobial activity of cinnamon and clove oils under modified atmosphere conditions. *International Journal of Food Microbiology* 107, 180–185.

Molyneux, R.J., Mahoney, N., Kim, J.H. and Campbell, B.C. (2007) Mycotoxins in edible tree nuts. *International Journal of Food Microbiology* 119, 72–78.

Montes-Belmont, R. and Carvajal, M. (1998) Control of *Aspergillus flavus* in maize with plant essential oils and their components. *Journal of Food Protection* 61, 616–619.

Mossini, S.A.G., De Oliveira, K.P. and Kemmelmeier, C. (2004) Inhibition of patulin production by *Penicillium expansum* cultured with neem (*Azadirachta indica*) leaf extracts. *Journal of Basic Microbiology* 44, 106–113.

Neri, F., Mari, M. and Brigati, S. (2006) Control of *Penicillium expansum* by plant volatile compounds. *Plant Pathology* 55, 100–105.

Nesci, A., Gsponer, N. and Etcheverry M. (2007) Natural maize phenolic acids for control of aflatoxigenic fungi on maize. *Journal of Food Science* 72, M180–M185.

Nielsen, P.V. and Rios, R. (2000) Inhibition of fungal growth on bread by volatile components from spices and herbs, and the possible application in active packaging, with special emphasis on mustard essential oil. *International Journal of Food Microbiology* 60, 219–229.

Owolade, O.F., Amusa, A.N. and Osikanlu,Y.O.K. (2000) Efficacy of certain indigenous plant extracts against seedborne infection or *Fusarium moniliforme* on maize (*Zea mays* L.) in South Western Nigeria. *Cereal Research Communications* 28, 323–327.

Reddy, K.R.N., Reddy, C.S. and Muralidharan, K. (2009) Potential of botanicals and biocontrol agents on growth and aflatoxin production by *Aspergillus flavus* infecting rice grains. *Food Control* 20, 173–178.

Rodriguez, A., Batlle, R. and Nerín, C. (2007) The use of natural essential oils as antimicrobial solutions in paper packaging. Part II. *Progress in Organic Coatings* 60, 33–38.

Sánchez, E., Heredia, N. and García, S. (2005) Inhibition of growth and mycotoxin production of *Aspergillus flavus* and *Aspergillus parasiticus* by extracts of *Agave* species. *International Journal of Food Microbiology* 98, 271–279.

Sandosskumar, R., Karthikeyan, M., Mathiyazhagan, S., Mohankumar, M., Chandrasekar, G. and Velazhahan, R. (2007) Inhibition of *Aspergillus flavus* growth and detoxification of aflatoxin B_1 by medicinal plant zimmu (*Allium sativum* L. x *Allium cepa* L.). *World Journal of Microbial Biotechnology* 23, 1007–1014.

Shukla, Y.N., Srivastava, A., Kumar, S. and Kumar S. (1999) Phytotoxic and antimicrobial constituents of *Argyreia speciosa* and *Oenothera biennis*. *Journal of Ethnopharmacology* 67, 241–245.

Shukla, R., Kumar, A., Prasad, C.S., Srivastava, B. and Dubey, N.K. (2008) Antimycotic and antiaflatoxigenic potency of *Adenocalymma alliaceum* Miers. on fungi causing biodeterioration of food commodities and raw herbal drugs. *International Biodeterioration & Biodegradation* 62, 348–351.

Sidhu, O.P., Chandra, H. and Behl, H.M. (2009) Occurrence of aflatoxins in mahua (*Madhuca indica* Gmel.) seeds: Synergistic effect of plant extracts on inhibition of *Aspergillus flavus* growth and aflatoxin production. *Food and Chemical Toxicology* 47, 774–777.

Singh, P., Srivastava, B., Kumar, A., Dubey, N.K. and Gupta, R. (2008a) Efficacy of essential oil of *Amomum subulatum* as a novel aflatoxin B_1 suppressor. *Journal of Herbs, Spices & Medicinal Plants* 14, 208–218.

Singh, P., Srivastava, B., Kumar, A., Kumar, R., Dubey, N.K. and Gupta, R. (2008b) Assessment of *Pelargonium graveolens* oils as a plant-based antimicrobial and aflatoxin suppressor in food preservation. *Journal of the Science of Food and Agriculture* 88, 2421–2425.

Soliman, K.M. and Badeaa, R.I. (2002) Effect of oil extracted from some medicinal plants on different mycotoxigenic fungi. *Food and Chemical Toxicology* 40, 1669–1675.

Somda, I., Leth, V. and Sereme, P. (2007) Antifungal effect of *Cymbopogon citratus*, *Eucalyptus camaldulensis* and *Azadirachta indica* oil extracts on sorghum seed-borne fungi. *Asian Journal of Plant Sciences* 6, 1182–1189.

Souza, A.E.F., Araujo, E. and Nascimento, L.C. (2007) Atividade antifúngica de extratos de Alho e Capim-Santo sobre o desenvolvimento de *Fusarium prolifera-tum* isolado de grãos de milho. *Fitopatologia Brasilera* 32, 465–471.

Srivastava, B., Singh, P., Shukla, R. and Dubey, N.K. (2008) A novel combination of the essential oils of *Cinnamomum camphora* and *Alpinia galanga* in checking aflatoxin B_1 production by a toxigenic strain of *Aspergillus flavus*. *World Journal of Microbiological Biotechnology* 24, 693–697.

Srivastava, B., Singh, P., Srivastava, A.K., Shukla, R. and Dubey, N.K. (2009) Efficacy of *Artabotrys odoratissimus* oils as a plant-based antimicrobial against storage fungi and aflatoxin B_1 secretion. *International Journal of Food Science and Technology Microbiological Biotechnology* 44, 1909–1915.

Tzortzakis, N.G. (2009) Impact of cinnamon oil-enrichment on microbial spoilage of fresh produce. *Innovative Food Science and Emerging Technologies* 10, 97–102.

Varma, J. and Dubey, N.K. (2001) Efficacy of essential oils of *Caesulia axillaris* and *Mentha arvensis* against some storage pests causing biodeterioration of food commodities. *International Journal of Food Microbiology* 68, 207–210.

Velluti, A., Sanchis, V., Ramos, A.J., Egido, J. and Marín, S. (2003) Inhibitory effect of cinnamon, clove, lemongrass, oregano and palmarose essential oils on growth and fumonisin B_1 production by *Fusarium proliferatum* in maize grain. *International Journal of Food Microbiology* 89, 145–154.

Velluti, A., Marin, S., Gonzalez, P., Ramos, A.J. and Sanchis, V. (2004a) Initial screening for inhibitory activity of essential oils on growth of *Fusarium verticillioides*,

F. proliferatum and *F. graminearum* on maize-based agar media. *Food Microbiology* 21, 649–656.

Velluti, A., Sanchis, V., Ramos, A.J., Turon, C. and Marín, S. (2004b) Impact of essential oils on growth rate, zearalenone and deoxynivalenol production by *Fusarium graminearum* under different temperature and water activity conditions in maize grain. *Journal of Applied Microbiology* 96, 716–724.

Venturini, M.E., Blanco, D. and Oria, R. (2002) *In vitro* antifungal activity of several antimicrobial compounds against *Penicillium expansum*. *Journal of Food Protection* 65, 834–839.

3

Natural Products from Plants: Commercial Prospects in Terms of Antimicrobial, Herbicidal and Bio-stimulatory Activities in an Integrated Pest Management System

J.C. Pretorius and E. van der Watt

Department of Soil, Crop and Climate Sciences, University of the Free State, Bloemfontein, South Africa

Abstract

The use of natural products developed from wild plants is gaining interest and momentum throughout the world in both developed and developing countries. In developing countries the use of natural plant extracts is simply the result of the inability of subsistence farmers to afford commercial synthetic pesticides. However, in developed countries this is largely due to consumer resistance towards synthetic chemicals, including antimicrobial, herbicidal and bio-stimulatory agents, believed to be potentially hazardous to the environment and human health. In this chapter an overview of the latter three pesticide groups is supplied in terms of its current 'natural product' status from an integrated pest management perspective as it is applied in the agricultural industry. First, some background is provided in order to cover the history of natural product development in these three pesticide categories. Second, a short synopsis of screening programmes that identified wild plants containing natural compounds which have the potential to be considered in natural product developing programmes is supplied. Lastly, the outcome of these development programmes that realized commercialized natural products is covered.

3.1 Introduction

Plant diseases cause large yield losses throughout the world and all important food crops are attacked with disastrous consequences for food security. In many cases, plant diseases may be successfully controlled with synthetic fungicides, but this is costly to African peasantry and often has disadvantages and side effects on the ecosystem (De Neergaard, 2001). It is, however, an established fact that the use of synthetic chemical pesticides provides many benefits to crop producers. These benefits include higher crop yields, improved crop quality and increased food production for an ever increasing world population. Despite this, synthetic pesticides may pose some hazards to the environment, especially when improperly used by farmers in developing countries who lack the technical skill of handling them, and who fail to adapt to this technology easily. This may result in undesirable residues left in food, water and the environment, toxicity to humans and animals, contamination of soils and groundwater and may lead to the development of crop pest populations that are resistant to treatment with agrochemicals.

Moreover, in Africa and the Near East obsolete pesticides have become a source of great environmental concern. Some stocks are more than 30 years old and are kept in poor conditions because of inadequate storage facilities and lack of staff trained in storage management. The Food Agricultural Organization (FAO) estimated that developing countries are holding stocks of more than 100,000 tonnes of obsolete pesticides, of which 20,000 tonnes are in Africa. Many of these chemicals are so toxic that a few grams could poison thousands of people or contaminate a large area. Most of these pesticides were left over from pesticide donations provided by foreign aid programmes. In the absence of environmentally sound disposal facilities, stocks are constantly increasing (Alemayehu, 1996). Obsolete pesticide stocks are, therefore, potential time bombs. Leakage, seepage and various accidents related to pesticides are quite common and widespread. Storage conditions rarely meet internationally accepted standards. Many pesticide containers deteriorate and leak their contents into the soil, contaminating groundwater and the environment. Most stores are in the centres of urban areas or close to public dwellings. According to the World Health Organization (WHO) there are 25 million cases of acute occupational pesticide poisoning in developing countries each year (Alemayehu, 1996).

As a result of the problems outlined above, farmers in developing countries and researchers alike are seeking less hazardous and cheaper alternatives to conventional synthetic pesticides. One such alternative is the use of natural products from plants to control plant diseases in crops as part of an organic approach to Integrated Pest Management (IPM) programmes. Justification for pursuing this alternative can be found in the following statement published a decade ago by the Environmental Protection Agency (EPA) regarding the advantages of natural products from plants in the control of plant diseases:

> Natural products from plants have a narrow target range and highly-specific mode of action; show limited field persistence; have a shorter shelf life and present no residual threats. They are often used as part of Integrated Pest Management (IPM) programmes; are generally safer to humans and the environment than conventional synthetic chemical pesticides and can easily be adopted by farmers in developing countries who traditionally use plant extracts for the treatment of human diseases.   (Deer, 1999)

A further reason for exploring the use of plant extracts or natural products as biological pesticides more extensively can be found in the plant itself. Plants have evolved highly specific chemical compounds that provide defence mechanisms against attack by disease-causing organisms, including fungal attack, microbial invasion and viral infection (Cowan, 1999). These bioactive substances occur in plants as secondary metabolites, and have provided a rich source of biologically active compounds that may be used as novel crop-protecting agents (Cox, 1990). In nature some wild plants have the potential to survive both harsh biotic and abiotic environmental conditions. This has initiated the postulate that such plants might be utilized as sources for the development of natural products to be applied in agriculture by man as natural herbicides, bactericides, fungicides or products with bio-stimulatory properties in crude or semi-purified form.

It is estimated that there are more than 250,000 higher plant species on earth (Cowan, 1999) offering a vast, virtually untapped, reservoir of bioactive chemical compounds with many potential uses, including their application as pharmaceuticals and agrochemicals. It is generally assumed that natural compounds from plants pose less risk to animals and humans and are more environmentally friendly than their synthetic counterparts (Johnson, 2001). As in pharmacology, biochemicals isolated from higher plants may contribute to the development of natural products for the agricultural industry in three different ways (Cox, 1990): (i) by acting as natural pesticides in an unmodified state (crude extracts); (ii) by providing the chemical 'building blocks' necessary to synthesize more complex compounds; and (iii) by introducing new modes of pesticidal action that may allow the complete synthesis of novel analogues in order to counter the problem of resistance to currently used synthetic products by bacterial and fungal pathogens.

However, it is quite common that a natural compound isolated from a plant may be of great biological interest but may not be sufficiently robust for use (Steglich et al., 1990). Subsequently, a need for the modification of natural products into synthetic analogues that will give the desired effect may still exist. One can, for example, isolate a natural compound with promising antimicrobial activity and, by introducing a stable chemical structure with higher activity synthetically, develop a commercial product (Steglich et al., 1990). The alternative that is now more vigorously being pursued in organic farming systems is the application of the plant material itself in a natural form. However, to date and despite statements such as these made almost two decades ago, the use of natural plant extracts as pesticides to control pathogens in crops, for example, are not widespread while synthetic chemical pesticides still remain the major tool in pest management systems. Nevertheless, some natural products have been commercialized recently and the use of a combination of synthetic chemicals and natural products in IPM programmes will probably become more popular in future.

Another related area of organic farming systems is the potential to apply natural plant extracts as either plant growth regulators or natural herbicides. A plant growth regulator is an organic compound, either natural or synthetic, that modifies or controls one or more specific physiological processes within a plant (Salisbury and Ross, 1992). If the compound is produced within the plant it is called a plant hormone e.g. auxins, gibberellins, cytokines, abscissic acid and ethylene. Two decades ago, Roberts and Hooley (1988) stated that the potential exists to apply a plant extract as a foliar spray in order to stimulate growth in crop plants and hence increase yields. According to the authors, a principal objective of the agricultural and horticultural industries is to manipulate plant growth and development in such a way that the quantity or quality of a crop is increased. After the late-1980s an elevated interest developed in terms of identifying natural plant compounds that possess the potential to manipulate plant growth and development over a short period, e.g. a growing season.

From a crop production perspective, the term 'integrated pest management', in the broad sense of the word, means to apply more than one method or product in order to control 'pests' that farmers have to deal with on a regular basis. The term 'pests', in the broad sense of the word, includes, among other things, insect and weed pests as well as viral, bacterial and fungal diseases. However, the term 'integrated pest management' has a different meaning when applied in either conventional cropping systems or organic agriculture. Pest control in conventional cropping systems is mostly driven by the large-scale use of many different synthetic chemicals and it cannot be denied that great success has been achieved in this way. Due to consumer resistance and pressure generated by the 'green revolution' towards the use of synthetic chemicals, a shift towards organic agriculture was inevitable. Restrictions in terms of synthetic chemical application in organic agriculture have, in turn, forced a shift towards the use of natural or organic products.

In conventional cropping systems integrated pest management is rather straightforward and entails the use of any registered product on the market that a farmer might choose, and this applies for any of the 'pests' mentioned earlier. The methods of application of these products are in many cases similar, e.g. foliar application, but might include integrated seed, soil and foliar treatments. IPM in organic agriculture is much more complex and controlled and may include sowing multiple crops, extended rotation cycles, mulching, specific soil cultivation methods (Dayan *et al.*, 2009) and the use of organically certified natural products.

Be that as it may, the reduction in the number of synthetic products as a result of more stringent pesticide registration procedures (Dayan *et al.*, 2009), such as the Food Quality Protection Act of 1996 in the United States, has opened the door for the vigorous pursuit of natural products from plants over the past two decades. In the meantime, many natural compounds from wild plants have been isolated, purified, identified and patented but, only a few products are commercially available.

In this chapter attention will be given to the integrated pest management concept, the rationale for considering natural compounds from plants and their potential to be applied as agrochemicals in an IPM system in the agricultural industry as well as selected areas where some progress have been made over the past three decades. In terms of the latter these will include natural compounds with application potential as antimicrobial, herbicidal and bio-stimulatory agents.

3.2 Rationale for Considering Natural Compounds from Wild Plants to be Developed as Commercial Products

Wild plants are a valuable source for the development of new natural products with the potential to be used for disease management in organic crop production systems (Duke *et al.*, 1995). Widespread public concern for long-term health and environmental effects of synthetic pesticides, especially in developed countries, has prompted a renewed effort to search for natural compounds of both plant and microbial origin that can be applied as alternatives to existing synthetic pesticides (Ushiki *et al.*, 1996). However, only a small proportion has been investigated for possible use in plant disease control in agriculture.

World agriculture still encounters huge crop losses annually due to plant diseases and the situation is more serious for most subsistence farmers in developing countries that depend on non-conventional disease management practices, often with doubtful results. In contrast, crop producers in developed countries rely heavily on synthetic fungicides and bactericides to control plant diseases. Despite their efficacy against plant diseases, synthetic chemicals are considered to cause environmental pollution and are potentially harmful to human health. According to the National Academy of Sciences (Wilson, 1997), the carcinogenic risk of fungicide residues in food is more than insecticides and herbicides put together.

The long-term effect of pesticides in contaminating the environment is of particular concern, together with the fact that frequent application of fungicides has resulted in fungal mutation and, subsequently, new resistant strains (Khun, 1989). Moreover, consumer resistance towards the use of synthetic chemicals has escalated, especially in developed countries, supplying a rationale for the application of natural product alternatives in the agricultural industry (Duke *et al.*, 1995). Currently, in many developed countries, the tendency to shift to organic farming systems has evolved under consumer pressure in an attempt to reduce the risk of pesticide application. As a result, research on the possible utilization of biological resources and its application potential in agriculture has become very relevant. A promising approach in this regard is the use of natural plant products as an alternative to synthetic chemicals due to their apparently less negative impact on the environment (Ganesan and Krishnaraju, 1995).

Conceptually and practically the use of natural plant products in agriculture is not new, but dates back to the time of Democratus (470 BC), where sprinkling of amurca, an olive residue, was recommended to control late blight disease (David, 1992). In short, plant extracts have been recognized to solve agricultural problems ever since man took to farming (Pillmoor, 1993). For example, the control of insect pests through the use of natural plant products, such as Pyrethrum extracted from the Pyrethrum plant,

Chrysanthemum cinerariefolium, and neem seed from *Azadirachta indica* (Richard, 2000), has a long history in plant protection.

Despite the fact that reports on the practical application of natural plant extracts towards the control of plant pathogens in modern agriculture are found less frequently in the literature (Menzies and Bélanger, 1996), a considerable number of reports on natural chemicals that are biologically active against various plant diseases is currently available. These natural compounds are usually secondary metabolites and are synthesized in plants as a result of biotic and abiotic interactions (Waterman and Mole, 1989; Helmut *et al.*, 1994). By means of bioassay guided screening, a number of these natural plant compounds, with antimicrobial activity, have been isolated and progress has also been made towards the identification and structural elucidation of these bioactive compounds (Grayer and Harborne, 1994). Although extractable secondary metabolites have long been considered as an important source of pharmaceuticals, the evaluation of their application potential in crop production systems has been largely neglected. A wide range of activities with both positive and negative effects, including the control of microorganism, plant growth regulation (Adam and Marquardt, 1986), the induction of plant resistance to various diseases (Daayf *et al.*, 1995; Schmitt *et al.*, 1996) and promotion of beneficial microorganisms in the soil rhizosphere (Williams, 1992) have been reported. Despite these efforts, the isolation of plant secondary metabolites has led to very few commercial successes in the agricultural industry and more specifically in crop management practices.

An accelerated search for alternative options to synthetic fungicides, based on natural products from plants, therefore, seems to be an important consideration in light of the current restrictions in pesticide use in both developed and developing countries. This especially applies to the search for environmentally friendly bioactive components with broad-spectrum antimicrobial activity (Benner, 1993). This probably needs serious consideration in developing countries where yield losses are high as a result of low-input production systems due to the unaffordability of synthetic fungicides to local farmers.

According to the Natural Antifungal Crop Protectants Research Agency (Hall, 2002) spoilage and plant pathogenic fungi are responsible for some 20% loss of the potential global plant production for food and non-food use. The very large amount of chemical crop protectants used to control these losses can be detrimental to both the environment and human health. Therefore research has been initiated to develop and implement non-synthetic crop protectants using natural antifungal agents (green chemicals) or antifungal metabolites from plants. These natural crop protectants will be designed for use on food or non-food crops vulnerable to fungal deterioration (Hall, 2002). From an agronomic perspective, a secondary aim of research on natural plant products is to cultivate bioactive plants, as alternative agricultural crops, to serve as sources for the bioactive compounds.

Despite the inherent potential of compounds from plants to be applied as natural products in agriculture, cognizance has to be taken of problems associated with its use in plant disease control systems that have been encountered in the past, and that have probably contributed to it not being regarded as a viable strategy in the agrochemical industry. First, claims have been made that the efficacy of plant extract compounds are not comparable to that of synthetic fungicides and lack consistency (Benner, 1993). Second, natural products have, in some cases, been reported to be phytotoxic to crop plants (Benner, 1993). Third, some natural compounds are unstable and can be broken down by UV-light or oxidation before the desired biological effects have been produced (Seddon and Schmidt, 1999). However, these arguments cannot be generally accepted as less than 10% of natural plants have been screened for their application potential and in most cases only towards one target pathogen (Hamburger and Hostettman, 1991). In light of the vast number of plant species known in the world today, the subsequent chemical diversity should allow for the identification of desired biologically active compounds with sufficient stability.

In contrast to the arguments against the use of natural plant compounds, other arguments towards its potential beneficial attributes should be considered. These include the possible reduction in the risk of fungicidal resistance, and the fact that they are potentially less toxic to humans and animals (Ganesan and Krishnaraju, 1995). Most importantly, it is envisaged that crude plant extracts might be more affordable to subsistence farmers as they are readily available and are probably cheaper to produce. Hence, attempts to develop plant-derived natural products and the consideration of its application potential in disease management systems in both developed and developing countries does not seem to be out of line.

In developing countries in particular, the consideration of applying natural plant products in their crude form should be high on the agenda. The approach has long been used in most traditional farming systems in many developing countries. Most African farmers possess substantial indigenous knowledge of insect and pathogen control. Although this knowledge is probably not scientifically based, some examples of natural insecticides applied in Ethiopian subsistence farming systems are worth mentioning. Extracts of chinaberry (*Melia azedarach*), pepper tree (*Schiunus molle*) and endod (*Phytolacca dodecandra*) are used to control insects in both organized agriculture and home gardens (Gebre-Amlak and Azerefegne, 1998). Crude extracts of all three plant species were reported by the authors to be effective against *Busseola fusca* (maize stalk borer) larvae. With regard to disease management, a crude extract of *Dolichos kilimandscharicus* L. (Bosha) has been used as a slurry to treat sorghum seed in the control of covered (*Sporisorium sorghi*; Ehrenberg) and loose (*Sphacelotheca cruenta*, Kuhn) kernel smuts in Ethiopia (Tegegne and Pretorius, 2007). Experimentally, treatment of sorghum seed with *D. kilimandscharicus*, in a powder form, provided excellent control of both pathogens and was as effective as the standard chemical, Thiram®. However, this has been practised on a small and isolated scale. It seems necessary to obtain a more scientific base through additional research in order to consider an expansion of these practices as well as to consider the economic potential of this approach.

3.3 Secondary Metabolites: Characteristics, Functions and Possible Applications in Agriculture

Except for primary metabolites such as carbohydrates, proteins, lipids and nucleic acids, plants also contain a large variety of secondary metabolites. Secondary metabolites are classified by different authors in different ways but for the purpose of this précis, a simple classification into three groups is supplied: (i) isoprenoid; (ii) aromatic; and (iii) alkaloid components (Stumpf and Conn, 1981; Dey and Harborne, 1989; Salisbury and Ross, 1992).

Isoprenoid components

This group of compounds is relatively diverse and three terms have been allocated to it, namely isoprenoids, terpenoids and terpenes. The term 'isoprenoid' is relatively descriptive in the sense that the common factor that relates these compounds is the 5C units, called isoprene units (Fig. 3.1). However, the term 'terpene' is probably used more often.

Under this group, known hormones such as gibberillic acid (GA), abscisic acid (ABA) and brassinosteroids as well as other components such as sterols, carotenoids, rubber and the phytol 'tail' of chlorophyll are classified. All of these consist of repeating isoprene units. Isoprene units join in different ways to form components with chain, ring or combined chain–ring structures (Fig. 3.2).

Terpenes are found abundantly in nature and are, as far as chemical structure is concerned, a diverse group of secondary metabolites. On the basis of the number of isoprene units, terpenes are divided into seven classes.

Hemiterpenes consist of one isoprene unit (C_5); they are not found in this form (freely) in nature, but can be found in the form of alcohols or acids (examples shown in Fig. 3.3).

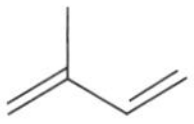

Fig. 3.1. An isoprene unit.

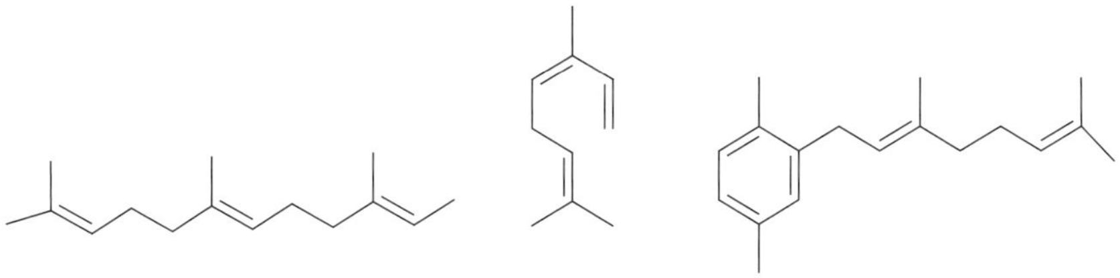

Repeating isoprene-units (chain)   Ring   Ring and chain combined

Fig. 3.2. Different ways in which isoprene units can combine.

Monoterpenes, consisting of two isoprene units (C_{10}), form the main component of essential oils and are therefore economically important due to their aroma (perfume), e.g. menthol, camphor, geraniol and pinene (Fig. 3.4). Their application potential in agriculture lies in the fact that specific monoterpenes are growth inhibitors of higher plants (e.g. menthol and pinene) and therefore have the potential to be developed as natural herbicides.

Sesquiterpenes consist of three isoprene units (C_{15}) (Fig. 3.5). One of the best known sesquiterpenes is absissic acid (ABA), which has a hormonal function. Plants showing natural resistance towards drought stress synthesize more ABA than drought-sensitive crops. Theoretically, the external application of ABA to crops has the potential to manipulate drought resistance in crops. Other sesquiterpenes have medicinal value e.g. as diuretics and as antimicrobials (e.g. kadinene from juniper berry).

Diterpenes consist of four isoprene units (C_{20}) and are the rarest types of terpenes in plants. Of the most important in this group is the hormone gibberellic acid (GA; Fig. 3.6). Although natural hormones are used in the horticultural and agricultural industries on a small scale as growth stimulants, it is expensive and not economically justifiable on a large scale.

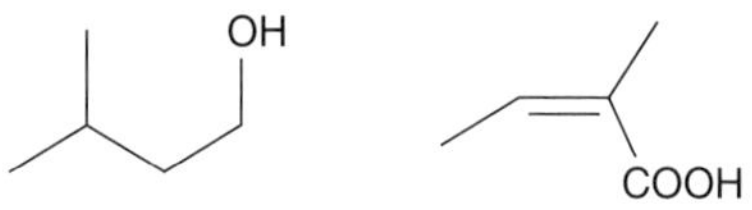

Iso-amyl alcohol   Angelic acid

Fig. 3.3. Hemiterpenes found in the form of alcohols or acids.

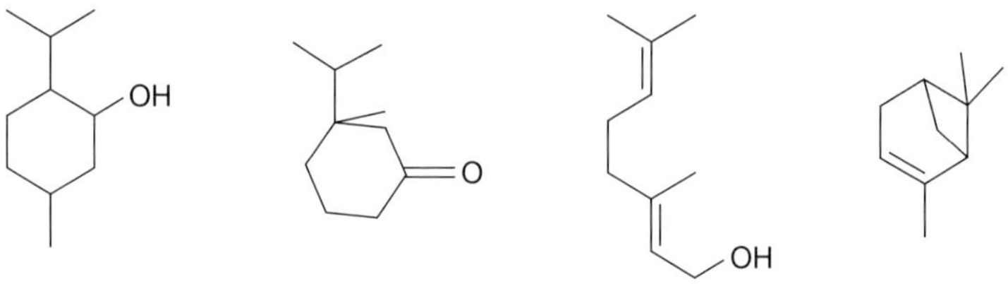

Menthol   Camphor   Geraniol   Pinene

Fig. 3.4. Examples of monoterpenes.

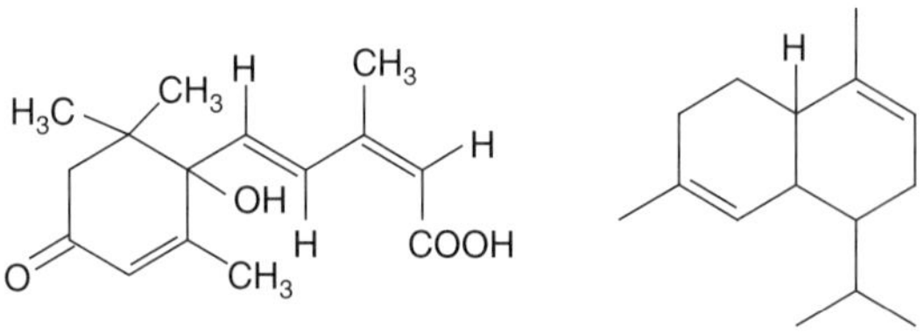

ABA   Kadinene

Fig. 3.5. Examples of sesquiterpenes.

Triterpenes consist of six isoprene units (C_{30}) and are divided into four groups, namely common triterpenes, steroids, saponins and cardiac glycosides. Common triterpenes are commonly found in plants as a wax layer on leaves and on some fruits e.g. limonene (Fig. 3.7).

Most steroids are hydrolysed on carbon 3 and are in fact all sterols. They play important roles either as hormones in plants (e.g. brassinosteroids) or vitamin precursors (e.g. 1-α-25-dihydroxy-vitamin D_3 glycoside; Fig. 3.8).

Brassinosteroids (e.g. brassinolide; Fig. 3.9) have growth-stimulating and yield-increasing properties and are regarded as the secondary metabolites with great application potential in both the agricultural and horticultural industries (see below).

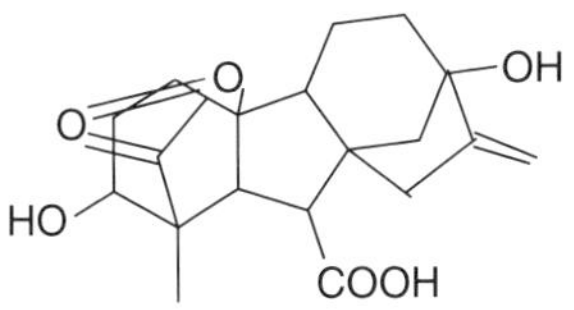

Fig. 3.6. Gibberellic acid.

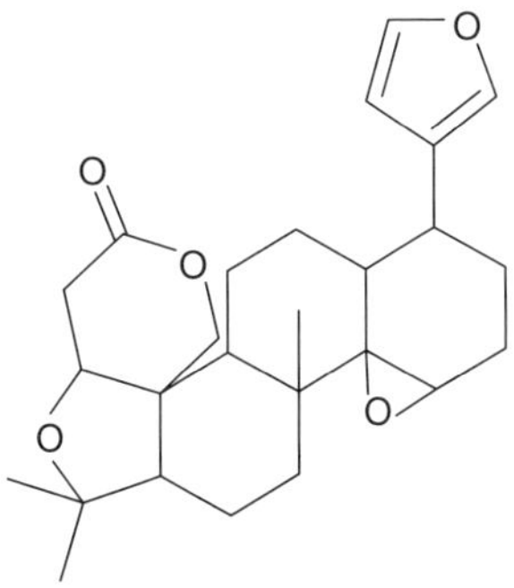

Fig. 3.7. Limonene.

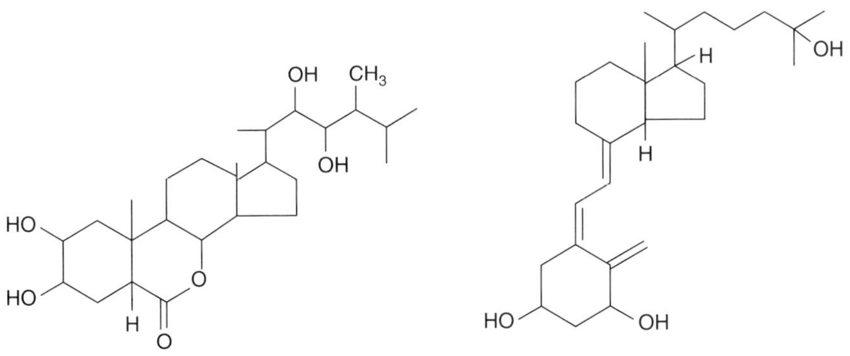

A brassinosteroid (brassinolide)   1-α-25-dihydroxy-vitamin D_3 glycoside

Fig. 3.8. Examples of triterpenes.

Other known sterols are cholesterol that acts as a precursor of other steroids in plants, and ergosterol that is rare in plants but abundant in fungi (Fig. 3.10). Usually, if the ergosterol concentration increases in plants, it is an indication of fungal infection.

Saponins are so-called triterpene-glycosides coupled to a sugar group. The molecule consists of two parts, namely the glycone (sugar) and aglycone (triterpene) moieties (Fig. 3.11). Saponins have soap properties, foam in water and have a bitter taste. They are also known for their medicinal properties in Asiatic communities.

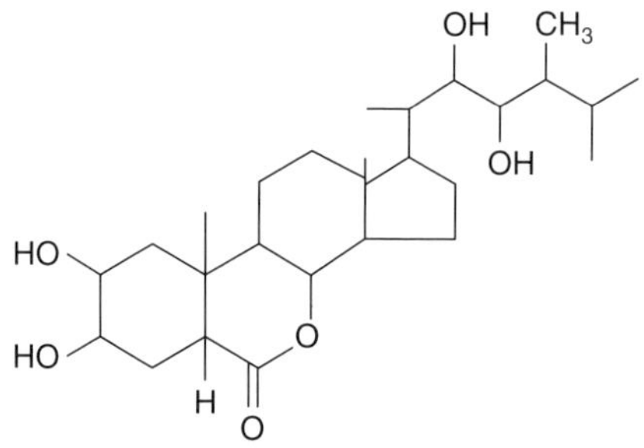

Fig. 3.9.   A typical brassinosteroid (brassinolide).

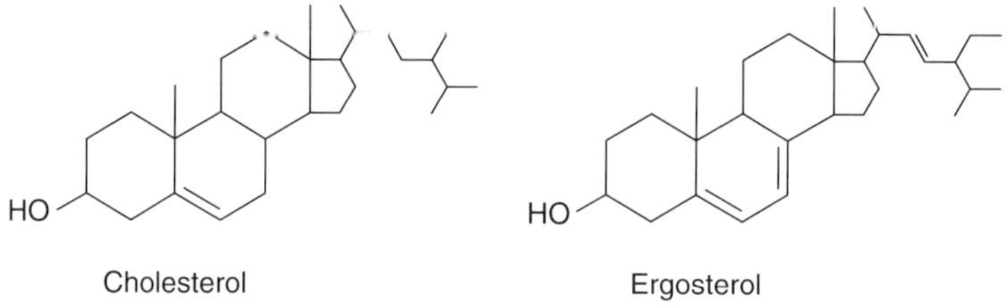

Cholesterol   Ergosterol

Fig. 3.10.   Two known sterols, cholesterol and ergosterol.

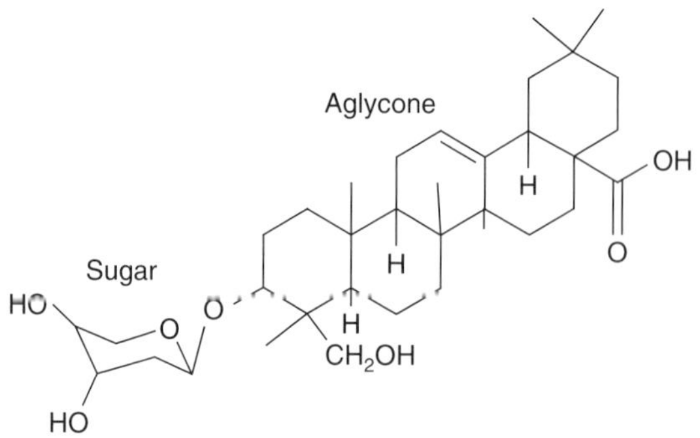

Fig. 3.11.   A typical saponin.

Tetraterpenes consist of eight isoprene units (C_{40}). The best known in this group are the carotenoids, e.g. β-carotene (Fig. 3.12). It is known as a colour pigment supplying colour to flowers, but also for its role in protecting chlorophyll against over exposure to light.

Aromatic components

All compounds in the aromatic group of secondary metabolites contain at least one aromatic ring (benzene ring; Fig. 3.13) in their structure, of which one or more hydroxyl groups are substituted. Thousands of aromatic compounds have already been identified in plants. The group can be divided into two sections, namely non-phenolic and phenolic aromatic compounds.

Non-phenolic aromatic compounds

Although most aromatic compounds contain one or more OH groups in the benzene ring, the OH groups are absent in non-phenolic aromatic compounds. For the sake of convenience, this group is further divided into two groups, namely non-phenolic amino acids and hormones, as well as tetrapyrroles. In the former group, the two amino acids phenyl alanine and tryptophan, also precursors of many phenolic compounds, and the hormone auxin (indole acetic acid or IAA) are best known (Fig. 3.14). As is the case for GA, auxin is also used as a growth stimulant in the horticultural industry, but on a small scale.

In the tetrapyrrole group of non-phenolic aromatic compounds, chlorophyll (Fig. 3.15) is best known and consists of four pyrrole rings forming a porphyrine 'head' attached to a phytol 'tail'. Another example is phytochrome that is a light-sensitive component, allowing plants to distinguish between different day lengths.

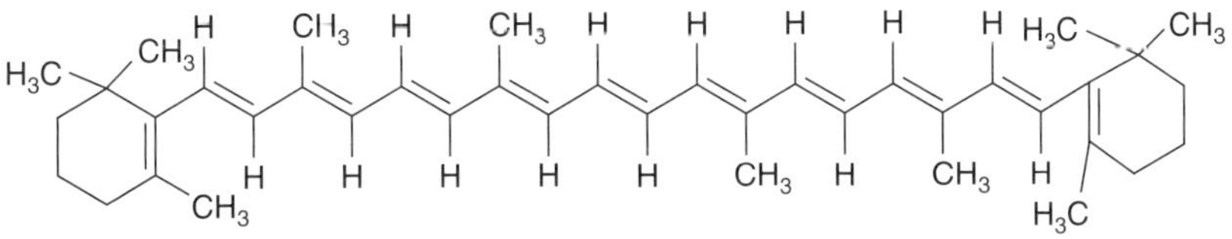

Fig. 3.12.   β-carotene.

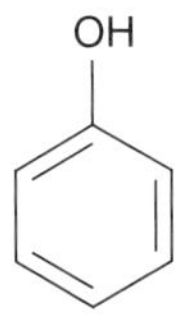

Fig. 3.13.   A typical aromatic ring (benzene ring).

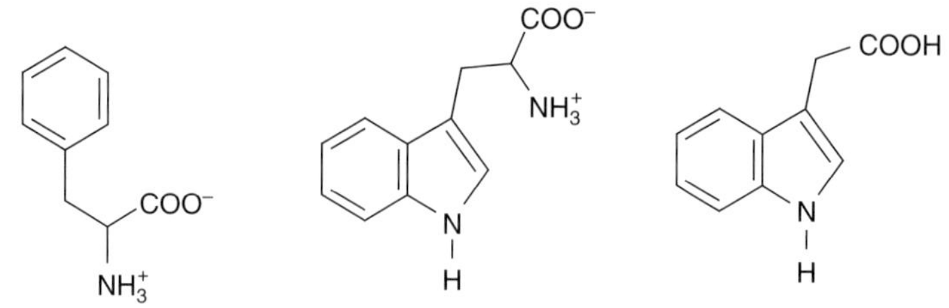

Phenyl alanine

Tryptophan

Auxin (IAA)
(indole acetic acid)

Fig. 3.14. Non-phenolic aromatic compounds.

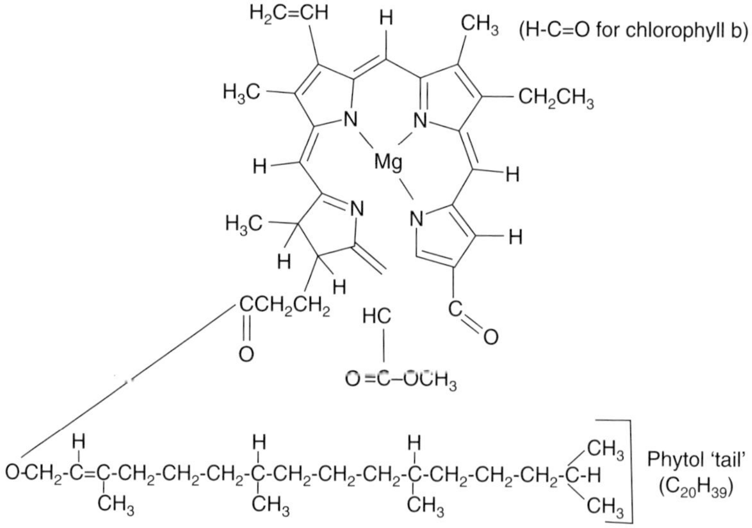

Chlorophyll a and b

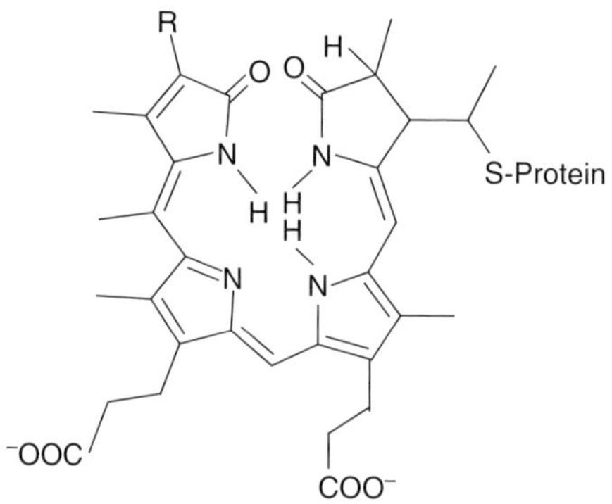

Phytochrome (P_f-form)

Fig. 3.15. Non-phenolic chlorophyll and phytochrome.

Phenolic aromatic compounds

Phenols are the most abundant aromatic and naturally occurring components of plants. The phenols are classified in different ways, depending on authors. We will only briefly discuss the simple phenols, phenyl propanoids, flavonoids, tannins and kinones.

Simple phenols are all monomeric (consisting of only one aromatic ring; Fig. 3.16). However, the most general simple phenols with growth-inhibiting allelopathic properties that have the potential to be developed as natural herbicides are vanillin, vanillic acid (a benzoic acid) and hydrokinone (Fig. 3.17). p-Hydroxy-benzoic acid and vanillic acid are the most common growth-inhibiting benzoic acids involved with allelopathy and are found in maize, wheat, sorghum and barley.

Phenol is probably the precursor of all other phenolic compounds found in plants. Hydrokinone, resorsinol and catechol are found in low concentrations in plants and their functions are not fully understood. They are mostly secreted by insects as a defence mechanism against other insects and animals. Salicylic acid, on the other hand, possesses anaesthetic properties and is the active ingredient of Aspirin®. In plants a number of functions have been identified for salicylic acid, namely induction of flowering as well as the induction of 'pathogenesis related' (PR) proteins (peroxidase, chitinase and β-1,3-glucanase) that increases a plant's resistance to fungal infections. When secreted by plants, it also has an allelopathic growth inhibiting effect on other plants in the environment.

Phenyl propanoids are synthesized from the aromatic amino acid phenyl alanine and contained in their structure is a 3-carbon side chain coupled to a phenol, e.g. coumaric acid and caffeic acid (Fig. 3.18).

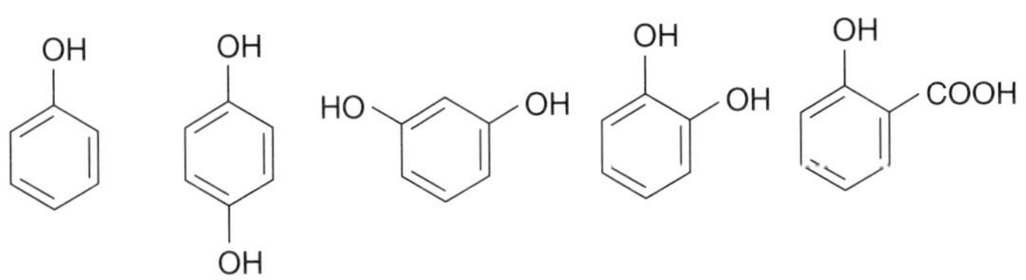

Phenol   Hydrokinone   Resorsinol   Catechol   Salicylic acid

Fig. 3.16. Simple phenols.

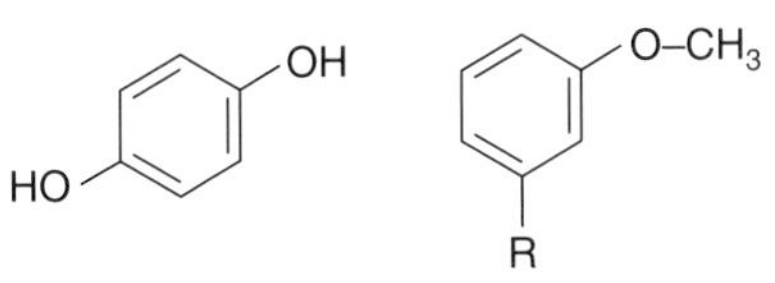

Hydrokinone   R = -COH=vanillin
                     R = -COOH=vanillic acid

Fig. 3.17. Simple phenols with great potential to be developed as natural products.

p-Coumaric acid is the direct precursor of other coumarins such as umbelliferone and scopoletin but coumarin itself is synthesized from cinnamic acid (Fig. 3.19).

The potential of phenyl propanoids to be developed into natural products for the agricultural industry is probably underestimated at present. For example, coumarin, isolated from *Mellilotus alba* (White Clover; a legume), is a strong inhibitor of seed germination and has the potential to be developed as a pre-emergence herbicide. However, depending on the concentration, it is also known to stimulate IAA activity and therefore growth. Its potential as bio-stimulant has not been researched to date. Coumarin can also combine with monosaccharide sugars such as glucose to form coumarin glycosides, e.g. umbelliferone glycoside (Fig. 3.20).

These glycosides inhibit the activity of the enzyme 6-phosphogluconate dehydrogenase, the regulatory enzyme of the oxidative pentose phosphate (OPP)-pathway, and therefore possess excellent potential to be developed as natural herbicides. Umbelliferone also has an inhibiting effect on the growth

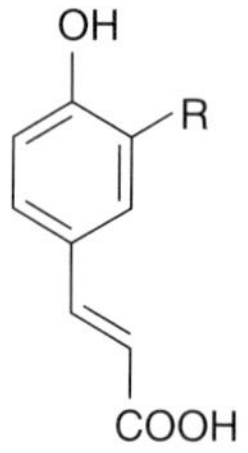

p-Coumaric acid (R=H)
Caffeic acid (R=OH)

Fig. 3.18. Two well-known phenyl propanoids.

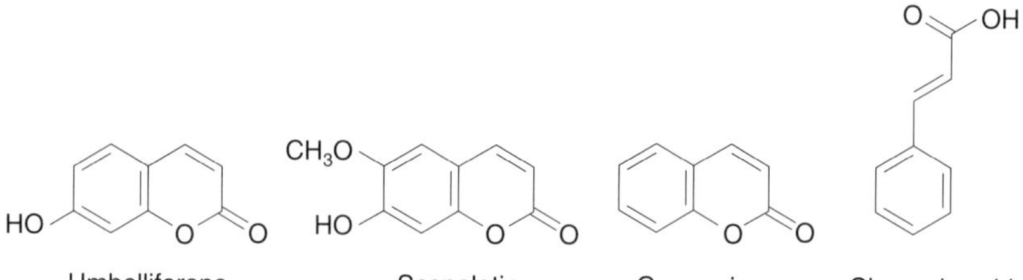

Umbelliferone   Scopoletin   Coumarin   Cinnamic acid

Fig. 3.19. Phenyl propanoids synthesized from the precursor, p-coumaric acid.

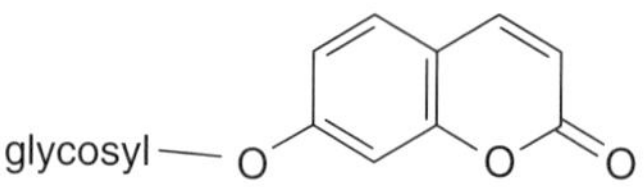

Fig. 3.20. Umbelliferone glycoside.

of certain seedlings and shows potential as an herbicide. Scopoletin has a stimulating effect on seed germination but an inhibiting effect on seedling growth, indicating a potential as pre-emergence herbicide.

Flavonoids are the largest group of natural phenolic compounds found in plants. More than 5000 different flavonoids have been described. All flavonoids have one thing in common, namely a 15-carbon 'skeleton' structured in three phenyl rings (A, B and C; Fig. 3.21). Other examples are shown in Fig. 3.22.

A large variety of biological activities have been associated with flavonoids, including antiviral, anti-inflammatory, antioxidant, antidiabetic, anticancer, insect-repelling and free-radical-scavenging activities. What makes this group extraordinary from an industrial perspective are the excessive quantities found in both edible and non-edible plants as well as their potential to be developed into natural products. It is estimated that about 2% of all carbon photosynthesized by plants is converted to flavonoids and this amounts to approximately 1×10^9 tons per annum. With this enormous amount of flavonoids available, and in light of the numerous bioactivities identified to date, the time is now ripe to consolidate our knowledge of the

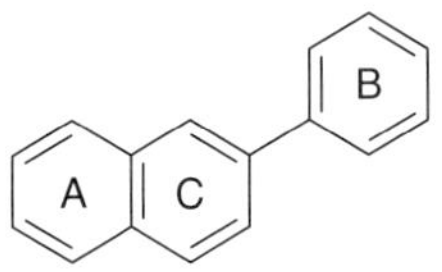

Fig. 3.21. Typical flavonoid structure.

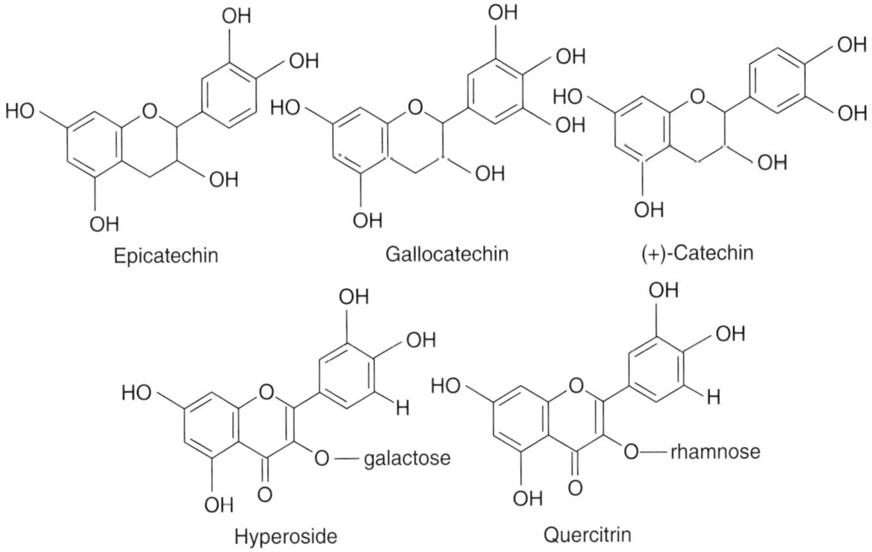

Fig. 3.22. Flavonoid examples.

group and to evaluate the application potential of these compounds in the agricultural industry (Pretorius, 2003 and references therein).

Tannins are generally found in the xylem and bark of vascular plants. Two types of tannins have been distinguished namely hydrolysable and condensed tannins. When hydrolysable tannins are broken down (hydrolysed), the well known gallic and digallic acids are formed (Fig. 3.23). Tannins combine with proteins to form insoluble co-polymers and this technique is used to tan animal skins. However, tannins are toxic to humans and animals and not suitable for natural product development (Borris, 1996).

Kinones are divided in three groups, benzokinones, naphtakinones and antrakinones (Fig. 3.24). Most kinones are coloured but do not contribute to colour in plants as they are found in the inside tissue layers. The best known example of benzokinone is ubikinone (Fig. 3.25), also known as Co-enzyme Q, which acts as an electron carrier during the light reaction of photosynthesis. Juglone, the well known naphtakinone, has a strong

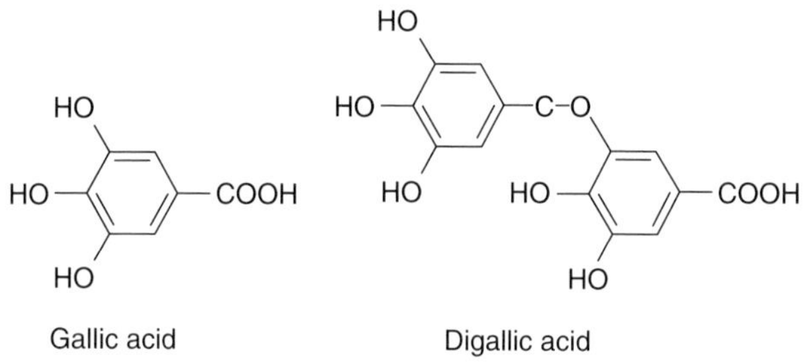

Gallic acid   Digallic acid

Fig. 3.23. Two hydrolysable tannins.

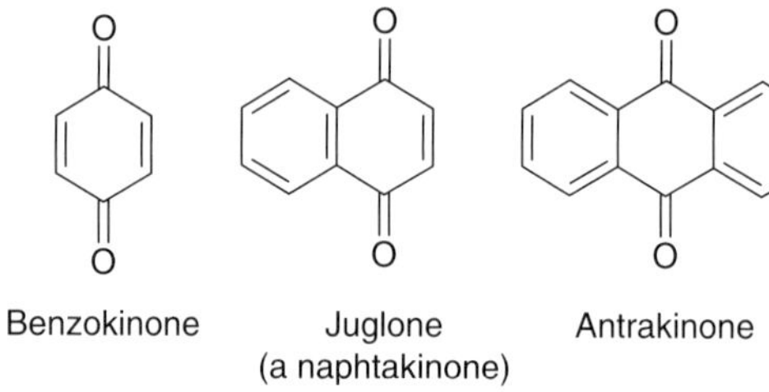

Benzokinone   Juglone   Antrakinone
                  (a naphtakinone)

Fig. 3.24. Three groups of kinones.

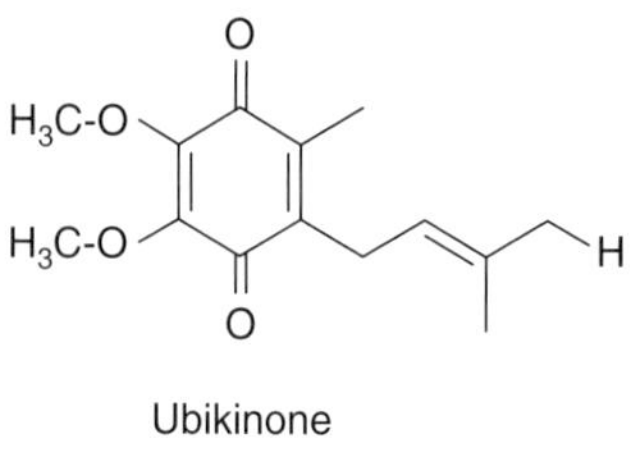

Ubikinone

Fig. 3.25. Ubikinone, the best known example of a benzokinone.

inhibiting effect on seed germination and seedling growth. Antrakinone is known as an insect repellent.

Alkaloids

More than 3000 alkaloids have been isolated from plants. The best known alkaloids are morphine (isolated from the poppy flower), nicotine (tobacco) and caffeine (coffee; Fig. 3.26).

Alkaloids are best known for their medicinal properties but also for their toxicity. Their functions in plants are not well known but are possibly linked to the natural resistance of plants against insects and pathogens. Alkaloids are also known to inhibit seed germination and seedling growth of certain plants. Because of their toxicity, chances are slim that alkaloids will be considered in natural product development programmes.

3.4 Antimicrobial Properties of Plant Extracts

Plants are under constant attack from various microorganisms and fungi. Their survival is testamentary to their ability to defend themselves chemically against these attacks by producing a myriad of secondary metabolites with antimicrobial properties (Cowan, 1999; Richard, 2001). Therefore, in addition to other functions, one of the most important functions of secondary metabolites in plants is antimicrobial activity against bacteria, fungi and viruses as well as acting as deterrents towards insects and predators (Lazarides, 1998; Minorsky, 2001). Failure of microorganisms to colonize wild plants has often been attributed to the presence of these inhibitory compounds within challenged tissues. However, monoculture crops have lost their ability to defend themselves against biotic stressors to a large extent.

Callow (1983) classified antimicrobial compounds (secondary metabolites) isolated from plants into two categories: (i) constitutive compounds, which are present in healthy plants; and (ii) induced compounds synthesized from remote precursors following infection. The term 'constitutive' includes compounds that are released from inactive precursors following tissue

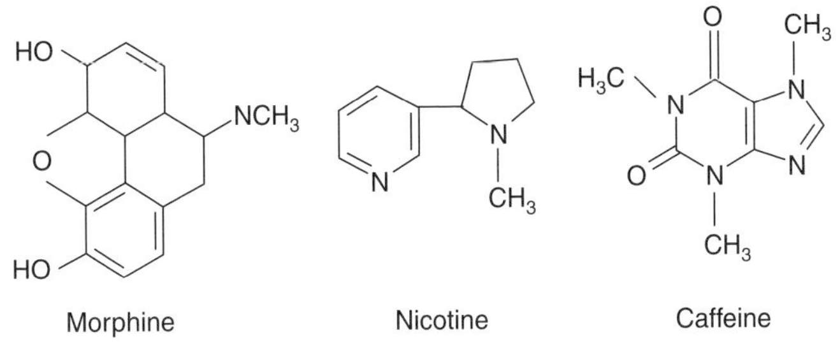

| Morphine | Nicotine | Caffeine |

Fig. 3.26. Examples of alkaloids.

damage, for example the release of toxic hydrogen cyanide from cyanogenic glycosides. Induced compounds include the types of secondary metabolites that are low molecular weight antimicrobial compounds and are synthesized by and accumulate in plants that have been exposed to microorganisms (e.g. phytoalexins).

Since plants have evolved highly elaborate chemical defences against attack, these have provided a rich source of biologically active compounds that may be used as novel crop-protecting agents. It follows, therefore, that secondary metabolites with antimicrobial activity purified and isolated from plant extracts, possess the potential to be developed into natural products. As a result, many organizations and institutions in different countries of the world are currently concentrating on natural product research. For example, the aim of the Agrochemical Discovery and Development Program of the National Centre for Natural Products Research in the USA is to identify lead compounds for the development of environmentally benign and toxicologically safe pest management agents (Borris, 1996). This programme is done in collaboration with scientists in the Natural Products Utilization Research Unit of the United States Development Agency (USDA) agricultural research service. Emphasis is on the discovery and development of compounds that are useful in the control of diseases affecting small niche crops. The research centre is devoted to improving agricultural productivity through the discovery, development and commercialization of agrochemicals derived from natural plant compounds (Johnson, 2001).

Moreover, an increasing number of scientists have become involved in intensive plant screening programmes for bioactivity and have contributed to the identification of plant species with the potential to be included in programmes for the development of natural products for the agricultural industry. It is therefore not surprising that a significant amount of evidence over the past three decades has demonstrated that plant extracts are active against plant pathogenic fungi (Lawson *et al.*, 1998), soil-borne fungi (Awuah, 1994; Bianchi *et al.*, 1997), bacteria (Leksomboon *et al.*, 2001) and nematodes (Oka *et al.*, 2000). In recent years research interest has turned towards isolating, purifying and identifying these active compounds that have application potential in plant disease control.

According to Grayer and Harborne (1994), secondary metabolites with antimicrobial properties include terpenoids (e.g. iridoids, sesquiterpenoids and saponins), nitrogen- and/or sulphur-containing compounds (e.g. alkaloids, amines and amides), aliphatics (especially long-chain alkanes and fatty acids) and aromatics (e.g. phenolics, flavonoids, bi-benzyls, xanthones and benzo-quinones). Most of these compounds have been studied for their antifungal and antibacterial activities and potential usefulness against plant pathogens *in vitro* (Verporte, 1998; Baldwin, 1999; Paul *et al.*, 2000) and some also *in vivo*. Many aromatic compounds, including simple and alkylated phenols, phenolic acids, phenylpropanoids, coumarins, flavonoids, isoflavonoids, quinones and xanthones, have been reported to show notable antifungal activities (Grayer and Harborne, 1994). For example, Bae *et al.* (1997) isolated flavonol diglycoside from leaves of *Phytolacca americana* L.

which exhibited significant antifungal activity against *Botrytis cinerea, Botry-osphaeria dothidea* and *Colletotrichum gloeosporioides* (*Glomerella cingulata*).

Flavonoid classes most often associated with antifungal activity are fla-vanones, flavonols, certain biflavones, chalcones and dihydrochalcones. For example, three glycosides isolated from *Terminalia alata* (*T. elliptica*) roots showed antifungal activity against the plant pathogen *Aspergillus niger* and the human pathogen *Candida albicans* at extremely low concentrations of between 25 and 32 ppm (Srivastava *et al.*, 2001). The three new glycosides identified were: compound 1, 3,3'-di-*O*-methylellagic acid 4-*O*-β-D-glucopyranosyl-(1,4)-β-D-glucopyranosyl-(1,2)-α-L-arabinopyranoside; compound 2, 5,7,2'-tri-*O*-methyl-flavanone 4'-*O*-α-L-rhamnopyranosyl-(1,4)-β-D-glucopyranoside and compound 3, 2-α,3-β,19-β,23-tetrahydroxyolean-12-en-28-oic acid 3-*O*-β-D-g-alacto-pyranosyl-(1,3)-β-D-glucopyrano-side-28-*O*-β-D-glucopyrano-side (Fig. 3.27). Compound 1 was a glycoside of an ellagic acid, whereas compounds 2 and 3 were a flavanone glycoside and a triterpene saponin, respectively.

Saponins are one of several groups of compounds that originate from tri-terpenoids and show a wide range of biological activities. For instance, fungi-cidal triterpenoid saponins, which also have molluscidal activity, have been isolated from roots of *Dolichos kilimandscharicus* (Marston *et al.*, 1988). The saponins from *Dolichos* were identified as the 3-*O*-β-D-glucopyranosides of hederagenin, bayogenin and medicagenic acid (Marston *et al.*, 1988). Saponins from *Mimusops elengi* and *M. littoralis* seeds were also reported to be active against *Phytophthora palmivora* and *Colletotrichum capsici* (Johri *et al.*, 1994).

The anti-infective potency of extracts from plants emphasizes the vast potential for natural compounds to be developed into commercial products (Duke, 1990). Although many of these plant compounds have been applied

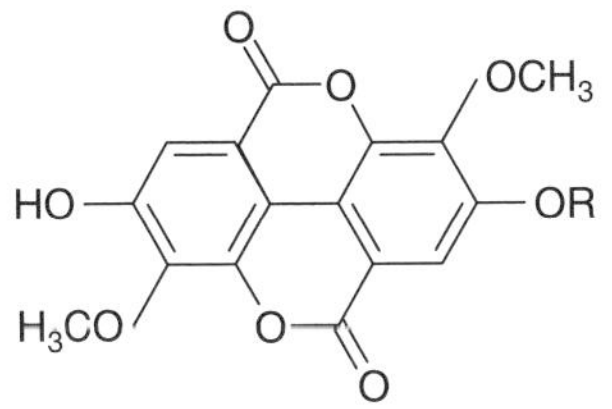

Compound 1 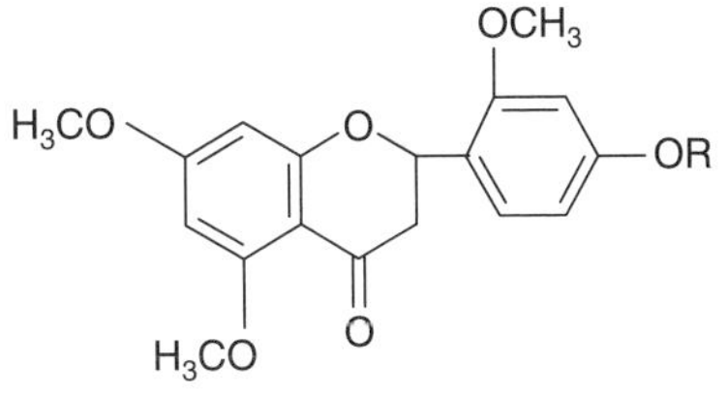   Compound 2

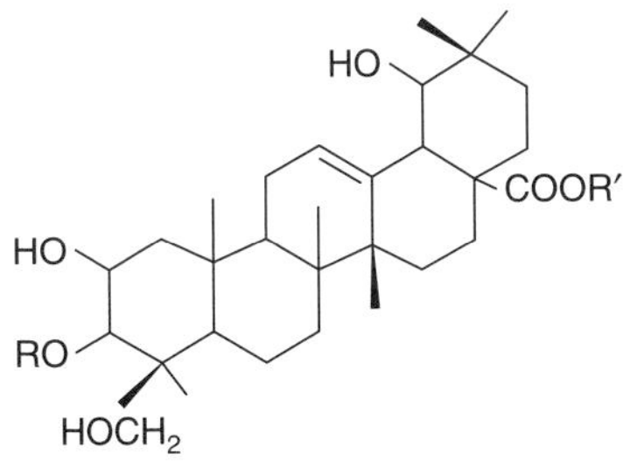

Compound 3

Fig. 3.27. Three new glycosides with antifungal properties isolated and identified from *Terminalia alata* (redrawn after Srivastava *et al.*, 2001). See text for names of compounds.

in the pharmaceutical industry (Naseby *et al.*, 2001) and folk medicine (Duncan *et al.*, 1999), their potential for plant disease management has not yet been fully realized. Crop production still depends heavily on synthetic-based products (Philip *et al.*, 1995) despite consumer resistance towards their possible residual effects. For this reason the switch to organic farming, including the use of natural products in disease management systems, has become a priority (Benner, 1993; Michael, 1999). The next section provides an overview of selected plants containing compounds with antimicrobial properties, identified during the past two decades, as well as the current status of natural products that have recently been commercialized.

Plants with antifungal properties

Two decades ago Naidu (1988) assayed young and mature leaf extracts of *Codiaeam variegatum* for antifungal activity. All extracts inhibited *Alternaria alternata* and *Fusarium oxysporum in vitro*, with the young leaves being more active against *A. alternata* and the old leaves more active against *F. oxysporum*. The active secondary metabolites from the leaves extracts were identified as phenolic compounds by chromatographic analysis. Phytochemical screening of the leaves also revealed other metabolites that may be responsible for antifungal activity.

In another study in the same year Bandara *et al.* (1988) reported that steam distillates of the leaves of *Croton aromaticus* and *C. lacciferus* and root extracts of *C. officinalis* inhibited mycelial growth of *Cladosporium cladosporioides in vitro*. Root extracts of *C. lacciferus* were moderately active while those of *C. aromaticus* were inactive. Of the six compounds isolated from root extracts showing antifungal activity, only 2,6-dimethoxybenzoquinone obtained from the chloroform extract of *C. lacciferus* was significantly active. Comparatively small quantities of this compound were required to inhibit growth of the pathogens *Botryodiplodia theobromae* and *Colletotrichum gloeosporioides* (*Glomerella cingulata*).

Also in the same year Gonzalez *et al.* (1988) extracted *Alnus acuminata* nodules with either 5% NaOH or water while constituents were separated by silica gel column chromatography. The following compounds were isolated: xylose, ribose, an aromatic carboxylic acid, a fatty acid, a phenolic biarylheptanoid and a flavonoid glycoside. The flavonoid glycoside was found to inhibit the growth of *Fusarium oxysporum* and *Pythium* species.

In the 1990s a number of contributions by natural product researchers were published. Only a few are mentioned here. Ajoene, a secondary metabolite derived from garlic (*Allium sativum*), was shown to inhibit spore germination of some fungi including *Alternaria solani* and other *Alternaria* spp., *Collectotrichum* spp., *Fusarium oxysporum* and other *Fusarium* spp. that cause serious diseases in some important crop plants in India (Singh *et al.*, 1990). In a study conducted by Hoffmann *et al.* (1992) a methanol extract of *Castela emoryi* was active as both a preventative and curative agent against grape downy mildew caused by *Plasmopara viticola*. An active secondary metabolite was identified as a glycoside, 15-glucopyranosyl-glaucarubolone. Saponins from *Mimusops elengi* and *M. littoralis* seeds and crude extract of *Ammi majus* were 86–100% effective against *Phytophthora palmivora in vitro* and the saponins were 100% effective against *Colletotrichum capsici* (Johri *et al.*, 1994). Encouraging results were obtained in 2 years of field trials using these products for the control of pathogens on *Piper betle*. No phytotoxicity was observed. Bae *et al.* (1997) isolated an antifungal secondary metabolite, flavonol diglycoside, from the leaves of *Phytolacca Americana L.* and identified the compound as kaemferol-3-O-β-D-apiofwanosyl-(1,2)-β-D-lucopyranoside by spectral analyses. The compound exhibited significant antifungal activity against *Botrytis cinerea*, *Botryosphaeria dothidea* and *Colletotrichum gloeosporioides* (*Glomerella cingulata*).

Earlier reports on the potential of compounds from plants to be considered in natural product research most probably played a role in the elevation of large-scale screening programmes that followed during the past decade. For example, Pretorius *et al.* (2002a) performed a wide search for South African plant species with fungitoxic properties against plant pathogens of economic importance in agriculture. For this study, 39 plant species, representing 20 families from the subclasses *Rosidae, Asteridae, Commelinidae* and *Liliidae*, were collected from the Blyde River Canyon Nature Reserve, Mpumalanga, South Africa. Crude extracts were prepared and bio-assayed, at equal concentrations, for their antifungal potential by determining the inhibitory effects on the mycelial growth of seven economically important plant pathogenic fungi. Statistically, significant differences between plants and plant parts were observed as well as the resistance of different fungi to treatment with different plant extracts. The most significant broad spectrum mycelial growth inhibition was obtained with extracts from two species of the subclass *Liliidae*, namely *Aristea ecklonii* and *Agapanthus inapertus*. The crude extract of *A. ecklonii* performed best of all extracts as it totally inhibited the mycelial growth of all seven of the plant pathogenic test organisms and outperformed the inhibition by a broad-spectrum synthetic fungicide (carbendazim/difenoconazole). Crude extracts of *A. inapertus* showed complete inhibition of four, and strong inhibition of the remaining three, plant pathogenic fungi.

In the same year Pretorius *et al.* (2002b) also performed an *in vivo* study on the control of black spot (Ascochyta blight) in pea leaves, caused by *Mycosphaerella pinodes*, by a crude bulb extract of *Eucomis autumnalis*. The fourth internode leaves were removed from 4-week-old pea (cv. Mohanderfer) plants, placed on moist filter paper in Petri dishes and inoculated with an *M. pinodes* spore suspension before and after treatment with the extract. The crude extract prevented *M. pinodes* spore infection of the leaves when the leaves were inoculated with spores both before and after treatment with the extract, confirming complete inhibition of spore germination. The crude *E. autumnalis* extract showed no phytotoxic effect on the leaves even at the highest concentration applied.

Equally promising was the results of a comprehensive study conducted by Chen *et al.* (2002) to determine the inhibitory effect of 58 plant extracts on spore germination and the effective control of grape downy mildew (*Plasmopara viticola*). Among the plant extracts, those of *Chloris virgata, Dalbergia hupeana, Pinus massoniana, Paeonia suffruticosa* and *Robinia pseudoacacia* inhibited spore germination of the pathogen significantly. An *in vivo* leaf disc test showed that the infected leaf discs, treated with these five plant extracts, exhibited no disease symptoms. Their effects were the same or better than that of the traditional fungicide, liquid Bordeaux.

Pandey *et al.* (2002), similarly, compared the antifungal potential of leaf extracts from 49 angiosperms, collected in Uttar Pradesh, India, with commercial fungicides, by screening them against *Helminthosoporium sativum* (*Cochliobolus sativus*). The leaf extract of *Mangifera indica* completely inhibited the mycelial growth of the test fungus while four plant species, *A. sativum, Azadirachta indica, Lawsonia inermis* and *Matricaria chamomila* (*Chamomilla recutita*) showed more than 90% inhibition. On assaying different parts of *Mangifera indica*, the leaf and seed extracts were found to possess the highest activity and, together with a leaf extract of *Matricaria chamomile*, performed better than the commercial fungicides.

An approach to screen for plants with antimicrobial activity against plant pathogens that often results in success is to exploit indigenous knowledge on medicinal plants and to screen these known plants for likely candidates. The rationale behind this approach is to screen traditional medicinal plants known in a specific area for their antimicrobial properties instead of randomly choosing potential candidates from the long list of currently known flowering plants, conifers, ferns or bryophytes. Rajiv *et al.* (2002) conducted a study to screen for the most effective extracts out of 15 medicinal plants against *Helminthosporium nodulosum* (*Cochliobolus nodulosus*) causing blight in finger millet. These included *Impatiens balsamina, Solanum nigrum, Tagetes erecta, A. sativum, A. indica, Datura metel, Emblica officinalis* (*Phyllanthus emblica*), *Eucalyptus citriodora, Euphorbia pulcherrima, Lantana camara, Mentha arvensis, Mimosa pudica, Nerium indicum* (*N. oleander*), *Ocimum sanctum* (*O. tenuiflorum*) and *Ricinus communis*. Extracts were sprayed on the potted finger millet plants at 15, 30, 45, 60 and 75 days after sowing. Crude extracts of *S. nigrum* and *I. balsamina* showed the highest mycelial growth inhibition, followed by *T. erecta*. Overall, the crude extract of *S. nigrum* recorded the best result *in vitro* but was found inferior to the *I. balsamina* extract in *in vivo* tests (Rajiv *et al.*, 2002).

Besides large-scale screening programmes, data from quite a number of smaller projects that included the screening of one or more plants were published during the past decade. In most cases these smaller projects included the isolation and identification of active compounds involved, even if only the chemical group level. Although large-scale screening programmes are important to at least identify plant orders, families, genera or species with promising potential, the more concerted approach is probably preferable. Only a few approaches are mentioned here.

In their study of antimicrobial properties of plant extracts, Orlikowski (2001a) used grapefruit extract (GE; Biosept 33 SL) to control *Phytophthora* spp. Amendment of peat with GE at a concentration of 165 µg cm^{-3} resulted in a drastic decrease in colony-forming units of the pathogen *Phytophthora cryptogea* and suppression of its development in potted gerbera (*Gerbera jamesonii*) and cypress (*Chamaecyparis lawsoniana*). About 40 µg GE ml^{-1} inhibited approximately 50% of the mycelial growth, whereas the pathogen did not develop at all in the presence of GE at a concentration of 1000 µg ml^{-1}. The antifungal property of GE against *P. cryptogea* was attributed to the presence of the active ingredient 7-geranoxycoumarin in GE (Orlikowski *et al.*, 2001b).

Kishore *et al.* (2002) reported on the antimicrobial activity of aqueous leaf extracts from *Lawsonia inermis* and *Datura metel* against *Mycosphaerella berkeleyi* causing late leaf spot in groundnuts (*Arachis hypogaea*). Field experiments were conducted at the International Crops Research Institute for the Semi-Arid Tropics (ICRISAT), Patancheru, India during 1999 and 2000 using a susceptible groundnut cultivar (TMV2). The *D. metel* extract continuously reduced disease progress up to 115 days after sowing while the severity of late leaf spot at harvest was significantly less than that of the controls. The *L. inermis* extract was slightly less effective, containing disease progress up to 95 days after sowing, but disease severity was also considerably less than that of the untreated and the positive chlorothalonil treated controls. Pod yields in plots sprayed with *L. inermis* and *D. metel* extracts were 20 and 48% higher, respectively, than in the control plots.

In the same year the aqueous extracts from bird cherry tree (*Padus avium* [*Prunus padus*]), aspen (*Populus tremula*), and celandine (*Chelidonium majus*) effectively suppressed the germination of *Puccinia triticina* (*Puccinia recondite*) uredospores (Karavaev *et al.*, 2002). Fungitoxic activity of the extracts was attributed to the high phenolic compound content and high peroxidase activity in the leaves of these plants.

Curir *et al.* (2003) investigated the phenol compositions of two cultivars of carnation (*Dianthus caryophyllus*) namely 'Gloriana' and 'Roland', which were partially and highly resistant, respectively, to *Fusarium oxysporum* subsp. *dianthi*. The aim was to determine if endogenous phenols could have an antifungal effect against the pathogen. Analyses were performed on healthy and *F. oxysporum*-inoculated tissues *in vitro* as well as on plants *in vivo*. Two benzoic acid derivatives, protocatechuic acid (3,4-dihydroxybenzoic acid) and vanillic acid (4-hydroxy-3-methoxybenzoic acid), were found within healthy and inoculated tissues of both cultivars, together with the flavonol glycoside peltatoside (3-[6-*O*-(α-L-arabinopyranosyl)-β-D-glucopyranosyl] quercetin). These molecules proved to be only slightly inhibitory towards the pathogen. 2,6-Dimethoxybenzoic acid was detected in small amounts only in the inoculated cultivar 'Gloriana', while the highly resistant cultivar 'Roland' showed the presence of the flavone datiscetin (3,5,7,2'-tetrahydroxyflavone). The latter compound exhibited an appreciable fungitoxic activity against *F. oxysporum* subsp. *dianthi*.

Natural plant extracts have in recent years also received attention as a significant and safe resource for the control of soil-borne pathogens due to the phasing out of methyl bromide, which has played a major role as a soil fumigant worldwide (Eshel *et al.*, 2000). Methyl bromide, the major fumigant used, is scheduled to be phased out, because it was defined by the Montreal Protocol of 1991 as a chemical that contributes to the depletion of the ozone layer (Bowers and Locke, 2000). Presently, pre-plant soil fumigation and fungicide applications are used to control wilt diseases. Subsequently, extracts from a number of plants have been identified as possible alternatives to methyl bromide against soil-borne pathogens.

Due to environmental and safety concerns associated with pesticides, as well as the need for a replacement of methyl bromide, Bowers and Locke (2000) investigated the effect of several formulated plant extracts and essential oils on soil populations of *F. oxysporum*. *Fusarium* wilts are some of the most widespread and destructive diseases of many major ornamental and horticultural crops (Bowers and Locke, 2000). Treatment of the soil with 10% aqueous emulsions of the formulated extracts of a chilli pepper extract and essential oil of mustard mixture, a cassia tree extract and clove oil reduced populations of *Fusarium* by 99.9, 96.1 and 97.5%, respectively, 3 days after soil treatment. The same formulations also suppressed disease development in the greenhouse and resulted in an 80–100% plant stand after 6 weeks. The observed reductions in the pathogen population in soil and increase in plant stand in the greenhouse indicated that these natural plant products may play important roles in future biologically based management strategies for the control of *Fusarium* wilt diseases.

Similar results were observed with neem tree (*Azadirachta indica*) extracts against soil-borne fungi (Ume *et al.*, 2001). It was shown that extracts from the leaves and seed kernels possess antifungal activity against *Sclerotium rolfsii* (*Corticium rolfsii*). Both leaf and seed extracts showed some effect against different growth stages of the fungus, but the effects were fungistatic rather than fungitoxic. The non-polar extracts of the seed kernels were reported to be more effective than those rich in polar terpenoids such as azadirachtin and an aqueous leaf extract was also more effective than the kernel-derived material. Neem is perhaps the most useful traditional medicinal plant in India and each part of the tree has some medicinal property that has made it commercially exploitable (Kausik *et al.*, 2002).

Amadioha (2002) evaluated the antifungal activities of different extracts of neem, both *in vitro* and *in vivo*. The oil extract from seeds as well as water and ethanol leaf extracts of the plant were effective in reducing the radial mycelial growth of *Cochliobolus miyabeanus* in culture and in controlling the spread of brown spot disease in rice. However, the oil extract was found to be the most effective, followed by the ethanol leaf extract, in inhibiting the growth of the pathogen *in vitro* and in controlling the development of the disease *in vivo*. The oil and ethanol extracts compared favourably with carbendazim (Bavistin) at 0.1% active ingredient and had the potential to control the brown spot disease of rice *in vivo*. Additionally, Bohra and Purohit (2002) studied the effects of the aqueous extracts of 17 plant species collected from Rajasthan, India on a toxigenic strain of *Aspergillus flavus*. The neem extract recorded the highest mycelial growth inhibition of the fungus.

The antimicrobial potential of plant extracts has also been reported against the highly resistant fungi that cause soil-borne damping-off disease in plants. The efficacy of crude, boiled water and acetone extracts of 17 plant species against *Rhizoctonia solani,* causing wet root rot disease in chickpeas, was evaluated under laboratory conditions (Kane *et al.,* 2002). The crude, boiled water and acetone extracts of *A. sativum,* the crude and boiled water extract of *Eucalyptus* sp., as well as the boiled water and acetone extracts of *Zingiber officinale* contributed to 100% inhibition of the mycelial growth of the pathogen. In the same year Prabha *et al.* (2002) reported on the antifungal properties of extracts from *Foeniculum vulgare, Coriandrum sativum, Trigonella foenum-graecum, Anethum graveolens* and *Cuminum cyminum* against three fungi, including *Fusarium oxysporum.* All extracts showed a relatively greater inhibitory effect on *F. oxysporum,* with stem extracts from *A. graveolens* exhibiting complete growth suppression, while the mycelial growth of the other two fungi was also inhibited significantly.

Om *et al.* (2001) also reported the effect of essential oils extracted from *Callistemon lanceolatus* (*Callistemon citrinus*), *Citrus medica, Eclipta alba, Hyptis suaveolens* and *Ocimum canum* (*O. americanum*) against *Rhizoctonia solani,* the cause of damping-off disease of tomato and chilli (*Capsicum annuum*). The essential oils of *Citrus medica, E. alba* and *O. canum* completely inhibited the growth of the fungus within 24 h. The essential oils of *C. lanceolatus* and *O. canum* controlled the damping-off disease of tomato by 57 and 71% and that of chilli by 40 and 83%, respectively. The same effect was also observed by applying crude, boiled water and acetone extracts from *A. sativum,* and a boiled water extract of *Eucalyptus* sp., as well as boiled water and acetone extracts of *Zingiber officinale* (Kane *et al.,* 2002).

In some cases plant extracts not only inhibit fungal mycelial growth, but also spore germination. Chen *et al.* (2002) reported that *Chloris virgata, Dalbergia hupeana, Pinus massoniana, Paeonia suffruticosa* and *Robinia pseudoacacia* extracts inhibited spore germination of grape downy mildew (*Plasmopara viticola*) and the effect was comparable to that of a traditional fungicide, liquid Bordeaux. Other work also demonstrated that leaf extracts of *Lawsonia inermis* and *Datura metel* possess antifungal activity against *Mycosphaerella berkeleyi,* causing late leaf spot in groundnuts (Kishore *et al.,* 2002). Particularly, the *D. metel* extract continuously reduced disease progress up to 115 days after sowing while the severity of late leaf spot at harvest was significantly less than that of the controls.

Plant seeds also contain compounds with antimicrobial properties. Seed extracts of 50 plant species, belonging to different families, were evaluated for their ability to inhibit the growth of *Trichoderma viride in vitro* (Bharathimatha *et al.,* 2002). Of the various seed extracts, that of *Harpullia cupanioides* (Roxb.), belonging to the family *Sapindaceae,* displayed very high antifungal activity. The seed extract of *H. cupanioides* strongly inhibited the growth of *Rhizoctonia solani, Curvularia lunata* (*Cochliobolus lunatus*), *Colletotrichum musae* and *Alternaria alternata* and retained its antifungal activity even after heating at 100°C for 10 min or autoclaving at 121°C for 20 min. Rodriguez and Montilla (2002) reported on the *in vitro* and *in vivo* antimicrobial effect of a *Citrus paradisi* seed extract (Citrex) on *F. oxysporum lycopersici* causing tomato wilt. Five treatments were evaluated: (i) immersion of plant roots in a solution of

Citrex before transplant; (ii) weekly application to the foliage; (iii) weekly application to the soil; (iv) weekly application to the foliage and to the soil; and (v) immersion of plant roots at transplant plus weekly application to the soil. The control was infested soil without application of the product. Treatments (i) and (iii) reduced wilting by 85%, indicating that it is possible to control soil-borne pathogens with the *C. paradisi* seed extract.

Extremely promising, from a natural product development perspective, is that many plant extracts compare favourably with commercially synthesized fungicides. For example, leaf extracts of *Azadirachta indica, Atropa belladonna, Calotropis procera, Ocimum basilicum, Eucalyptus amygdalina, Ailanturs excelsa (Ailanthus excelsa)* and *Lantana camara,* at different concentrations, were compared to the fungicides Bavistin (carbendazim), Dithane M-45 (mancozeb), captan, thiram and Topsin M (thiophanate-methyl) at standard concentrations against *F. oxysporum* inducing fenugreek wilt under greenhouse conditions. Seeds of fenugreek were separately soaked overnight in each leaf extract as well as the fungicide solutions (Gupta and Bansal, 2003). All the leaf extracts, except that of *Ailanturs excelsa,* significantly inhibited the mycelial growth of *Fusarium* compared to the untreated control. Maximum germination of fenugreek seeds was observed with Bavistin (93.33%), followed by *Atropa belladonna* (90.66%), *Azadirachta indica* (87.99%), *L. camara* (87.99%), *C. procera* (85.99%), *O. basilicum* (85.33%) and *E. amygdalina* (82.40%). Dithane M-45 (80.00%), Captan (86.66%), thiram (86.66%) and Topsin M (86.66%) were on par with the leaf extracts.

An approach adopted by Solunke *et al.* (2001), following the increasing availability of data in the literature on plants with antifungal activity, was to evaluate the potential of plant extracts to be applied in IPM programmes. The authors conducted a study to manage sclerotium rot (*Sclerotium rolfsii* [*Athelia rolfsii*]) of potato using a commercial fungicide, carbendazim, separately and together with plant extracts. Sensitivity of fungal isolates (SRP-1, SRP-2, SRP-3 and SRP-4) against carbendazim was determined beforehand. Subsequently, aqueous extracts of *Azadirachta indica, Allium cepa, Glossocardia bosvallea* and *Vinca rosea (Catharanthus roseus)* were mixed with carbendazim in solution and used to treat potato slices. Based on the minimum inhibitory concentration of carbendazim against the four isolates, SRP-4 appeared to be tolerant while SRP-1 was sensitive. The percentage control efficacy (PCE) of carbendazim alone and in mixture with plant extracts were tested against the tolerant isolate, SRP-4. Using carbendazim in combination with plant extracts increased its PCE. Application of carbendazim along with *A. indica* and *G. bosvallea* recorded a PCE of 100%. Carbendazim with *V. rosea* and with *A. cepa* recorded PCEs of 90.77 and 80.53%, respectively. The results showed that it is possible to reduce the selection pressure of carbendazim when combined with plant extracts.

Finally, an aspect that needs special reference is that the antimicrobial efficacy of plant extracts compare favourably with that of commercial fungicides in controlling plant diseases. In fact, research into antimicrobial activities of plant extracts has offered some potential solution to disease control in the absence of effective chemical control. For example, few commercial fungicides have been effective in inhibiting teliospore germination of *Tilletia indica*, the causal agent of Karnal bunt of wheat. This disease is becoming more widespread, in part, because of the lack of effective chemical control. Various extracts of native plants from Sonora, Mexico were evaluated to determine their antifungal activity against *T. indica*. Dichloromethane and methanol extracts were incubated with the fungus to measure the inhibition of mycelial growth. Dichloromethane extracts from *Chenopodium ambrosiodes* and *Encelia farinosa* reduced radial mycelial growth significantly, but total inhibition occurred at a relatively high concentration of 500 mg ml^{-1} of the dichloromethane extract from *Larrea tridentata*. Teliospores subjected to treatment with the latter plant extract showed no viability when transferred to fresh culture media, confirming the extract's potential to be applied as controlling agent for *T. indica* (Rivera *et al.*, 2001).

Plants with antibacterial properties

Already in the 1990s, the application of natural plant extracts under both greenhouse (Blaeser and Steiner, 1999; Nikolov *et al.*, 1999; Scholz *et al.*, 1999) and postharvest (El-Ghaouth *et al.*, 1995; Bhaskara *et al.*, 1998) conditions against various pathogens confirmed the potential of natural compounds from plants to be developed into commercial products. Although the number of natural products applied in biological control systems is currently a relatively small percentage of the total world market, it is predicted to increase substantially. This implicates the possible decrease of pesticide application in agriculture in the future. In turn, this approach could create a significant market opportunity for alternative products to be applied on a larger scale in organic crop production systems.

A vast number of reports on large-scale screening programmes to evaluate wild plants for antibacterial activity are available in the literature; we refer to only a few. The antibacterial and antifungal activities of 38 plants belonging to 17 families were tested by Ghosh *et al.* (2000). The solvent extracts of different morphological parts of these plants were tested against 14 bacterial and 18 fungal strains. Out of these, *Alpinia mutica*, *Cephalandra indica*, *Croton bonplandianum*, *Curcuma amada*, *Holarrhena antidysenterica*, *Moringa oleifera* and *Zingiber spectabile* were found to contain antimicrobial properties (Ghosh *et al.*, 2000). Khan and Omoloso (2002) studied the antibacterial and antifungal activities of methanol extracts of *Harpullia petiolaris* leaves, stems, root barks and heartwoods (collected from Papua New Guinea). Antibacterial activities were found to be highest in fractions of root bark, petrol and dichloromethane fractions of stem bark, the petrol fraction of heartwood and the butanol fraction of leaves. Antifungal activity was only observed in the petrol fractions of the root bark and stem heartwood.

Ethanol and water extracts of leaves, flowers, shoots, bark and fruits of 30 herbal and woody plant species were tested for *in vitro* growth inhibition of *Erwinia amylovora* by Krupinski and Sobiczewski (2001) using an agar diffusion method. Active extracts were found in 23 species, while in 13 of these the active substances were found for the first time. The highest growth inhibition of this bacterium was recorded for extracts of *Aloe arborescens*, *Juglans regia*, *Rhus typhina* (*R. hirta*), *Salvia officinalis* and *Satureja hortensis*. In almost all cases ethanol appeared to be a better solvent of active plant substances against *E. amylovora* than water.

A similar approach was followed by Morais *et al.* (2002) in screening crude extracts of 45 known medicinal plants against bacterial pathogens of tomato. The antibacterial activity of crude extracts was tested against *Xanthomonas campestris* pv. *vesicatoria*, *Ralstonia solanacearum* and *Clavibacter michiganense* subsp. *michiganense*. Some assays were also performed to verify the capability of these plant extracts to show antibiosis. Five of the 45 extracts showed significant activity against the test bacteria, confirming the potential of using either these extracts or active substances contained in them as natural products under field conditions. In the same year, Devanath *et al.* (2002) compared the antibacterial effect of extracts from three medicinal plants (*Psidium guajava*, *Aloe vera* [*A. barbadensis*] and *Datura stramonium*) to two standard fungicides (Streptocycline and Blitox or copper oxychloride) against *Rhizoctonia solanacearum in vitro*. The aqueous extract of *Aloe vera* was most effective in suppressing the growth of *R. solanacearum* followed by the extract of *Psidium guajava*.

Zeller *et al.* (2002) reported on the effect of an extract from *Hedera helix* against fire blight on pome fruits, caused by *Erwinia amylovora*. The highest growth inhibition of this bacterium was recorded for extracts of *Aloe arborescens*, *Juglans regia*, *Rhus typhina* (*R. hirta*), *Salvia officinalis* and *Satureja hortensis*. In almost all cases ethanol appeared to be a better solvent of active plant substances against *E. amylovora* than water. Also using the bacterium *E. amylovora* as a test organism, Jin and Sato (2003) searched for secondary metabolites in aqueous extracts from succulent young shoots of pear (*Pyrus*). Aqueous extracts of the tissue of succulent young shoots exhibited strong antibacterial activity against *E. amylovora*. The active compound was isolated from the extract by steam distillation *in vacuo*, purified through charcoal powder column chromatography and identified as benzoquinone (2,5--cyclohexadiene-1,4-dione) by NMR spectra, mass spectra and HPLC analysis. In the same year, Pretorius *et al.* (2002a) demonstrated the broad-spectrum antibacterial activity of crude extracts from *Acacia erioloba*, *Senna italica* and *Buddleja saligna* against the plant pathogens *Agrobacterium tumefaciens*, *Clavibacter michiganense* pv. *michiganense*, *Erwinia carotovora* pv. *carotovora*, *Pseudomonas solanacearum* and *Xanthomonas campestris* pv. *phaseoli*.

Additionally, plant extracts with antibacterial properties can provide an alternative to certain antibiotics. Zeller *et al.* (2002) reported on such alternatives to the antibiotic streptomycin for the control of fire blight on pome fruits, caused by *E. amylovora*, which is of great economic importance for German and European fruit producers. An antagonistic preparation, BIOPRO, showed a control efficacy of up to 60% and the plant extract from *Hedera helix* revealed a high efficacy in the field in combination with a low concentrated copper compound and a metal salt. The control of fire blight in this manner was comparable to that by the antibiotic streptomycin under artificial and natural infection conditions. The latter is important in light of the fact that the use of copper-containing antimicrobials is phasing out in Europe due to recent legislation, while the use of antibiotics, such as streptomycin, is highly opposed.

Current status of antimicrobial products from plant extracts

Extracts from various plant species appear to have promising potential for their application as natural products in the agricultural industry against both plant pathogenic bacteria and fungi and to be integrated in plant disease management systems. However, to date, the number of commercially developed natural plant products that emerged from screening programmes is alarmingly low. The natural plant product Milsana®, extracted from the giant knotweed (*Reynoutria sacchalinensis*), is probably best known (Daayf *et al.*, 1995). The product has been reported to control powdery mildew, caused by *Sphaerotheca fuliginea*, in the long English cucumber under greenhouse conditions, and also showed broad spectrum activity against powdery mildew of tomato, apple and begonia as well as downy mildew of grapevine and rust of bean (Daayf *et al.*, 1995). It was concluded by Petsikos *et al.* (2002) that Milsana® can substantially contribute to the management of cucumber powdery mildew in organic or integrated farming systems.

A volatile natural product, Carvone™, derived from dill and caraway seed, has been developed to inhibit the growth of storage pathogens and to suppress sprouting of potatoes in the warehouse (Moezelaar *et al.*, 1999). Carvone™ is currently marketed as Talent® in the Netherlands. Additionally, Fungastop™ and Armorex™ II, two natural products developed in the USA (Soil Technologies Corp.), are commercially available for the control of various plant diseases in agriculture.

Dayan *et al.* (2009) supplied an excellent summary of new natural antimicrobial products that were recently commercialized for use in either conventional or organic agricultural or both industries. To name a few, these include Bla-S™ (against rice blast disease in eastern Asia), Kasugamin™ (against rice blast and other crop diseases in Japan), Mildiomycin™ (against powdery mildews mainly in Japan), Delvolan™ (against fungal diseases of ornamentals) and Validacin™ (used for *Rhizoctonia* spp. control on a variety of crops). All of these products are fermentation secondary products from Actinomycetes, mostly *Streptomyces* spp., commercialized for use as agricultural fungicides in Japan, and to a lesser extent in other parts of the world.

Several plant essential oils are marketed as fungicides for organic farmers (Dayan *et al.*, 2009). These include E-Rase™ from jojoba (*Simmondsia californica*) oil, Sporan™ from rosemary (*Rosemarinus officianalis*) oil, Promax™ from thyme (*T. vulgaris*) oil, Trilogy™ from neem (*A. indica*) oil and GC-3™ being a mixture of cottonseed (*Gossypium hirsutum*) oil and garlic (*A. sativum*) extract. According to Dayan *et al.* (2009), few scientific papers deal with these products or the actual active components, and their modes of action against individual plant pathogens are largely unknown.

A fairly new approach to microbial disease control is via systemic acquired resistance (SAR) or, in other words, the activation of natural defence mechanisms within the plant via elicitors (Dayan *et al.*, 2009). Harpin protein that induces SAR in plants is sold as Messenger™ or ProAct™. Brassinosteroids that induce the synthesis of PR proteins that resist fungal infection in crops, are sold as ComCat® (Roth *et al.*, 2000). The latter product seems to be used more often as a bio-stimulatory agent; an aspect that will be dealt with later in this chapter.

These past successes and the public's current concern over the impact of synthetic pesticides on the environment ensures a continued, if not an increased, interest in searching nature for environmentally friendlier pest management tools (Dayan *et al.*, 2009).

3.5 Natural Bio-herbicides

The herbicide component of all pesticides sold increased from less than 20% in the 1950s to almost 50% of the market in the 1980s (Jutsum, 1988) and this has since risen to over 60%. Today, herbicides account for more than half of the volume of all agricultural pesticides applied in the developed world and their use has also contributed to the concern of consumers towards the potential health and environmental impact they may have (Dayan *et al.*, 2009). However, almost three decades ago, McWhorter and Chandler (1982) estimated that weeds contribute to a significant 12% loss in worldwide crop production, emphasizing the need to control weeds. These two perspectives, namely the need to control weeds and the negative attitude of consumers, have in part fuelled the current elevated interest in the discovery of natural products to be applied as an environmentally friendly method of weed control.

However, the complexity of the environment in terms of weed control and especially the possible side effects of herbicides or biological control agents on non-targeted plants, calls for a comprehensive study on control strategies, whether biological, chemical, cultural or a combination of these in an integrated weed management system. The biological approach to weed control can be separated into three distinct strategies: conservation, inundative and classical depending on the setting (e.g. range, row crops, urban), the extent of the infestation and the biology of the system (McFadyen, 1998; Goeden and Andrés, 1999). According to the authors, conservation biocontrol refers to situations where a natural biocontrol agent is already present and is able to control the weed but requires assistance in the form of cultural practices or integrated management decisions that enable the agent to thrive. Although conservation strategies are less common in the biological approach to weed control than in the biocontrol of insects, examples do exist (Goeden and Andrés, 1999).

Inundative biocontrol involves the release of large numbers of a biological control agent at a time when weed populations are expected to escape control or exceed a critical economic or competitive threshold. When the agent is already present at a level that does not provide adequate, continuous control of the target, this strategy is known as augmentation biocontrol (Goeden and Andrés, 1999). In general, inundative biocontrol relies on the released organisms themselves to control the target without any expectation of continued control by future generations of the agent (Elzen and King, 1999). In an inundative strategy the agent may need to be released or applied several times during a single crop cycle in the event of re-growth or re-emergence of the target weed. Thus, inundative strategies typically apply to relatively high-input systems.

Classical biocontrol, sometimes referred to as inoculative biocontrol, involves the practice of identifying co-evolved natural enemies from the native range of a target weed species and releasing them into the invaded range to reduce the presence of the weed to acceptable levels (McFadyen,1998). In classical biocontrol, the agents are expected to reproduce and proliferate on the target weed and disseminate throughout its invaded range, reaching an ecological equilibrium with the target weed and providing continuous, perpetual control. Successful control depends almost wholly on damage caused by the descendants of the released individuals rather than by the released individuals themselves (Elzen and King, 1999). Classical biocontrol is generally practiced in low-input systems. Biologically based weed control can also take the form of weed-resistant properties in crop plants, akin to host-plant resistance versus insect and pathogen pests.

Allelopathy, the production by a plant of secondary metabolites that inhibit growth of nearby plants, is a phenomenon that has been studied for its potential utility in weed control for many years (Rector, 2008). This implies the use of natural compounds from plants, referred to as allelochemicals, in natural product development programmes. The term allelopathy is derived from two Greek words 'allelos' and 'patos' meaning 'to suffer on each other' (Delabays and Mermillod, 2002). According to the authors, allelopathy is defined as a chemical process whereby certain plants release natural compounds into the environment that can either stimulate or inhibit the growth and development of surrounding plants. In nature the latter is more likely, as a form of competition for growing space, where allelochemicals are released from the roots of one plant and absorbed by the roots of surrounding plants preventing them from growing in the same area. There are several ways in which an allelopathic plant can release its protective chemicals. These include: (i) volatilization, where a chemical is released in the form of a gas through the leaves and, on absorption, sensitive surrounding plants are stunted or die; (ii) leaching of allelochemicals from decomposing abscised leaves preventing surrounding plants from establishing in the same area; and (iii) exudation of allelochemicals into the soil through the roots, preventing surrounding plants from prospering (Delabays and Mermillod, 2002).

Despite its long history, little progress has been made in incorporating allelopathy into mainstream weed management programmes due to a failure to provide adequate weed control while maintaining other agronomic qualities of the crop (Belz, 2007). There is a strong possibility that formulation has played a key role in the failure of bio-herbicides in the market and their efficacy in the field. This is understandable when the bottom line criterion for any natural product developed from plant extracts, including natural herbicides, is considered. This criterion is that the product must be effective, safe and have consistent results and must have an adequate shelf life of at least 1 year.

Despite the apparent failure to develop bio-herbicides from plants or other biological sources at present (see current status later in this chapter), the need to do so is increasing due to weeds becoming more resistant towards the commercially available synthetic herbicides (Dudai *et al.*, 1999; Duke *et al.*, 2000; Tworkoski, 2002). For example, in New Zealand the problems posed by invasive weeds are among the most severe in the world with an estimated cost of NZ$100 million each year (Juliena *et al.*, 2007). Of all the weeds currently identified in New Zealand, only a third has been targeted by biological control. The latter emphasizes the need to integrate either biological control agents or chemical bio-herbicides or both into a broader management programme for weeds.

Plants with herbicidal properties

The greatest potential for developing natural herbicides probably lies in the plant kingdom, a largely untapped reservoir of natural compounds with allelopathic herbicidal properties. Although little success in terms of natural product development has been realized, extensive research has been done in the past to screen for the growth inhibitory activity of extracts from many plant species (Wu *et al.*, 2002). Only a few recent examples are supplied to emphasize the principle.

Nie *et al.* (2002) reported that extracts of *Wedelia chinensis* reduced seed germination, inhibited seedling growth, resulted in yellowing leaves and reduced resistance to disease in weeds such as *Cyperus difformis*, *Paspalum thunbergii*, *Alternanthera sessilis* and *Cynodon dactylon* at relative low concentration of 0.4 g FW (fresh weight) ml^{-1} water. This study confirmed the potential of *W. chinensis* extracts sprayed before crop emergence to control the germination of weed seeds. Another study by Randhawa *et al.* (2002) showed that a sorghum extract reduced seed germination and seedling growth of the weed *Trianthema portulacastrum* substantially at high concentrations (75–100%) but promoted shoot length of the weed at low concentrations (25%).

Similar contrasting results were reported for the effects of extracts from eight lucerne cultivars on seed germination as well as on root and hypocotyl development of lettuce seedlings (Tran and Tsuzuki, 2002). Extracts from some lucerne cultivars had a stimulatory effect in terms of seed germination as well as root and hypocotyl growth, whereas others showed the direct opposite effect, confirming that crop plants can also be affected by plant extracts aimed at controlling weed growth. Singh *et al.* (2003) confirmed this phenomenon by showing that aqueous leaf leachates of *Eucalyptus citriodora* inhibited the germination and seedling growth of all test crops (*Vigna radiata*, *V. mungo* and *Arachis hypogaea*) investigated. Further, in their study of bioactivity of plant extracts, Deena *et al.* (2003) demonstrated the inhibitory effect of leaf, stem and root leachates from *Andrographis paniculata* on germination and seedling growth in rice. From this it became clear that the bioactivities of plant extracts are unpredictable and may give different and often contrasting results with regard to inhibition or promotion of growth and development in other plants. The reaction of crops to treatment with plant extracts may depend on the interaction between different types of plant species or even on the concentration of the extracts (Channal *et al.*, 2002a).

Research into the allelopathic activity of plant extracts has resulted in the identification of active ingredients responsible for both inhibition or stimulation of either seed germination or seedling growth. In this respect, Chung *et al.* (2002) reported on the inhibitory effect of ferulic, p-hydroxybenzoic, p-coumaric and m-coumaric acids isolated from three rice cultivars on the growth of barnyard grass. This suggested that these compounds may be, at least, a key factor in rice allelopathy on barnyard grass, and the information presented may contribute to the development of natural herbicides.

Similar active allelochemicals were isolated by Sasikumar *et al.* (2002). In their study on the allelopathic effects of *Parthenium hysterophorus* leachates on cowpea, pigeonpea, greengram, blackgram and horsegram, the authors reported significant seed germination inhibition for all test crops. Gas chromatographic analysis showed the presence of phenolic acids, namely caffeic, p-coumaric, ferulic, p-hydroxybenzoic and vanillic acids in the leachates from different plant parts (leaf, stem, flower and root) of *P. hysterophorus*. A mixture of allelopathic compounds in bioassays significantly inhibited the germination and vigour index of all test crops. However, leachates from flowers had no inhibitory effect on the germination of blackgram and greengram seeds.

Kato and Kawabata (2002) isolated a growth-inhibiting compound from the acetone extract of 30-day-old lemon balm (*Melissa offinalis*) shoots by means of silica gel column chromatography. This uncharacterized compound inhibited the growth of cress seeds at concentrations higher than 0.3 µg ml^{-1}. Iqba *et al.* (2002) showed that living buckwheat reduced weed biomass compared to plots without buckwheat. A laboratory study revealed that root exudates from buckwheat (collected from Aomori, Japan) suppressed root and shoot growth of the weeds *Trifolium repens*, *Brassica juncea*, *Amaranthus palmeri*, *Echinochloa crus-galli* and *Digitaria ciliaris* but also that of lettuce, and reduced weed dry weight. Fagomine, 4-piperidone and 2-piperidinemethanol were isolated from a chloroform extract and identified as the active ingredients.

Several allelochemicals have also been characterized from *Helianthus annuus* that inhibit seed germination and seedling growth of *Amaranthus albus*, *Amaranthus viridis*, *Agropyron repens* (*Elymus repens*), *Ambrosia artemisiifolia*, *Avena fatua*, *Celosia cristata* (*C. argentea* var. *cristata*), *Chenopodium album*, *Chloris barbara* (*Chloris barbata*), *Cynodon dactylon*, *Digitaria sanguinalis*, *Dactyloctenium aegyptium*, *D. ciliaris*, *E. crus-galli*, *Flaveria australasica*, *P. hysterophorus*, *Portulaca oleracea*, *Sida spinosa*, *Trianthema portulacastrum* and *Veronica persica* (Macias *et al.*, 2002). The inhibitory effects of this crop may be utilized for weed management to attain reduced herbicide usage in sustainable agricultural systems (Azania et *al.*, 2003).

Previous studies by Salamci *et al.* (2007) confirmed that the essential oils of *T. aucheranum* and *T. chiliophylllum*, characterized by the relatively high content of 1,8-cineole, camphor, borneol and α-terpineol, exhibited potent inhibitory effects on seed germination and seedling growth of *A. retroflexus*. All commercialized essential oils act as non-selective, contact herbicides that can provide good but transient weed control. The use of essential oils for weed control in organic agriculture seems promising, but these natural herbicides all act very rapidly and their efficacy is limited by the fact that they probably volatize relatively quickly (Dayan *et al.*, 2009). Other oils such as pine and clover oils have also been used in organic farming systems, but with limited success because the relatively high rate of use required for control makes it expensive compared to only one treatment of glyphosate with the same or even better control of weeds (Dayan *et al.*, 2009).

Current status of natural herbicides

Although several natural products from various sources are probably currently under development and targeted against specific weeds across the globe, very little literature is available on the use and environmental impact of natural products in organic agriculture and little success can be reported at present. According to Babua *et al.* (2003), reasons why so many promising bio-herbicides have failed to reach the commercial phase probably include: (i) the economics of patenting and registration; (ii) inadequate or non-commercial market sizes; and (iii) technological constraints, especially relating to formulation chemistry. Discovery and development of a synthetic chemical herbicide can easily cost US$30 million (Heiny and Templeton, 1993) or more. It is generally expected that the development of a natural product herbicide should cost less. Indeed, two notable successes include COLLEGO™ and BIOMAL™, developed from exotic pathogens, where the development costs were US$1.5 million and US$2.6 million respectively. The costs involved with the development of these two examples of natural products can probably not be accepted as a rule and there are no guarantees that prospective entrepreneurs will experience the same. However, from an ecological perspective natural products are much more favoured (Heiny and Templeton, 1993) as a result of added advantages over classical synthetic weed control agents, including their narrow host ranges that reduce their impact on non-targeted plants in the environment (Rector, 2008), and this alone serves as a driving force for natural product developers to continue with this entrepreneurial business enterprise. Additionally, the quickly expanding organic agriculture industry does not allow synthetic pesticides, including herbicides (Anonymous, 2009), and this provides more momentum for the natural product industry.

In an excellent recent review by Dayan *et al.* (2009), a number of natural compounds from which bio-herbicides have been commercialized for weed control in organic agriculture have been listed. A few examples include corn gluten meal (e.g. WeedBan™, Corn Weed Blocker™ and Bioscape Bioweed™), a mixture of essential oils and other organic compounds (e.g. Burnout™, Bioorganic™, Weed Zap™ and GreenMatch™) as well as pelargonic acid mixed with related short-chain fatty acids and paraffinic petroleum oil (e.g. Scythe™). However, despite the examples of natural products mentioned above, a tripeptide obtained from the fermentation culture of the actinomycete *Streptomyces hygroscopis* and registered as Bialaphos™ is regarded by Dayan *et al.* (2009) to be the only true commercialized natural product herbicide to date. It is a pro-herbicide that is metabolized into the active ingredient L-phosphinothricin (Fig. 3.28) in the treated plant.

Bialaphos™ and phosphinothricin inhibit the enzyme glutamine synthetase which is necessary for the production of glutamine and for ammonia detoxification. The elevated ammonia levels in tissues of treated plants stops photosynthesis and results in plant death. Both Bialaphos™ and phosphinothricin are broad-spectrum post-emergence herbicides that can be used for total vegetation control in many agricultural settings or in non-cultivated areas and to desiccate crops before harvest (Dayan *et al.*, 2009).

The list of commercialized bio-herbicides has expanded substantially during the past two decades. However, their mere existence must be considered in view of the following very valid comments by Dayan *et al.* (2009): (i) as opposed to traditional synthetic herbicides, none of the natural herbicidal compounds allowed for use in organic agriculture are very active and they must, therefore, be applied in relatively large quantities that may lead to undesirable effects on the environment and the soil fauna and microbes, which is in direct opposition with the philosophical positions and purpose of those who practice organic agriculture; (ii) the use of organic weed management tools may be enhanced in the context of an integrated pest management programme that includes sowing multiple crops, extended rotation cycles, mulching, and soil cultivation and cover; (iii) existing natural herbicides show very little crop selectivity and still require laborious application methods to ensure they do not come in contact with the desired crop; and (iv) organic weed management methods may be possible in small-scale farming and high-value crops but do not seem feasible in the production of the agronomic crops such as grains grown in large-scale farming enterprises.

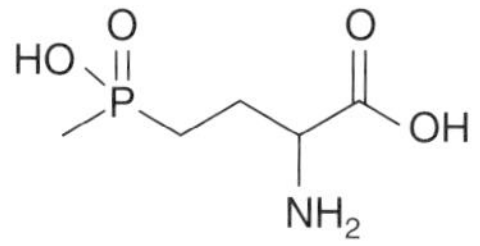

Fig. 3.28. L-phosphinothricin (redrawn from Dayan *et al.*, 2009).

3.6 Natural Compounds from Plants with Bio-stimulatory Potential

Allelochemicals found in plants are probably all secondary metabolites that are distinctive from primary metabolites in that they are generally non-essential for the basic metabolic processes such as respiration and photosynthesis (Richard, 2001). They are numerous and widespread, especially in higher plants (Pillmoor, 1993), and often present in small quantities (1–5%) as compared to primary metabolites (carbohydrates, proteins and lipids). Approximately 88,000–100,000 secondary metabolites have been identified in all plant forms, showing both structural and activity diversity (Verpoorte, 1998). Ecologically, these chemicals play essential roles in attracting pollinators, as adaptations to environmental stresses and serve as chemical defences against insects and higher predators as well as microorganisms (Rechcigl and Rechcigl, 2000). Although the purpose of the production of secondary metabolites in plants has long been argued among researchers, it is now universally accepted that they are produced as a result of abiotic (Beart *et al.*, 1985) and biotic stresses (Bourgaud, *et al.*, 2001), probably as part of a plant defence arsenal. Besides the role secondary metabolites play in plant metabolism, the growth promotion or inhibitory properties of certain natural compounds from plants have been extensively researched (Wu *et al.*, 2002).

Current status of plant products with bio-stimulatory potential

Already, at the end of the millennium, two successful natural products developed in Moldavia (formerly part of the Soviet Union) are Moldstim™ and Pavstim™, extracted from hot peppers (*Capsicum annum* L.) and leaves of *Digitalis purpurea* L., respectively (Waller, 1999). Both products have been used on a large scale as plant-growth regulators and for disease control. These developments are excellent examples of how natural plant resources can be exploited and applied in agriculture.

However, from an agricultural perspective, plant extracts containing growth-promoting substances have always been of interest to the research community in terms of the role they could play in addressing future food security issues. The ideal breakthrough would be to identify a plant or plants that contain bio-stimulatory substances promoting growth and resistance to pathogens, as well as yields in agricultural and horticultural crops. At this point it seems appropriate to consider recent discoveries of bio-stimulatory compounds from plants that have the potential to adhere to the 'ideal breakthrough' criterion in terms of their application potential as natural products in the agricultural industry.

Extracts from numerous plant species, with bio-stimulatory properties, were identified and evaluated for their commercial potential. Channal *et al.* (2002b) reported on seed germination as well as seedling growth enhancement of sunflower and soybean by leaf extracts from three tree species

(*Tectona grandis*, *Tamarindus indica* and *Samanea saman*). Terefa (2002) reported similar effects for *P. hysterophorus* extracts on tef (*Eragrostis tef*) while Neelam *et al.* (2002) demonstrated similar effects for *Leucaena leucocephala* extracts on wheat (*Triticum aestivum*). However, none of these studies revealed that treatment with the different plant extracts had any effect on the final yields of the crops under investigation.

In this regard, a report by Ferreira and Lourens (2002) demonstrating the effect of a liquid seaweed extract (now trading as a natural product under the name Kelpak™) on improving the yield of canola must be regarded as significant. Kelpak™ applied singly or in combination with the herbicide Clopyralid® at various growth stages of canola (*Brassica napus*) was assessed in a field experiment conducted in South Africa during 1998–1999. Foliar application of 2 l Kelpak ha^{-1}, applied at the four-leaf growth stage, significantly increased the yield of the crop. The active compounds in Kelpak™ are auxins and cytokinin.

In the same year, a study directed towards identifying bio-stimulatory properties in plant extracts was performed by Cruz *et al.* (2002a). The authors treated the roots of bean, maize and tomato with an aqueous leachate of *Callicarpa acuminata* and followed the *in vitro* effects on radicle growth, protein expression, catalase activity, free radical production and membrane lipid peroxidation in the roots. The aqueous extract of *C. acuminata* inhibited the radicle growth of tomato but had no effect on root growth of maize or beans. However, expression of various proteins in the roots of all treated plants was observed. In treated bean roots the expression of an 11.3 kDa protein by the leachate, showing a 99% similarity with subunits of an α-amylase inhibitor found in other beans, was induced. In treated tomato an induced 27.5 kDa protein showed 95% similarity to glutathione-S-transferases of other Solanaceae species. Spectrophotometric analysis and native gels revealed that catalase activity was increased twofold in tomato roots and slightly in bean roots, while no significant changes were observed in treated maize roots. Luminol chemiluminescence levels, a measure of free radicals, increased fourfold in treated tomato roots and twofold in treated bean roots. Oxidative membrane damage in treated roots, measured by lipid peroxidation rates revealed an almost threefold increase in peroxidation in tomato while no effect was observed in maize or beans (Cruz *et al.*, 2002a).

The significance of this study lies in the fact that various metabolic events can be manipulated in plants by treatment with certain plant extracts. What has to be established by researchers is whether these altered metabolic events contribute towards positive or negative physiological changes within the treated plants. The rationale for this type of research lies in the search for natural compounds to be applied in sustainable yield-improving, as well as weed-, pest- and disease-controlling, management systems (Singh *et al.*, 2001). According to the authors, natural compounds isolated from some plants show strong bio-herbicidal activity at high concentrations but at low concentrations these extracts can promote crop seed germination and seedling growth, hence showing a potential to be applied as bio-stimulatory agents or growth-promoting substances in agriculture. There is general consensus amongst scientists that research in this regard should concentrate on both the inhibitory and stimulatory effect of plant extracts on seed germination, seedling growth and the physiology of other test plants in order to verify the action at hand (Khan *et al.*, 2001; Ameena and George, 2002; Cruz *et al.*, 2002b; Duary, 2002; Obaid and Qasem, 2002).

Probably the most effective compounds to enhance crop yield, crop efficiency and seed vigour have been identified as brassinosteroids (BRs; Mandava, 1979; 1988), first extracted from rape (*Brassica napus* L.) pollen (Adam and Marquard, 1986). In a recent mini-report, Zullo and Adam (2002) confirmed the prospective agricultural uses of BRs. The assumption of their application potential was made from data collected over the past three decades and only a few examples are presented here. Yield increases that were in most instances significant were reported, as cited by Zullo and Adam (2002), in beans and lettuce (Meudt *et al.*, 1983), rice (Lim, 1987), maize (Lim and Han, 1988), wheat (Takematsu *et al.*, 1988), chickpea (Ramos, 1995) and tomato (Mori *et al.*, 1986). Many other examples of BR use for increasing crop yield can be found in the literature (Kamuro and Takatsuto, 1999; Khripach *et al.*, 1999; Khripach *et al.*, 2000).

Besides their yield-improving effects, BRs have been shown to increase plant growth in crops (Rao *et al.*, 2002) and especially root growth (Müssig *et al.*, 2003), to increase resistance in crops towards low-temperature injury (Kamuro and Takatsuto, 1999) and to increase resistance of potato to infections by *Phytophthora infestans* and *Fusarium sulfureum* (Kazakova *et al.*, 1991). Although many BRs, such as 24-epibrassinolide, are commercially available and employed in some countries, more accurate studies on dosage, method and time of application, its suitability for the plant or cultivar, and association with other phytohormones are needed, because many of the results were obtained by experiments performed in greenhouses or small fields (Zullo and Adam, 2002).

A report on a prototype bio-stimulatory natural product developed from a BR-containing extract of *Lychnis viscaria* came from Roth *et al.* (2000). In 2003, after 12 years of intensive research under laboratory, greenhouse and field conditions at the University of the Free State, South Africa, a product was listed in Germany as a plant-strengthening agent under the trade name ComCat™ and commercialized by a German company, Agraforum AG. Foliar applications of ComCat™ have been demonstrated to enhance consistently root development in seedlings and final yields in a number of vegetable, fruit and row crops, as well as to induce resistance in crops towards abiotic and biotic stress conditions, and the mechanisms of action were elucidated on both a metabolic and genetic level (unpublished results, Pretorius, J.C. and van der Watt, E., University of the Free State, South Africa). Recently, significant yield increases in tomato, preharvest treated with ComCat™, were reported by Workneh *et al.* (2009). The authors also claimed more than 70% shelf life extension and higher marketability in tomato fruit harvested from plants treated with ComCat™ during the vegetative growth phase compared to the untreated control under ambient storage conditions. Preharvest ComCat™ treated tomatoes contained lower total soluble sugar levels at

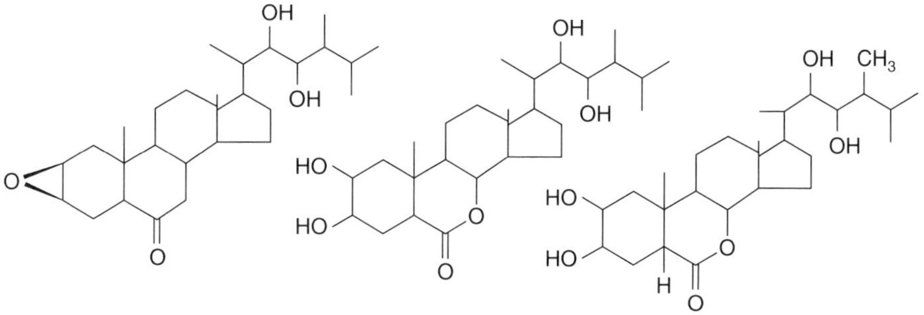

Fig. 3.29. Brassinosteroid active compounds contained in the plant-strengthening agent ComCat®.

harvest and showed better keeping quality in terms of physiological weight loss and juice content compared to untreated controls. In light of the diverse positive effects of ComCat™ on agricultural and horticultural crops, the product largely adheres to the 'ideal candidate' criterion stated earlier. Three BRs have been identified as the main active components of ComCat™ and these include 24-epi-secasterone, 24-epicastasterone and brassinolide (Fig. 3.29).

The future of commercialized bio-stimulants seems positive in light of the elevated costs of fertilizer. In this regard, research in terms of the use of bio-stimulants in combination with fertilizer levels lower than the recommended standard for different crops seems to be important in an attempt to lower the input costs that have become a gloomy issue for farmers recently.

3.7 Conclusions

Discovery programmes by the agrochemical industry are mostly driven by large-scale synthetic programmes followed by screening to identify potential new bio-pesticides, including antimicrobials and herbicides. Most companies have a more modest effort to evaluate natural products from outside sources and, to a lesser extent, from in-house isolation efforts. Although the literature is replete with reports of the isolation and characterization of phytotoxins from many sources, and many of these compounds have been patented for potential use, the use of natural or natural-product-derived herbicides in conventional agriculture is limited (Dayan *et al.*, 2009).

Since the Second World War, traditional agricultural practices have included the use of synthetic chemicals for the management of plant pathogens, pests and weeds. This has, without any doubt, increased crop production but with some deterioration of the environment and human health (Cutler, 1999). Research indicates that even if one never uses pesticides, one can still be exposed to them by being a consumer of commodities that others have treated with pesticides, e.g. through food.

In addition to the target pathogen, pesticides may kill various beneficial organisms and their toxic forms can persist in the soil. The increasing incidence of resistance among pathogens towards synthetic chemicals is also a cause for serious concern. The above is not only of major concern to the developed countries, where consumer preferences are for organically produced foods, but also in the developing world, such as Africa, where synthetic pesticides are too expensive for subsistence farming. Because of these problems there is a need to find alternatives to synthetic pesticides.

Among the various alternatives, natural plant products that are biodegradable and eco-friendly are receiving the attention of scientists worldwide. Such products derived from higher plants and microbes are relatively bioefficacious, economical and environmentally safe and can be ideal candidates for use as agrochemicals (Macias *et al.,* 2002). Additionally, the manufacturers of natural bio-stimulants applied in agriculture claim increased production, profit increases, cutting of operating costs and reduced fertilizer costs with no detrimental effect to the environment (Chen *et al.,* 2002). A number of plants showing the potential to act as donor plants for these natural products have been outlined in this chapter.

References

Adam, G. and Marquardt, V. (1986) Brassinosteroids. *Phytochemistry* 25, 1787–1799.

Alemayehu, W. (1996) Unused pesticides in developing countries: 100 000 tonnes threaten health and environment. News & Highlights. Food and Agriculture Organization of the United Nations Page.

Amadioha, A.C. (2002) Fungitoxic effects of extracts of *Azadirachta indica* against *Cochliobolus miyabeanus* causing brown spot disease of rice. *Archives of Phytopathology and Plant Protection* 35, 37–42.

Ameena, M. and George, S. (2002) Allelopathic influence of purple nut sedge (*Cyperus rotundus* L.) on germination and growth of vegetables. *Allelopathy Journal* 10, 147–152.

Anonymous. (2009) Organic farming. <http://www.epa.gov/pesticides/regulating/laws/fqpa/>.

Arras, G., Piga, A. and D'Hallewin, G. (1993) The use of *Thymus capitatus* essential oil under vacuum conditions to control *Penicillium digitatum* development on citrus fruit. *Acta Horticulturae* 344, 147–153.

Awuah, R.T. (1994) *In vivo* use of extracts from *Ocimum gratissimum* and *Cymbopogon citrates* against *Phytophthora palmivora* causing blackpod disease of cocoa. *Annals of Applied Biology* 124, 173–178.

Azania, A., Azania, C., Alves, P., Palaniraj, R., Kadian, H.S., Sati, S.C., Rawat, L.S., Dahiya, D.S. and Narwal, S.S. (2003) Allelopathic plants: Sunflower (*Helianthus annuus* L.). *Allelopathy Journal* 11, 1–20.

Babua, R.M., Sajeenaa, A., Seetharamana, K., Vidhyasekarana, P., Rangasamyb, P., Prakashc, M.S., Rajab, A.S and Bijib, K.R. (2008) Molecular biology approaches to control of intractable weeds: New strategies and complements to existing biological practices *Plant Science* 175, 437–448.

Bae, E.Y., Shin, E., Lee, D.H., Koh, Y.J., Kim, J.H., Bae, E.Y., Shin, E.J., Lee, D.H., Koh, Y.J. and Kim, J.H. (1997) Antifungal kaempferol-3-O-β-D-apiofuranosyl-(1,2)-β-D-glucopyranoside from leaves of *Phytolacca americana* L. *Korean Journal of Plant Pathology* 13, 371–376.

Baldwin, I.T. (1999) The Jasmonate cascade and the complexity of induced defense against herbivore attack. In: Michael, W. (ed.) *Function of Plant Secondary Metabolites and their Exploitation in Biotechnology.* CRC Press, Columbus, USA, pp. 155–179.

Bandara, B.M.R., Wimalasiri, W.R., Adikaram, N.K.B., Sinnathamby, B. and Balasubramaniam, S. (1988) Antifungal properties of *Croton aromaticus, C. lacciferus* and *C. officinalis* including the isolation of a fungicidal constituent. *Ceylon Journal of Science: Biological Sciences* 20, 11–17.

Beart, J.E., Terence, H.L. and Edwin, H. (1985) Plant polyphenols-secondary metabolism and chemical defense: some observations. *Phytochemistry* 24, 33–38.

Belz, R.G. (2007) Allelopathy in crop/weed interactions – an update, *Pest Management Science* 63, 308–326.

Benner, J.P. (1993) Pesticidal compounds from higher plants. *Pesticide Science* 39, 95–102.

Bharathimatha, C., Sabitha, D., Velazhahan, R. and Doraiswamy, S. (2002) Inhibition of fungal plant pathogens by seed proteins of *Harpullia cupanioides* (Roxb.). *Acta Phytopathologica et Entomologica Hungarica* 37, 75–82.

Bhaskara, R.M.V., Angers, P., Gosselin, A. and Arul, J. (1998) Characterisation and use of essential oil from *Thymus vulgaris* against *Botrytis cinerea* and *Rhizopus stolonifer* in strawberry fruits. *Phytochemistry* 47, 1515–1520.

Bianchi, A., Zambonelli, A., Zechini D.A. and Bellesia, F. (1997) Ultrastructural studies of the effects of *Allium sativum* on phytopathogenic fungi *in vitro*. *Plant Disease* 81, 1241–1246.

Bohra, N.K. and Purohit, D.K. (2002) Effect of some aqueous plant extracts on toxigenic strain of *Aspergillus flavus*. *Advances in Plant Sciences* 15, 103–106.

Borris, R.P. (1996) Natural products research: Perspectives from a major pharmaceutical company. *Journal of Ethnopharmacology* 51, 29–38.

Bourgaud, F., Gravot, A., Milesi, S. and Gontier, E. (2001) Production of plant secondary metabolites: A historical perspective. *Plant Science* 161, 839–851.

Bowers, J.H. and Locke, J.C. (2000) Effect of botanical extracts on the population density of *Fusarium oxysporum* in soil and control of Fusarium wilt in the greenhouse. *Plant Disease* 84, 300–305.

Callow, J.A. (1983) *Biochemical Plant Pathology*. John Wiley and Sons, New York, p. 484.

Channal, H.T., Kurdikeri, M.B., Hunshal, C.S., Sarangamath, P.A. and Patil, S.A. (2002a) Allelopathic influence of tree leaf extracts on greengram and pigeonpea. *Karnataka Journal of Agricultural Sciences* 15, 375–378.

Channal, H.T., Kurdikeri, M.B., Hunshal, C.S., Sarangamath, P.A., Patil, S.A. and Shekhargouda, M. (2002b) Allelopathic effect of some tree species on sunflower and soybean. *Karnataka Journal of Agricultural Sciences* 15, 279–283.

Chen, J. Dai, G., Gu, Z., Miao, Y., Chen, J., Dai, G., Gu, Z. and Miao, Y. (2002) Inhibition effect of 58 plant extracts against grape downy mildew (*Plasmopara viticola*). *Natural Product Research and Development* 14, 9–13.

Chung, I.M., Kim, K.H., Ahn, J.K., Chun, S.C., Kim, C.S., Kim, J.T. and Kim, S.H. (2002) Screening of allelochemicals on barnyard grass (*Echinochloa crus-galli*) and identification of potentially allelopathic compounds from rice (*Oryza sativa*) variety hull extracts. *Crop Protection* 21, 913–920.

Cowan, M.M. (1999) Plant products as antimicrobial agents. *Clinical Microbiology Reviews* 12, 564–582.

Cox, P.A. (1990) *Ethnopharmacology and the search for new drugs. Bioactive compounds from Plants*. Wiley, Chichester, (Ciba Foundation Symposium), UK, pp. 40–55.

Cruz, O.R., Ayala, C.G. and Anaya, A.L. (2002a) Allelochemical stress produced by the aqueous leachate of *Callicarpa acuminata*: effects on roots of bean, maize and tomato. *Physiologia Plantarum* 116, 20–27.

Cruz, M.E.S., Schwan-Estrada, K.R.F., Nozaki, M.H., Batista, M.A., Stangarlin, J.R., Ming, L.C., Craker, L.E., Scheffer, M.C. and Chaves, F.C.M. (2002b) Allelopathy of the aqueous extract of medicinal plants on *Picao preto* seeds germination. Proceedings of the First Latin American Symposium on the Production of Medicinal, Aromatic and Condiments Plants, Sao Pedro, Sao Paulo, Brazil, 30 July to 4 August 2000. *Acta-Horticulturae* 569, 235–238.

Curir, P., Dolci, M., Dolci, P., Lanzotti, V., Coomann, L.D.E. and De-Coomann, L. (2003) Fungitoxic phenols from carnation (*Dianthus caryophyllus*) effective against *Fusarium oxysporum f. sp. dianthi*. *Phytochemical Analysis* 14, 8–12.

Cutler, H.G. (1999) Biologically Active Natural Products: Agrochemicals. In: Macias, F.A., Galindo, J.C.B., Molinillo, J.M.G. and Cutler, H.G. (eds) *Recent Advances in Allelopathy. Vol. 1. A Science for the Future*. Servicio de Publicationes, Universidad de Cadiz, Spain, pp. 397–414.

Daayf, F., Schmitt, A. and Bélanger, R.R. (1995) The effect of plant extracts of *Reynoutria sachalinensis* on powdery mildew development and leaf physiology of long English Cucumber. *Plant Disease* 79, 577–580.

David, H.T. (1992) *Sustainable Practices for Plant Disease Management in Traditional Farming Systems*. Oxford & IBH Publishing Co., New Delhi, India, pp. 143.

Dayan, F.E, Cantrell, C.L. and Duke, S.O. (2009) Natural products in crop protection. *Bioorganic and Medicinal Chemistry*, 17, 4022–4034. [doi:10.1016/j.bmc.2009.01.046]

De Neergaard, E. (2001) Systemic acquired resistance: An eco-friendly strategy for managing diseases in rice and pearl millet. Enhanced Research Capacity. <http://www.plbio.kvl.dk/staffpresent/personer/han_jor/ENRECA.htm>

Deena, S., Rao, Y.B.N. and Singh, D. (2003) Allelopathic evaluation of *Andrographis paniculata* aqueous leachates on rice (*Oryza sativa* L.). *Allelopathy Journal* 11, 71–76.

Deer, H.M. (1999) Pesticide application training workshops: Biopesticides and their active ingredients. Utah Pesticide and Toxic News, Volume XVII, Number 11 Nov 1999. <http://extension.usu.edu/files/agpubs/nov99.htm>

Delabays, N. and Mermillod, G. (2002) The phenomenon of allelopathy: first field assessments. *Revue Suisse d'Agriculture* 34, 231–237.

Devanath, H.K., Pathank, J.J. and Bora, L.C. (2002) *In vitro* sensitivity of *Ralstonia solanacearum*, causing bacterial wilt of ginger towards antagonists, plant extracts and chemicals. *Journal of Interacademica* 6, 250–253.

Dey, P.M. and Harborne, J.B. (1989) *Methods in Plant Biochemistry Vol. 1*. Academic Press. London, pp. 552.

Duary, B. (2002) Effect of leaf extract of sesame (*Sesamum indicum* L.) on germination and seedling growth of blackgram (*Vigna mungo* L.) and rice (*Oryza sativa* L.). *Allelopathy Journal* 10, 153–156.

Dudai, N., Poljakoff-Mayber, A., Mayer, A.M., Putievsky, E. and Lerner, H.R. (1999) Essential oils as allelochemicals and their potential use as bioherbicides. *Journal of Chemical Ecology* 25, 1079–1089.

Duke, S.O. (1990) Natural pesticides from plants. In: Janick, J. and Simon, J.E. (eds). *Advances in New Crops*. Timber Press, Portland, USA, pp. 511–517.

Duke, S.O., Abbas, J.K. and Einhellig, F.A. (1995) Natural products with potential use as herbicides. In: Inderjit, A. and Dakshini, K.M.M. (eds). *Allelopathy: Organisms, Processes and Applications*. American Chemical Society, Washington, USA, pp. 348–362.

Duke, S.O., Dayan, F.E., Romagni, J.G. and Rimando, A.M. (2000) Natural products as sources of herbicides: Current status and future trends. *Weed Research* 40, 99–111.

Duncan, A.C., Jager, A.K. and Van Staden, J. (1999) Screening of Zulu Medicinal-Plants for Angiotensin-Converting Enzyme (ACE) Inhibitors. *Journal of Ethnopharmacology* 68, 63–70.

El-Ghaouth, A., Wilson, C.L. and Wisniewski, M.E. (1995) Sugar analogs as potential fungicides for post harvest pathogens of apple and peach. *Plant Disease* 79, 254–258.

Elzen, G.W. and King, E.G. (1999) Periodic release and manipulation of natural enemies. In: Bellows, T.S. and Fisher, T.W. (eds) *Handbook of Biocontrol*. Academic Press, San Diego, USA, pp. 253–270.

Eshel, D., Gamliel, A., Grinstein, A., Diprimo, P. and Katan, J. (2000) Combined soil treatments and sequence of application in improving the control of soil borne pathogens. *Phytopathology* 90, 751–757.

Ferreira, M.I. and Lourens, A.F. (2002) The efficacy of liquid seaweed extract on the yield of canola plants. *South African Journal of Plant and Soil* 19, 159–161.

Ganesan, T. and Krishnaraju, J. (1995) Antifungal properties of wild plant II. *Advances in Plant Sciences* 8, 194–196.

Gebre-Amlak, A. and Azerefegne, F. (1998) Insecticidal activity of chinaberry, endod and pepper tree against maize stalk borer (*Lepidoptera*: *Noctuidae*) in Southern Ethiopia. *International Journal of Pest Management* 35, 143–145.

Ghosh, S.K., Bhaskar-Sanyal, Satyabrata-Ghosh, Subhendu-Gupta, Sanyal, B., Ghosh, S. and Gupta, S. (2000) Screening of some angiospermic plants for antimicrobial activity. *Journal of Mycopathological Research* 38, 19–22.

Goeden, R.D. and Andrés, L.A. (1999) Biocontrol of weeds in terrestrial ecosystems and aquatic environments, In: Bellows, T.S. and Fisher, T.W. (eds) *Handbook of Biocontrol*. Academic Press, San Diego, USA, pp. 871–890.

Gonzalez, J., Suarez, M., Granda, D.E., Orozco-De, A.M. and De-Granda, E. (1988) Antifungal constituents in root nodules of *Alnus acuminata*. *Agronomia Colombiana* 5, 83–85.

Grayer, R.J. and Harborne, J.B. (1994) A survey of antifungal compounds from higher plants. *Phytochemistry* 37, 19–43.

Gupta, R.K. and Bansal, R.K. (2003) Comparative efficacy of plant leaf extracts and fungicides against *Fusarium oxysporum Schlecht* inducing fenugreek wilt under pot house condition. *Annals of Applied Biology* 19, 35–37.

Hall, K. (2002) Production, Processing and Practical Application of Natural Antifungal Crop Protectants: Production of Natural Antifungal Crop Protectants. <http://www.nf-2000.org/secure/Fair/S388. htm>.

Hamburger, M. and Hostettmann, K. (1991) Bioactivity in plants: The link between phytochemistry and medicine. *Phytochemistry* 30, 3864–3874.

Heiny, D.K. and Templeton, G.E. (1993) Economic comparisons of mycoherbicides to conventional herbicides. In: Altman, J. (ed.) *Pesticide Interactions in Crop Production*. CRC Press, Boca Raton, FL, pp. 395–408.

Helmut, K., Theo, S., Jim, L., Michael, O. and John R. (1994) Activation of systemic acquired disease resistance in plants. *Journal of Plant Pathology* 100, 359–369.

Hoffmann, J.J., Jolad, S.D., Hutter, L.K., Mclaughlin, S.P., Savage, S.D., Cunningham, S.D., Genet, J.L. and Ramsey, G.R. (1992) Glaucarubolone glucoside, a potential fungicidal agent for the control of grape downy mildew. *Journal of Agricultural and Food Chemistry* 40, 1056–1057.

Iqba, Z, Hiradate, S., Noda, A., Isojima, S. and Fujii, Y. (2002) Allelopathy of buckwheat: assessment of allelopathic potential of extract of aerial parts of buckwheat and identification of fagomine and other related alkaloids as allelochemicals. *Weed Biology and Management* 2, 110–115.

Jin, S. and Sato, N. (2003) Benzoquinone, the substance essential for antibacterial activity in aqueous extracts from succulent young shoots of the pear *Pyrus* spp. *Phytochemistry* 62, 101–107.

Johnson, R. 2001. National Centre for Natural Products Research. <http://www.ole-miss.edu/depts/usda/>

Johri, J.K., Balasubrahmanyam, V.R., Misra, G. and Nigam, S.K. (1994) Botanicals for management of betelvine disease. *National Academy Science Letters* 17, 7–8.

Juliena, M.H., Scottb, J.K., Orapac, W and Paynterd, Q. (2007) History, opportunities and challenges for biological control in Australia, New Zealand and the Pacific islands. *Crop Protection* 26, 255–265.

Jutsum, A.R. (1988) Commercial application of biological control: Status and prospects. *Philosophical Transactions of the Royal Society of London. Series B. Biological Sciences* 318, 357–370.

Kamuro, Y., Takatsuto, S. (1999) Practical applications of brassinosteroids in agricultural fields. In: Sakurai, A., Yokota, T. and Clouse, S.D. (eds) *Brassinosteroids – Steroidal Plant Hormones*, Springer, Tokyo, Japan, pp. 223–241.

Kane, P.V, Kshirsargar, C.R, Jadhav, A.C. and Pawar, N.B. (2002) *In vitro* evaluation of some plant extracts against *Rhizoctonia solani* from chickpea. *Journal of Maharashtra Agricultural Universities* 27, 101–102.

Karavaev, V.A., Solntsev, M.K., Kuznetsov, A.M., Polyakova, I.B., Frantsev, V.V., Yurina, E.V., Yurina, T.P., Taborsky, V., Polak, J., Lebeda, A. and Kudela, V. (2002) Plant extracts as the source of physiologically active compounds suppressing the development of pathogenic fungi. *Plant Protection Science* 38, 200–204.

Kato, N.H. and Kawabata, K. (2002) Isolation of allelopathic substances in lemon balm shoots. *Environment Control in Biology* 40, 389–393.

Kausik, B., Ishita, C., Banerjee, R.K., Uday, B., Biswas, K., Chattopadhyay, I. and Bandyopadhyay, U. (2002) Biological activities and medicinal properties of neem (*Azadirachta indica*). *Current Science* 82, 1336–1345.

Kazakova, V.N., Karsunkina, N.P. and Sukhova, L.S. (1991) Effect of brassinolide and fusicoccin on potato productivity and tuber resistance to fungal diseases under storage. *Izvestiia Timiryazevskoi Sel´skokhoziaistvennoi Akademii* 94(8), 85021.

Khan, M.R. and Omoloso, A.D. (2002) Antibacterial, antifungal activity of *Harpullia petiolaris*. *Fitoterapia* 73, 331–335.

Khan, P.A., Mughal, A.H. and Khan, M.A. (2001) Allelopathic effects of leaf extract of *Populus deltoides* M. on germination and seedling growth of some vegetables. *Range Management and Agroforestry* 22, 231–236.

Khripach, V.A., Zhabinskii, V.N. and de Groot A.E. (1999) *Brassinosteroids – a New Class of Plant Hormones*. Academic Press, San Diego, USA, pp. 325–346.

Khripach, V.A., Zhabinskii, V. and De Groot, A. (2000) Twenty Years of Brassinosteroids: Steroidal Plant Hormones Warrant Better Crops for the XXI Century *Annals of Botany* 86, 441–447.

Khun, P.J. (1989) The discovery and development of fungicide-Does biochemistry have a role? *Pesticide Science* 14, 272–293.

Kishore, G.K., Pande, S. and Rao, J.N. (2002) Field evaluation of plant extracts for the control of late leaf spot in groundnut. *International Arachis Newsletter* 22, 46–48.

Krupinski, G. and Sobiczewski, P. (2001) The influence of plant extracts on growth of *Erwinia amylovora* – the causal agent of fire blight. *Acta Agrobotanica* 54, 81–91.

Lawson, M. and Kennedy, R. (1998) Evaluation of garlic oil and other chemicals for control of downy mildew (*Peronospora parasitica*) in organic production of brassica. *Annals of Applied Biology* 132, 14–15.

Lazarides, L. (1998) Secondary Plant Metabolites. <http://www.waterfall 2000.com/ a-z/secondarypl.htm>

Leksomboon, C., Thaveechai, N., Kositratana, W., Chalida, L., Niphone, T. and Wichai, K. (2001) Potential of plant extracts for controlling citrus canker of lime. *Kasetsart Journal Natural Sciences* 35, 392–396.

Lim, U.K. (1987) Effect of brassinolide treatment on shoot growth, photosynthesis, respiration and photorespiration of rice seedlings. *Agricultural Research of Seoul National University* 12, 9–14.

Lim, U.K. and Han, S.S. (1988) The effect of plant growth regulating brassinosteroid on early state and yield of corn. *Agricultural Research of Seoul National University* 13, 1–14.

Macias, F.A., Varela, R.M., Torres, A., Galindo, J.L.G., Molinillo, J.M.G., Inderjit and Mallik, A.U. (2002) Allelochemicals from sunflowers: chemistry, bioactivity and applications. *Chemical Ecology of Plants: Allelopathy in Aquatic and Terrestrial Ecosystems*, 73–87.

Mandava, N.B. (1979) Natural products in plant growth regulation. In: Mandava, N.B. (ed.) *Plant Growth Substances*. ACS Symposium series III, American Chemical Society, Washington, USA, pp. 135–213.

Mandava, N.B. (1988) Plant growth-promoting brassinosteroids, *Annual Reviews of Plant Physiology Plant Molecular Biology* 39, 23–52.

Marston. A., Gafner, F., Dossaji. S.F. and Hostettmann. (1988) Fungicidal and molluscicidal saponins from *Dolichos kilimandscharicus*. *Phytochemistry* 27, 1325–1326.

McFadyen, R.E.C. (1998) Biocontrol of weeds, *Annual Reviews in Entomology* 43, 369–393.

McWhorter, C.G., Chandler, J.M. (1982) Conventional weed control technology. In: Charudattan, R. and Walker, H.L. (eds) *Biological Control of Weeds with Plant Pathogens.* Wiley, New York, USA, pp. 5–27.

Menzies, J.G. and Bélanger, R.R. (1996) Recent advances in cultural management of diseases of green house crops. *Canadian Journal of Plant Pathology* 18, 186–193.

Meudt W.J., Thompson M.J. and Bennett, H.W. (1983) Investigations on the mechanism of brassinosteroid response III: Techniques for potential enhancement of crop production. In: *Proceedings of the 10th Annual Meeting of the Plant Growth Regulators Society of America.* Madison, USA, pp. 312–318.

Michael, W. (1999) Plant–microbe interactions and secondary metabolite with antiviral, antibacterial and antifungal properties. In: Michael, W. (ed.) *Function of Plant Secondary Metabolites and their Exploitation in Biotechnology.* CRC Press, Columbus, USA, pp. 187–273.

Minorsky, P.V. (2001) Natural Products (Secondary metabolites). American Society of Plant Biologists. <http://www.aspb.org/publications/biotext/sumrys/ch24.cfm>.

Moezelaar, R., Braam, C., Zomer, J., Gorris, L.G.M., and Smid, E.J. (1999) Volatile plant Metabolites for postharvest crop protection. In: Lyr, H., Russell, P.E, Dehne, H-W. and Sisler, H.D. (eds) *Modern Fungicides and Antifungal Compounds II.* Intercept Limited, pp. 453– 467.

Morais, L.D.E., Carmo, M.D.O., Viegas, E.C., Teixeira, D.F., Barreto, A.S. Pizarro, A.P.B., Gilbert, B., De Morais, L.A.S., Do Carmo, M.G.F., Ming, L.C., Craker, L.E., Scheffer, M.C. and Chaves, F.C.M. (2002) Evaluation of antimicrobial activity of extracts of medicinal plants on three tomato phytopathogens. *Acta Horticulturae* 569, 87–90.

Mori, K., Takematsu, T., Sakakibara, M. and Oshio, H. (1986) Homobrassinolide, its production and use. US Patent 4,604,240.

Müssig, C., Shin, G-H and Altmann, T. (2003) Brassinosteroids promote root growth in Arabidopsis. *Plant Physiology*, 133, 1261–1271.

Naidu, G.P. (1988) Antifungal activity in *Codiaeum variegatum* leaf extract. *Current Science* 57, 502–504.

Naseby, D.C., Way, J.A., Bainton, N.J. and Lynch, J.M. (2001) Biocontrol of *Pythium* in the pea rhizosphere by antifungal metabolite producing and non-producing *Pseudomonas* strains. *Journal of Applied Microbiology* 90, 421–429.

Neelam, K., Bisaria, A.K. and Khare, N. (2002) The allelopathic effect on *Triticum aestivum* of different extracts of *Leucaena leucocephala. Indian Journal of Agroforestry* 4, 63–65.

Nie, C., Wen, Y., Li, H., Chen, l., Hong, M., Huang, J., Nie, C., Wen, Y., Li, H., Chen, L.Q., Hong, M.Q. and Huang, J.H. (2002) Study on allelopathic effects of *Wedelia chinensis* on some weeds in South China. *Weed Science China* 2, 15.

Nikolov, A. and Boneva, I. (1999) Screening tests of fungicides from plant origin towards powdery mildews from roses and cucumbers. *Bulgarian Journal of Agricultural Science* 5, 975–978.

Obaid, K.A. and Qasem, J.R. (2002) Inhibitory effects of *Cardaria draba* and *Salvia syriaca* extracts to certain vegetable crops. *Dirasat Agricultural Sciences* 29, 247–259.

Oka, Y., Nacar, S., Putievsky, E. Ravid, U., Yaniv, Z. and Spiegel, Y. (2000) Nematicidal activity of essential oils and their components against the root-knot nematode. *Phytopathology* 90, 710–715.

Om, P., Pandey, V.N., Pant, D.C. and Prakash, O. (2001) Fungitoxic properties of some essential oils from higher plants. *Madras Agricultural Journal* 88, 73–77.

Orlikowski, L.B. (2001a) Effect of grapefruit extract on development of *Phytophthora cryptogea* and control of foot rot of gerbera. *Journal of Plant Protection Research* 41, 288–294.

Orlikowski, L.B., Skrzypczak, C. and Harmaj, I. (2001b) Biological activity of grapefruit extract in the control of *Fusarium oxysporum. Journal of Plant Protection Research* 41, 420–427.

Pandey, M.K., Singh, A.K. and Singh, R.B. (2002) Mycotoxic potential of some higher plants. *Plant Disease Research* 17, 51–56.

Paul, R., Birch, J. and Sophien, K. (2000) Studying interaction transcriptomes: Coordinated analyses of gene expression during plant-microorganism interactions. *New technologies for life sciences: A Trends Guide.* <http://trends.com>. pp. 77–82.

Petsikos, P.N, Shmitt, A., Markellou, E., Kalamarakis, A.E, Tzempelikou, K., Siranidou, E. and Konstantinidou, D.S. 2002. Management of cucumber powdery mildew by new formulations of Reynoutria sachalinensis (F. Schmidt) Nakai extract. *Zeitschrift fur Pflanzenkrankheiten und Pflanzenschutz* 109, 478–490.

Philip, E.R., Richard, J.M. and Ken, W. (1995) Control of fungi pathogenic to plants. In: Hunter, P.A., Darby, G.K. and Russell, N.J. (eds) *Fifty years of Antimicrobials: Past Perspectives and Future Trends.* Cambridge University Press, pp. 85–110.

Pillmoor, J.B. (1993) Natural products as a source of agrochemical and leads for chemical synthesis. *Pesticide Science* 39, 131–140.

Prabha, P., Bohra, A. and Purohit, P. (2002) Antifungal activity of various spice plants against phytopathogenic fungi. *Advances in Plant Sciences* 15, 615–617.

Pretorius, J.C., Zietsman, P.C. and Eksteen, D. (2002a) Fungitoxic properties of selected South African plant species against plant pathogens of economic importance in agriculture. *Annals of Applied Biology* 141, 117–124.

Pretorius, J.C, Craven, P. and Van der watt, E. (2002b) *In vivo* control of *Mycosphaerella pinodes* on pea leaves by a crude bulb extract of *Eucomis autumnalis. Annals of Applied Biology* 141, 125–131.

Pretorius, J.C. (2003) Flavonoids: A review of its commercial application potential as anti-infective agents. *Current Medical Chemistry: Anti Infective Agents.* 2, 335–353.

Rajiv, K., Jha, D.K., Dubey, S.C. and Kumar, R. (2002) Evaluation of plant extracts against blight of finger millet. *Journal of Research, Birsa Agricultural University* 14, 101–102.

Ramos, M.T.B. (1995) Efeito da aplicação de 24-epibrassinolídio sobre o rendimento e a omposição química de sementes de grão-de-bico (*Cicer arietinum* L.). MSc thesis.Araraquara, Universidade Estadual Paulista Júlio de Mesquita Filho.

Randhawa, M.A., Cheema, Z.A. and Ali, M.A. (2002) Allelopathic effect of sorghum water extract on the germination and seedling growth of *Trianthema portulacastrum. International Journal of Agriculture and Biology* 4, 383–384.

Rao, S.S-R., Vardhini, B.V., Sujatha, E. and Anuradha, S. (2002) Brassinosteroids – A new class of phytohormones. *Current Science* 82 (10), 1239–1245.

Rechcigl, J.E. and Rechcigl, N.A. (2000) *Biological and Biotechnological Control of Insect Pests.* Lewis, New York, pp. 101–121.

Rector, B.G. (2008) Molecular biology approaches to control of intractable weeds: New strategies and complements to existing biological practices *Plant Science* 175, 437–448.

Richard, A.W. (2000) Botanical insecticides, soaps and oils. In: Rechcigl, J.E. and Rechcigl, N.A. (eds) *Biological and Biotechnological Control of Insect Pests.* Lewis Publishers, CRC Press, Columbus, USA, pp. 101–121.

Richard, A.D. (2001) Natural products and plant disease resistance. *Nature* 411, 843–847.

Rivera, C.G., Martinez, T.M.A., Vallejo, C.S., Alvarez, M.G., Vargas, A.D.C., Moya, S.P., Primo, Y.E., Del, C. and Vargas, A.I. (2001) *In vitro* inhibition of mycelial growth of *Tilletia indica* by extracts of native plants from Sonora, Mexico. *Revista Mexicana de Fitopatologia* 19, 214–217.

Roberts, J.A. and Hooley, R. (1988) *Plant Growth Regulators.* Chapman & Hall, New York, pp. 164–174.

Rodriguez, D.A. and Montilla, J.O. (2002) Decrease in wilt caused by *Fusarium* on tomato by means of extract of *Citrus paradisi*. *Manejo Integrado de Plagas* 63, 46–50.

Roth, U., Friebe, A., Schnabl, H. (2000) Resistance induction in plants by a brassinosteroid-containing extract of *Lychnis viscaria* L. *Zeitschrift fur Naturforschung.Section C, Biosciences* 55, 552–559.

Salamci, E., Kordali, S., Kotan, R., Cakir, A. and Kaya, Y. (2007) Chemical compositions, antimicrobial and herbicidal effects of essential oils isolated from *Turkish Tanacetum aucheranum* and *Tanacetum chiliophyllum* var. *chiliophyllum*. *Biochemical Systematics and Ecology* 35, 569–581.

Salisbury, F.B. and Ross, C.W. (1992) *Plant Physiology*, 4th Edition, Wadsworth Publishing Company, Belmont, California, pp. 357–407.

Sasikumar, K., Parthiban, K.T., Kalaiselvi, T., and Jagatram, M. (2002) Allelopathic effects of *Parthenium hysterophorus* on cowpea, pigeonpea, greengram, blackgram and horsegram. *Allelopathy Journal* 10, 45–52.

Schmitt, A., Eisemann, S., Strathmann, S., Emslie, K.A. and Sedon, B. (1996) *The use of Reynoutria sachalinensis extracts for induced resistance in integrated disease control: Effects on Botrytis cinerea*. Programme and book of abstracts of the XIth International *Botrytis* Symposium, 23–27 June, Wageningen, The Netherlands, pp. 69.

Scholz, K., Vogt, M. and Kunz, B. (1999) Application plant extracts for controlling fungal infestation of grains and seeds during storage, In: Lyr, H., Russell, P.E., Dehne, H-W. and Sisler, H.D. (eds) *Modern Fungicides and Antifungal Compounds II*. Intercept Limited, pp. 429–435.

Seddon, B. and Schmitt, A. (1999) Integrated biological control of fungal plant pathogens using natural products. In: Lyr, H., Russell, P.E., Dehne, H-W. and Sisler, H.D. (eds) *Modern Fungicides and Antifungal Compounds II*. Intercept Limited, pp. 423–428.

Singh, H.P., Batish, D.R., Kohli, R.K. and Kaur, S. (2001) Crop allelopathy and its role in ecological agriculture. *Crop Production* 4, 121.

Singh, N.B., Ranjana-Singh and Singh, R. (2003) Effect of leaf leachate of Eucalyptus on germination, growth and metabolism of green gram, black gram and peanut. *Allelopathy Journal* 11, 43–52.

Singh, U.P., Pandey, V.N., Wagner, K.G. and Singh, K.P. (1990) Antifungal activity of ajoene, a constituent of garlic (Allium sativum). *Canadian Journal of Botany* 68, 1354–1356.

Solunke, B.S., Kareppa, B.M. and Gangawane, L.V. (2001) Integrated management of *Sclerotium* rot of potato using carbendazim and plant extracts. *Indian Journal of Plant Protection* 29, 142–143.

Srivastava, S.K., Srivastava, S.D., Chouksey, B.K. (2001) New antifungal constituents from *Terminalia alata*. *Fitoterapia* 72, 106–112.

Steglich, W., Steffan, B., Eizenhöfer, T., Fugmann, B., Herrmann, R. and Klamann, J.D. (1990) Some problems in the structural elucidation of fungal metabolites. In: *Bioactive Compounds from Plants*, Wiley, Chichester (Ciba Foundation Symposium 154), UK, pp. 56–65.

Stumpf, P.K. and Conn, E.E. (1981) *The Biochemistry of Plants. A Comprehensive Treatise, Vol. 7*. Academic Press, New York, pp. 34.

Takematsu, T., Ikekawa, N., Shida, A. (1988) Increasing the yield of cereals by means of brassinolide derivatives. US Patent 4,767,442.

Tegegne, G. and Pretorius, J.C. (2007) *In vitro* and *in vivo* antifungal activity of crude extracts and powdered dry material from Ethiopian wild plants against economically important plant pathogens. *BioControl* 52, 877–888.

Terefa, T. (2002) Allelopathic effects of *Parthenium hysterophorus* extracts on seed germination and seedling growth of *Eragrostis* tef. *Journal of Agronomy and Crop Science* 188, 306–310.

Tran, D.X. and Tsuzuki, E. (2002) Varietal differences in allelopathic potential of alfalfa. *Journal of Agronomy and Crop Science* 188, 2–7.

Tworkoski, T. (2002) Herbicide effects of essential oils. *Weed Science* 50, 425–431.

Ume, K., Iftikhar, A., Malik, S.A., Strang, R.H.C., Cole, M. and Strang, R. (2001) Efficacy of neem products against *Sclerotium rolfsii* infecting groundnut: *The science and application of neem*. Glasgow, UK, pp. 33–37.

Ushiki, J. Hayakawa, Y. and Tadano, T. (1996) Medicinal plants for suppressing soilborne plant diseases. I. Screening for medicinal plants with antimicrobial activity in roots. *Soil Science and Plant Nutrition* 42, 423–426.

Verpoorte, R. (1998) Exploration of nature's chemodiversity: the role of secondary metabolites as leads in drug development. *Drug Discovery Today* 3, 232–238.

Waller, G.R. (1999) Recent advances in saponins used in foods, agriculture, and medicine. In: Cutler, G. and Cutler, S.J. (eds) *Biologically Active Natural Products: Agrochemicals*. CRC Press, Columbus, USA, pp. 243–274.

Waterman, P.G. and Mole, S. (1989) Extrinsic factors influencing production of secondary metabolites in plants. In: Bernays, E.A. (ed.) *Insect-Plant Interactions*, Vol. II. CRC Press, Boca Raton, FL., pp. 107–134.

Williams, R.J. (1992) Management of weeds in the year 2000. In: Kadir, A-A.S.A. and Barlow, H.S. (eds) Pest management and the environment in 2000. C.A.B International, Wallingford, Oxon, UK, pp. 257–280.

Wilson, C.L. (1997) Rapid evaluation of plant extracts and essential oils for antifungal activity against *Botrytis cenerea*. *Plant Disease* 81, 204–210.

Workneh, T.S., Osthoff, G. and Steyn, M.S. (2009) Integrated agrotechnology with preharvest ComCat® treatment, modified atmosphere packaging and forced ventilation evaporative cooling of tomatoes. *African Journal of Biotechnology* 8, 860–872.

Wu, H., Pratley, J., Lemerle, D., Haig, T. and An, M. (2002) Screening methods for the evaluation of crop allelopathic potential. *Botanical Reviews* 67, 403–415.

Zeller, W., Laux, P., Hale, C. and Mitchell, R. (eds) (2002) *Newest Results on the Biocontrol of Fire Blight in Germany*. Proceedings of the IXth International Workshop on Fire Blight, Napier, New Zealand, 8–12 October, *Acta Horticulturae* 590, 243–2465.

Zullo, M.A.T and Adam, G. (2002) Brassinosteroid phytohormones – structure, bioactivity and applications. *Brazilian Journal of Plant Physiology* 14, 143–181.

4 Antimicrobials of Plant Origin to Prevent the Biodeterioration of Grains

K.A. Raveesha

Department of Studies in Botany, University of Mysore, Manasagangotri, India

Abstract

A significant portion of stored food becomes unfit for human consumption due to the biodeterioration of grains. The incessant and indiscriminate use of synthetic chemicals in crop protection has been one of the major factors in polluting soil and water bodies. Thus, there is a need to search for effective, efficient and eco-friendly alternative methods for preventing the biodeterioration of grains during storage. Antimicrobials of plant origin are an important alternative, which could be better exploited to prevent grain biodeterioration. Fungi are significant destroyers of food so the biodeterioration of grains can be prevented by inhibiting fungal growth. Of late, many plant extracts have been screened for antifungal activity. *Decalepis hamiltonii* Wight & Arn. (Asclepiadaceae) and *Psoralea corylifolia* L. (Leguminosae) have revealed highly significant antifungal activity. The antifungal active compounds from these plants have been isolated by antifungal activity guided assays and characterized using NMR, IR and mass spectral studies. The biomolecules responsible for the activity were identified as 2H-Furo [2,3-II]-1-benzopyran-2-one in *P. corylifolia* and 2-hydroxy-4-methoxybenzaldehyde in *D. hamiltonii*. *In vitro* and *in vivo* evaluations of these biomolecules have shown promising inhibitory activity of important biodeterioration-causing fungi. Comparative evaluation of the bioactive compounds with those of the routinely used chemical fungicides is highly encouraging. Results suggest the potential of these biomolecules for commercial exploitation to develop eco-friendly herbal remedies for preventing the biodeterioration of grains during storage.

4.1 Introduction

Storage of food is a necessity to ensure the availability of food throughout the year. Stored foods are prone to postharvest loss in quality and quantity due to infestation by different groups of organisms. Biodeterioration may be defined as the quality and quantity loss of stored food caused by organisms. The organisms broadly responsible for such deterioration are microbes, insects and rodents. Among microbes, fungi are significant destroyers of food, particularly species of *Aspergillus* and *Penicillium*. Grain produced is not a grain until it is consumed without quality loss (Neergaard, 1977). In spite of the developments in the methods of food preservation, nearly 30% of the food produced is lost during storage, due to poor handling and storage practices.

The incessant and indiscriminate use of chemical pesticides in agriculture has resulted in drug resistance, residual toxicity, and has caused soil and water pollution. This has necessitated the search for natural eco-friendly alternatives, which are biodegradable and non-toxic to non-target species. Plants are a repository of novel biomolecules with various biological activities. Among the 250,000–500,000 species of plants available on earth, the number of plants screened phytochemically is a small percentage, among these the number of plants screened for various biological activities is infinitesimally small. In recent years there have been serious efforts to screen plants for various biological activities and the scientific endeavour is focused mainly on utilizing plants for human health and least towards plants for plant health. Considering the fact that large amounts of synthetic pesticides known to cause environmental pollution are used in present-day agriculture, there is an urgent need to search for natural eco-friendly alternatives. Antimicrobials of plant origin appear to be one of the safest and eco-friendly alternatives. Hence, there is a need to search for antimicrobials of plant origin to manage field crop loss in general and to prevent biodeterioration of grains during storage in particular.

Clinical microbiologists have two reasons to be interested in antimicrobials of plant origin. First, it is very likely that these phytochemicals or biomolecules will find their way into the arsenal of prescribed antimicrobial drugs. Second, the public is becoming increasingly aware of the problems with and side effects from the over prescription and traditional use of antibiotics (Cowan, 1999).

Agricultural microbiologists have many more reasons to be interested in antimicrobials of plant origin: (i) it is very likely that these phytochemicals may find their way into the array of antimicrobial drugs prescribed; (ii) it is known that the effective life span of any antimicrobial is limited, hence newer antimicrobials are necessary; (iii) it may help to overcome the rising incidence of drug resistance amongst pathogenic microbes and the mechanism of action could be different; (iv) the phytochemicals present in plant extracts may alleviate the side effects that are often associated with synthetic antimicrobials; (v) phytomedicines usually have multiple effects on the plant body and their actions are often growth promoting, more systemic and beyond the symptomatic treatments of disease; and (vi) it may help in developing cost effective remedies that are affordable to the population (Cowan, 1999; Doughari, 2006).

Finding healing powers in plants for human health needs is an ancient idea and practice. Searching herbal remedies for crop protection in general and prevention of biodeterioration in particular is recently gaining importance. Agricultural microbiologists around the world are evaluating a number of plant extracts for antimicrobial activity against important plant pathogens *in vitro*. Some attempts are also being made to identify the potential of plant extracts to prevent biodeterioration of grains during storage.

The body of literature available in the area of antimicrobials of plant origin is voluminous and concerns different areas of study. In the present context, the literature reviewed concerns mainly the antifungal agents of plant origin with specific reference to important biodeterioration-causing fungi. The body of literature available may be broadly categorized under the following headings:

- *In vitro* evaluation of plant extracts for antifungal activity against important biodeterioration-causing fungi.
- *In vitro* evaluation of the active compounds for antifungal activity against important biodeterioration-causing fungi.
- Isolation and identification of the antifungal active compounds from plants.

4.2 *In Vitro* Evaluation of Plant Extracts for Antifungal Activity against Important Biodeterioration-causing Fungi

Screening plant extracts for antimicrobial activity is the first step towards the isolation of the active component. Water and/or different organic solvents, singly or in combination, are generally used in the preparation extracts.

Aqueous extracts

Antifungal activity assays of leaf extract obtained from *Ocimum sanctum, Adhatoda vasica, Emblica officinalis, Saussurea lappa, Glycyrrhiza glabra, Zingiber officinale, Piper longum, Piper nigrum, Onosma bracteatum, Tinospora cordifolia, Fagonia cretica* and *Terminalia chebula* were tested against *Aspergillus flavus* at 10% and 50% concentration. *P. longum* and *Z. officinale* showed significant antifungal activity (Farooq and Pathak, 1998). Crude extracts of 40 Iranian and Canadian plants were tested for antifungal activity against several species of *Aspergillus*. Of these, 26 plants (65%) showed activity and the spectrum of activity was wide in case of *Diplotaenia damavandica, Heracleum pessicum, Sanguisorba minor* and *Zataria multiflora* (Sardari *et al.*,1998).

Sinha and Saxena (1999) conducted an antifungal activity assay of leaf extract of *Allium cepa* and *Allium sativum* (garlic) against *Aspergillus niger* and reported that the garlic was more effective in inhibiting the germination of spores and mycelial growth of the fungus. Mahmoud (1999) evaluated the antifungal activity of five different concentrations of aqueous extract of *Lupinus albus, Ammivis naga* and *Xanthium pungens* against *A. flavus* and production of aflatoxins and reported the inhibition of mycelial growth of *A. flavus* at 2, 4, 6, 8 and 10 mg/ml.

Aqueous extracts of 50 plants belonging to 27 families were screened for antifungal activity against *A. flavus* and *A. niger* and it was found that only four plants, namely *Trachysper neumammi, Allium sativum, Syzygium aromaticum* and *Plectranthus rugosus* were effective against both species of *Aspergillus* (Singh and Singh, 2000). Aqueous extract of *Terminalia australis* has been reported to be effective in inhibiting *Aspergillus* strains by Carpano *et al.* (2003).

Aqueous extracts of *Aloe barbadensis, Datura stramonium, Zingiber officinale, Murraya koenigii* and *Azadirachta indica* were evaluated against species of *Aspergillus* by employing the poisoned food technique and seed inoculation method. All the plant extracts showed significant inhibitory activity. *D. stramonium* and *A. indica* showed higher activity (Sharma *et al.*, 2003). The extracts of *Toona ciliata* (stem bark) and *Amoora rohituka* (stem bark) exhibited significant *in vitro* antifungal activities against *A. flavus* at 20 and 30% concentration (Chowdhury *et al.*, 2003).

Solvent extracts

Welsh onion ethanol extract was tested against *A. flavus* and *A. parasiticus* for mycelial growth inhibitory activity and aflatoxins production (Fan and Chen, 1999). Sharma *et al.* (2002) evaluated an alcoholic extract of *Semecarpus anacardium* L. against *Aspergillus fumigatus* at different concentrations (20–400 µg/ml) and observed complete inhibition at 40 µg/ml.

Petroleum ether and methanolic extract of *Eupatorium ayapana* were tested for antifungal activity at 250, 500, 750 and 1000 µg/ml against *A. niger* and *A. flavus*. Petroleum ether extract showed higher antifungal activity than the methanolic extract (Gupta *et al.*, 2002). An ethyl acetate soluble fraction of acidified aqueous mother liquor and buffer soluble fraction of neutral mother liquor of stem bark of *Alangium salvifolium* were tested for antifungal activity against species of *Aspergillus* by Katyayani *et al.*, (2002) and reported significant activity at 10 mg/ml concentration.

The methanolic extract of stem bark of *Ailanthus excelsa* partitioned with chloroform recorded significant antifungal activity against *A. niger, A. fumigatus, Penicillium flequentence*, and *Penicillium notatum* at 300, 90, 70, and 140 µg/ml concentrations, respectively (Joshi *et al.*, 2003). Moderate antifungal activity of ethanolic extracts of the trunk bark of *Zanthoxylum fagara, Z. elephantiasis* and *Z. martinicense* was observed against *A. niger*, and *A. flavus* at 500 and 1000 µg/disc (Hurtado *et al.*, 2003). Ethanolic extract of seeds of *Piper guineense* and other solvent fractions obtained by column chromatography tested against *A. flavus*, revealed highly significant antifungal effect (Ngane *et al.*, 2003). Methanol and methyl chloride extracts of 20 Indonesian plants were tested for antifungal activity against *A. fumigatus* at 10 mg/ml by the poisoned food technique. Extracts of six plants, viz. *Terminalia catappa, Swietenia mahagoni, Phyllanthus acuminatus, Ipomoea* spp., *Tylophora asthmatica* and *Hyptis brevipes* recorded significant antifungal activity (Goun *et al.*, 2003).

Extracts of aerial parts of *Achillea clavennae, Achillea holosericea, Achillea lingulata* and *Achillea millefolium* (hexane:ether:methanol = 1:1:1) were tested for antifungal activity by disc diffusion assay against *A. niger*. All four species exhibited antifungal activity against tested strains (Stojanovic *et al.*, 2005).

Organic solvent leaf extracts of two Moroccan *Cistus* L. species *Cistus villosus* L. and *Cistus monspeliensis* L. (Cistaceae) used in traditional medicine were tested for their antifungal properties against *A. fumigatus*. The extracts differed in their antifungal activities. *C. villosus* extracts exhibited higher activity than *C. monspeliensis* (Bouamama *et al.*, 2006).

Antifungal activity of *Trapa natans* L. fruit rind, extracted in different solvents with increasing polarity was observed and 1,4-dioxan, chloroform, acetone, dimethylformamide, ethanol, and water did not reveal any activity against *A. niger* (Parekh and Chanda, 2007).

4.3 *In Vitro* Evaluation of the Active Components for Antifungal Activity against Important Biodeterioration-causing Fungi

Plants possess unlimited ability to produce secondary metabolites and more than 12,000 of them have been isolated, which is probably less than 10% of the total. Useful antimicrobial phytochemicals are phenolics and poly phenols, quinones, flavones, flavonoids and flavanols, tannins, coumarins, terpenoids and essential oils, alkaloids, and lectins and polypeptides. Cowan (1999) has critically reviewed the antimicrobials of plant origin and has listed the plants with antimicrobial activity.

Many secondary metabolites have been identified and many more are yet to be discovered. There is growing evidence that most of these compounds are involved in the interaction of plants with other species, primarily in the defence of the plant from plant pests. Secondary compounds represent a large reservoir of chemical structures with biological activity (Duke, 1990). This resource is largely untapped for use as pesticides.

Considering the advantages of organic pesticides over synthetic pesticides, nowadays, attempts are made by many workers to screen plants for antimicrobial activity, isolate and characterize the bioactive compounds from different parts. Such attempts to isolate antimicrobials of plant origin and test their efficacy against different storage fungi to prevent biodeterioration of grains are, however, very few.

The high terpene hydrocarbon content in the oils of chamomile, lavender, eucalyptus, and geranium are responsible for their antifungal activity against *A. niger* and *Aspergillus ochraceous* (Lis-Balchin *et al.*, 1998). Antimicrobial compounds, such as 3-0-methylveracevine, 3-0-[3-(2,2,2,-trifluoroethoxy)]-5-(methoxybenzoyl) veracevine 3-0-(3,5-diiodobenzoyl) veracevine, 3-0-(3-thienoyl) veracevine, isolated from the seeds of *Schoenocaulon officinale* showed highly significant antifungal activity against *A. flavus* and *A. niger* (Oros and Ujvary, 1999).

Mathekga *et al.* (2000) isolated an acylated form of a phloroglucinol from *Helichrysum caespititium*. The structural elucidation revealed that the compound was 2-methyl-4-2'-4'(6'-trihydroxy-3'-(2-methylpropanoylphenyl)-2-0-enylacetate). This compound completely inhibited the mycelial growth of *A. niger* and *A. flavus* at 1.0 μg/ml. Bioactive compounds 12β-hydroxysandarocopimar-15-one and 2-propionoxyo-β-chesorcylic acid isolated from methanolic extract of leaves of *Trichilia heudelotti* showed highly significant antifungal activity against *A. niger* at 100 μg/ml (Aladesanmi and Odediran, 2000).

Investigations on the antifungal active methanol fraction of the root of *Epinetrum villosum* (Exell) Troupin (Menispermaceae) led to the isolation of the bisbenzylisoquinoline alkaloid cocsoline, which displayed significant antifungal activity with a minimum inhibitory concentration (MIC) for *A. flavus* and *A. niger* of 31.25 g/ml (Otshudi *et al.*, 2005).

Phytochemical analysis of the leaves of *Vernonia amygdalina* yielded two known sesquiterpene lactones, vernolide and vernodalol, that exhibited significant antifungal activity. Vernolides exhibited high activity and the 50% lethal concentration (LC_{50}) values ranged from 0.2 to 0.4 mg/ml for *P. notatum*, *A. flavus*, *A. niger* and *Mucor hiemalis*. Vernodalol showed moderate inhibitory activity against *A. flavus*, *A. niger* and *P. notatum* with LC_{50} values of 0.3, 0.4 and 0.5 mg/ml, respectively (Erasto *et al.*, 2006).

4.4 Isolation and Identification of the Antifungal Active Component from Plants

The isolation and identification of antimicrobials of plant origin involve the following stepwise approach:

- Selection of the plant, followed by the selection of the plant part.
- Selection of appropriate solvent for extraction.
- Separation and purification of the active component.
- Characterization and structural elucidation of the active component.
- *In vitro* evaluation of the active component for antimicrobial activity against test fungi.
- *In vivo* evaluation of the active component for antifungal activity.

Employing bioactivity-directed fractionation and isolation (BDFI) for a specific antimicrobial activity against biodeterioration-causing fungi is advantageous. Two plants have been extensively studied using this method, namely *Psoralea corylifolia* and *Daecalepis hamiltonii*. The active component responsible for antifungal activity has been isolated, characterized and the activity has been demonstrated *in vitro*. Further investigations have been done to prove its efficacy to prevent biodeterioration *in vivo*.

Potential of 2H-furo[2,3-H]-1-benzopyran-2-one isolated from *Psoralea corylifolia* L. to prevent the biodeterioration of maize

Psoralea corylifolia L. (Leguminosae) is an annual herb (Fig. 4.1) reported to have medicinal properties. In traditional medicine, the seeds (Fig. 4.2) are used in the treatment of psoriasis, leucoderma and inflammatory disease of

Fig. 4.1. *Psoralea corylifolia* L. (plant).

Fig. 4.2. *Psoralea corylifolia* L. (seeds).

the skin. An antifungal activity guided assay of different solvent extracts revealed that petroleum ether and methanol extracts were highly active. Hence a combination of petroleum ether and methanol was employed for the isolation of the antimicrobial active component from the seeds of *P. corylifolia* (Kiran and Raveesha, 2004).

Isolation and characterization of the antifungal active component

Powdered seeds of *P. corylifolia* were refluxed in a petroleum ether and methanol mixture [9:1(v/v)] in a Soxhlet apparatus for 8 h at 60°C. An excess of solvent was removed by distillation under reduced pressure. The concentrated extract was cooled for 48 h at 5°C to obtain the pure compound as transparent, colourless, rectangular prism-shaped crystals (Fig. 4.3). The average yield of the compound was 5 mg per 25 g of seed. The melting point was 138°C. The purity of the compound was confirmed using thin-layer chromatography (TLC) and the R_f value was 0.47.

The bioactive compound was subjected to infra red (IR), ^{1}H-NMR, ^{13}C-NMR (Al-Fatimi *et al.*, 2006) and gas chromatography–mass spectral (GC–MS) analyses (Yanez *et al.*, 2005) for structural elucidation.

The IR spectrum showed an absorption band in the region of 1652.9 cm^{-1} for C–O-stretching. Further absorption bands at 1550.7 cm^{-1} and 1454 cm^{-1} were due to the presence of a coumarin ring oxygen and a furan ring oxygen, respectively.

In ^{1}H-NMR spectra, the signal due to C_3–H and C_4–H of coumarin appeared at δ 6.41 as a doublet and at δ 7.81 as a singlet. The aromatic protons

Fig. 4.3.   Stereo micrograph of the active compound [2H-furo[2,3-H]-1-benzopyran-2-one] isolated from seeds of *P. corylifolia*, showing transparent, colourless, rectangular, prism-shaped crystals.

(C_7 + C_8) are mingled together and appeared at δ 7.7 as a multiplet. The signal due to the C_3' and C_2 protons appeared at 6.84 and 7.49 as a doublet, respectively.

^{13}C-NMR data of the bioactive compound showed peaks at δ 146.85(C_2), 143.8 (C_4), 119.77(C_3), 115.5(C_7), 114.76(C_8), 2 106.3(C_3'), and 99.86 (C_2'), which are inconsistent with structure.

GC–MS analysis showed a molecular ion peak at M/z 186.17 consistent with the molecular formula $C_{11}H_6O_3$. The peak at M/z 158 was due to the formation of the coumarin cation. The recorded chromatogram of the plot matched with the chromatogram of an already known compound, 2H-furo[2,3-H]-1-benzopyran-2-one. Figure 4.4 presents the molecular structure of the bioactive compound.

In vitro *evaluation of the antifungal active component*

Ten species of *Aspergillus*, viz. *A. flavus*, *A. niger*, *A. terreus*, *A. tamarii*, *A. flavus oryzae*, *A. fumigatus*, *A. candidus*, *A. ochraceous*, *A. flavipes* and *A. flavus columnaris*, and two species of *Penicillium*, viz., *P. chrysogenum* and *P. notatum*, isolated from maize seeds and known to cause biodeterioration of grains served as test fungi for the antifungal activity assay using the poisoned food technique.

The poisoned food technique is as follows. Malt extract salt agar (MESA) medium amended with different concentrations of the bioactive compound were prepared and poured into sterile Petri plates and allowed to cool and solidify. Mycelium discs (5 mm diameter) of 7-day-old cultures of species of *Aspergillus* and *Penicillium* were placed at the centre of the plates and incubated at 25 ± 1°C for 7 days. The MESA medium without the bioactive compound served as a control. The colony diameter was measured. Similarly the fungicides Captan ($C_9H_8Cl_3NO_2S$) and Thiram ($C_6H_{12}N_2S_4$) were also tested against all the test fungi at the recommended dose of 2000 ppm concentration for comparative evaluation. The percentage inhibition of mycelial growth if any was determined by the formula PI = C − T/C × 100; where C is the diameter of control colony and T is the diameter of treated colony (Pinto *et al.*, 1998). MICs for each of the test fungi were determined on the basis of the concentration needed to inhibit totally the test fungi (Fig. 4.5).

The total inhibition of *A. flavus* was observed at 100 ppm. *A. niger* and *A. fumigatus* were totally inhibited at 500 and 600 ppm, respectively. *A. flavus oryzae* and *A. flavus columnaris* were totally inhibited at 700 ppm. *A. ochraceous* and *A. flavipes* were totally inhibited at 900 ppm. Total inhibition of

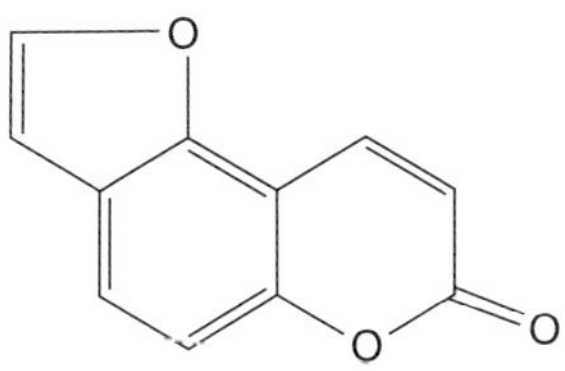

Fig. 4.4. Molecular structure of the bioactive compound, 2H-furo [2,3-H]-1-benzopyran-2-one, isolated from seeds of *P. corylifolia*.

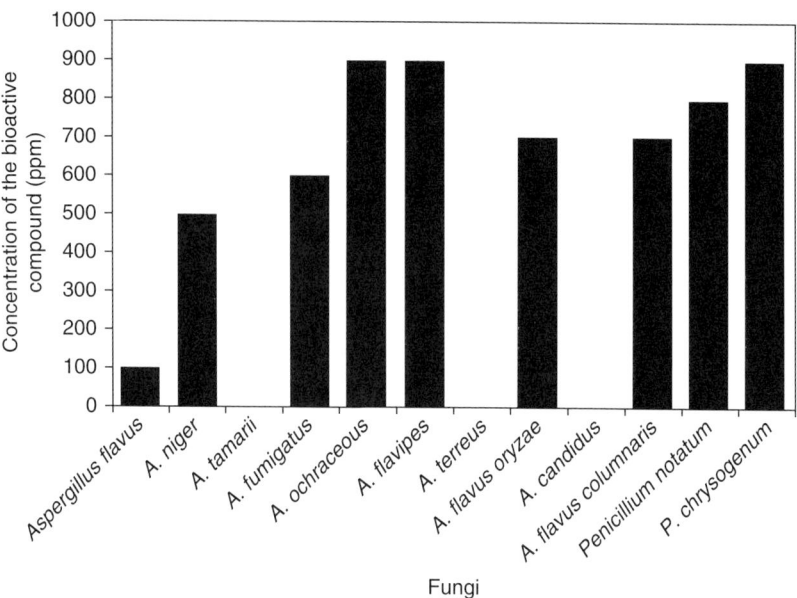

Fig. 4.5. Minimal inhibitory concentration (MIC) of the bioactive compound [2H-furo[2,3-H]-1-benzopyran-2-one] isolated from seeds of *P. corylifolia* L. against species of *Aspergillus* and *Penicillium*.

A. tamarii, *A. terreus* and *A. candidus* was not observed even at 1000 ppm. *Penicillium notatum* and *P. chrysogenum* were totally inhibited at 800 ppm and 900 ppm concentration, respectively.

The concentration of the bioactive compound needed for total inhibition of *Aspergillus* and *Penicillium* species was much lower than the recommended dose of the test fungicides.

In vivo *evaluation of the active component to prevent the biodeterioration of maize*

Maize grains naturally infected with diverse species of *Aspergillus* and *Penicillium* were treated with 250, 500, 1000 and 1500 ppm concentration of the bioactive compound. Untreated seeds served as a control. The treated and untreated seeds were stored at room temperature (30 ± 2°C) for 4 months and the moisture content of the seed sample was maintained at 13.5% (Janardhana *et al.*, 1999). Samples were drawn at regular intervals of 30, 60, 90 and 120 days from each treatment and subjected to seed mycoflora analysis employing the standard method (ISTA, 1999), determination of protein content (Lowry *et al.*, 1951) and carbohydrate content (Dubios *et al.*, 1956) to assess the level of biodeterioration.

Results revealed a significant reduction in the seed mycoflora in all the treatments with a total elimination of seed-borne fungi including species of *Aspergillus* and *Penicillium* in the grains treated with 1000 ppm and 1500 ppm concentration of the bioactive compound in all the storage periods tested. No change in protein and carbohydrate content was observed in the grains treated with 1000 and 1500 ppm even after 4 months of storage, suggesting that the active compound, 2H-furo[2,3-H]-1-benzopyran-2-one isolated from seeds of *P. corylifolia* L. could be exploited to prevent the biodeterioration of grains.

Potential of 2-hydroxy-4-methoxybenzaldehyde isolated from *Decalepis hamiltonii* Wight & Arn. to prevent biodeterioration of paddy

Decalepis hamiltonii Wight & Arn., a member of the family Asclepiadaceae, is an important medicinal plant widely used in traditional medicine. The rhizome (Fig. 4.6) of this plant is largely used in South India for pickling.

An antifungal activity assay of different solvent extracts of the fresh rhizome of this plant revealed highly significant activity in the petroleum ether extract against seed-borne fungal pathogens (Mohana *et al.*, 2006). Further experimentation conducted to isolate and characterize the antifungal active principle from the petroleum phenolic fraction using chloroform as a

Fig. 4.6. Rhizome of *Decalepis hamiltonii* Wight & Arn.

solvent for TLC revealed the presence of seven bands. An antifungal activity assay of each of these bands revealed that band five with an R_f value of 0.77 showed significant antifungal activity, whereas the other bands did not show any antifungal activity. Light blue fluorescence at 365 nm was observed.

The active principle was isolated and subjected to ^{1}H-NMR, ^{13}C-NMR and MS analysis to confirm the identity of the compound. The ^{1}H-NMR analysis of the compound showed NMR peaks at δ 3.85 (s,-0CH3), 6.52 (dd, J = 2 HZ; 3-H), 6.55(d, J = 7 Hz, 5-H), 7.40 (d, J = 7 Hz; 6-H), 9.70 (s, CHO), 11.6 (s,-OH) functional groups. ^{13}C-NMR analysis of the compounds showed eight carbon signals 135.6 (1-CH), 108.7 (3-CH), 167.2 (C of carbonyl), 101.05 (5-CH), 164.8(2-C), 115.5(C), 194.7 (6-CH), and 56.09 (CH3) and its identity was confirmed by MS analysis [m/z (% abundance): 57(48), 95(46), 108(24), 121(20), 151(100), 152(70)]. The strong molecular ion peak (m/z, 152) and stronger M-1 ion peak (m/z, 151) observed were characteristic of an aromatic aldehyde. The melting point of the active compound is 46°C. Results revealed that the active compound responsible for the activity was 2-hydroxy-4-methoxybenzaldehyde (Fig. 4.7), reported in the literature by Nagaraju *et al.* (2001).

The active component isolated was subjected to an *in vitro* antifungal activity assay by the poisoned food technique and *in vivo* experiments to evaluate the potential of this active compound to prevent the biodeterioration of paddy during storage.

An *in vitro* antifungal activity assay against important seed-borne fungi associated with paddy known to cause biodeterioration during storage revealed a highly significant inhibitory activity against the test fungi. Complete inhibition of the test fungi was observed at 650 μg/ml of the compound (Table 4.1). Further *in vivo* experimentation done to assess the efficacy of the active principle to prevent biodeterioration of paddy up to 90 days of storage revealed a high potency of the compound to prevent biodeterioration of grains. The percentage incidence of seed-borne fungal species decreased significantly in the samples treated with 1 g/kg of the active compound. Comparative efficacy studies revealed that the concentration of the active compound needed to prevent growth of fungi responsible for biodeterioration of paddy was much less than that of thiram (2 g/kg) (Table 4.2).

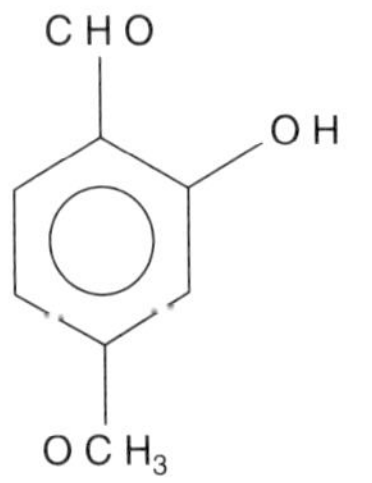

Fig. 4.7. Molecular structure of the bioactive compound, 2-hydroxy-4-methoxybenzaldehyde isolated from the rhizome of *Decalepis hamiltonii* Wight & Arn.

Table 4.1. The antifungal activity of 2-hydroxy-4-methoxybenzaldehyde isolated from *D. hamiltonii* and that of thiram against phytopathogenic fungi isolated from paddy.

Concentration of active compound (µg/ml)	Mycelium growth inhibition (%) of seed-borne pathogenic fungi of paddy					
	Alternaria alternata	*Drechslera tetramera*	*Fusarium oxysporum*	*Fusarium proliferatum*	*Pyricularia oryzae*	*Trichoconis padwickii*
40	3.25 ± 0.3	3.34 ± 0.2	11.47 ± 0.3	12.72 ± 0.4	0.00 ± 0.0	2.10 ± 0.5
60	8.09 ± 0.5	6.6 ± 0.6	24.33 ± 0.3	23.35 ± 0.1	5.66 ± 0.4	3.61 ± 0.5
80	11.33 ± 0.4	19.72 ± 0.4	39.31 ± 0.4	42.30 ± 0.4	10.34 ± 0.2	8.84 ± 0.6
10	16.71 ± 0.5	24.20 ± 0.6	53.05 ± 0.3	51.63 ± 0.2	14.36 ± 0.4	11.88 ± 0.4
100	21.82 ± 0.7	34.60 ± 0.5	56.90 ± 0.5	57.58 ± 0.4	17.77 ± 0.4	12.64 ± 0.5
150	34.34 ± 0.7	45.89 ± 0.6	68.53 ± 0.3	68.06 ± 0.5	22.64 ± 0.2	24.67 ± 0.6
200	43.07 ± 0.6	67.10 ± 0.3	71.26 ± 1.2	79.06 ± 0.3	26.60 ± 0.3	26.55 ± 0.5
250	59.94 ± 0.7	86.69 ± 0.4	83.14 ± 0.4	86.21 ± 0.3	42.76 ± 0.7	31.41 ± 0.7
300	64.78 ± 0.7	91.35 ± 0.5	90.07 ± 0.4	92.78 ± 0.6	56.57 ± 0.6	37.18 ± 0.5
350	76.14 ± 0.5	100.0 ± 0.0	98.33 ± 0.7	100.0 ± 0.0	63.54 ± 2.3	41.75 ± 1.0
400	91.49 ± 1.1	100.0 ± 0.0	100.0 ± 0.0	100.0 ± 0.0	67.94 ± 0.4	49.56 ± 0.5
450	100.0 ± 0.0	100.0 ± 0.0	100.0 ± 0.0	100.0 ± 0.0	78.19 ± 0.5	65.00 ± 0.6
500	100.0 ± 0.0	100.0 ± 0.0	100.0 ± 0.0	100.0 ± 0.0	85.88 ± 0.7	76.19 ± 0.6
550	100.0 ± 0.0	100.0 ± 0.0	100.0 ± 0.0	100.0 ± 0.0	89.79 ± 0.6	90.08 ± 0.8
600	100.0 ± 0.0	100.0 ± 0.0	100.0 ± 0.0	100.0 ± 0.0	97.57 ± 0.3	100.0 ± 0.0
650	100.0 ± 0.0	100.0 ± 0.0	100.0 ± 0.0	100.0 ± 0.0	100.0 ± 0.0	100.0 ± 0.0
Thiram	100.0 ± 0.0	100.0 ± 0.0	100.0 ± 0.0	100.0 ± 0.0	100.0 ± 0.0	100.0 ± 0.0

Data given are mean of four replicates ± standard error.
Analysis of variance (ANOVA) d.f. = 15 at $P < 0.0001$.

Table 4.2. Effect of 2-hydroxy-4-methoxy-benzaldehyde isolated from *D. hamiltonii* and thiram on seed-borne fungi of paddy stored up to 90 days.

| Storage periods (days) | Incidence of seed-borne fungal species in paddy seeds (%) | | | | | | | | | | | | | | | |
| | Untreated | | | | 0.5 g/kg active compound | | | | 1 g/kg active compound | | | | Thiram (2 g/kg) | | | |
	0	30	60	90	0	30	60	90	0	30	60	90	0	30	60	90
Alternaria spp.	22 ± 0.5	26 ± 0.5	28 ± 0.3	27 ± 0.8	4 ± 0.5	12 ± 0.6	14 ± 0.6	19 ± 0.6	0.0	0.0	7 ± 0.4	11 ± 0.7	0.0	0.0	8 ± 0.6	11 ± 0.7
Aspergillus spp.	42 ± 0.8	51 ± 1.2	64 ± 0.5	72 ± 0.4	9 ± 0.3	14 ± 0.3	26 ± 0.4	35 ± 0.5	0.0	5 ± 0.2	12 ± 0.6	17 ± 1.1	0.0	5 ± 0.6	23 ± 0.5	32 ± 0.5
Curvularia spp.	28 ± 0.6	27 ± 0.5	31 ± 0.5	39 ± 0.8	0.0	8 ± 0.8	13 ± 0.3	18 ± 0.8	0.0	0.0	4 ± 0.3	6 ± 0.4	0.0	0.0	0.0	3 ± 0.3
Drechslera spp.	41 ± 0.8	40 ± 0.8	44 ± 0.4	48 ± 0.7	0.0	5 ± 0.5	9 ± 0.5	14 ± 0.8	0.0	0.0	0.0	4 ± 0.3	0.0	0.0	0.0	1 ± 0.2
Fusarium spp.	36 ± 0.3	42 ± 0.7	49 ± 0.9	54 ± 0.8	0.0	6 ± 0.3	10 ± 0.5	15 ± 0.8	0.0	0.0	6 ± 0.3	13 ± 0.6	0.0	0.0	3 ± 0.4	8 ± 0.5
Penicillium spp.	21 ± 0.4	34 ± 0.8	42 ± 0.5	58 ± 0.9	4 ± 0.3	11 ± 0.5	19 ± 0.4	28 ± 0.5	0.0	0.0	8 ± 0.3	18 ± 0.5	0.0	8 ± 0.6	15 ± 0.6	22 ± 0.5
Pyricularia spp.	22 ± 0.5	26 ± 0.8	27 ± 0.8	25 ± 0.4	0.0	6 ± 0.5	8 ± 0.8	12 ± 0.5	0.0	0.0	2 ± 0.9	4 ± 0.8	0.0	0.0	0.0	2 ± 0.3
Trichoconis padwickii	23 ± 0.7	25 ± 0.5	21 ± 0.8	19 ± 0.5	0.0	0.0	4 ± 0.5	8 ± 0.7	0.0	0.0	0.0	2 ± 0.3	0.0	0.0	0.0	1 ± 0.6
Trichothecium spp.	13 ± 0.4	16 ± 0.8	21 ± 0.5	26 ± 0.5	0.0	0.0	5 ± 0.3	10 ± 0.8	0.0	0.0	0.0	0.0	0.0	0.0	0.0	0.0

Per cent incidence is based on four replicates with 100 seeds each, $F = 321.64$; $P < 0.001$.

The evaluation of the bioactive compound to understand the impact on the nutritional quality of the paddy grains stored up to 90 days revealed that 1 g/kg treatment is effective in preventing biodeterioration of paddy grains without affecting the nutritional quality in terms of dry matter loss, lipid, carbohydrate and protein content (Fig. 4.8a–d), suggesting that 2-hydroxy-4-methoxybenzaldehyde is an important antimicrobial of plant origin that could be exploited to prevent biodeterioration of grains during storage (Mohana *et al.*, 2009).

Even though a few antimicrobial active components have been isolated and characterized by earlier workers, evaluations of these to assess their potency to prevent biodeterioration of grains are lacking. The active compounds isolated from the seeds of *P. corylifolia* L. and from the rhizome of *D. hamiltonii* Wight & Arn. have been successively evaluated to assess their potency to

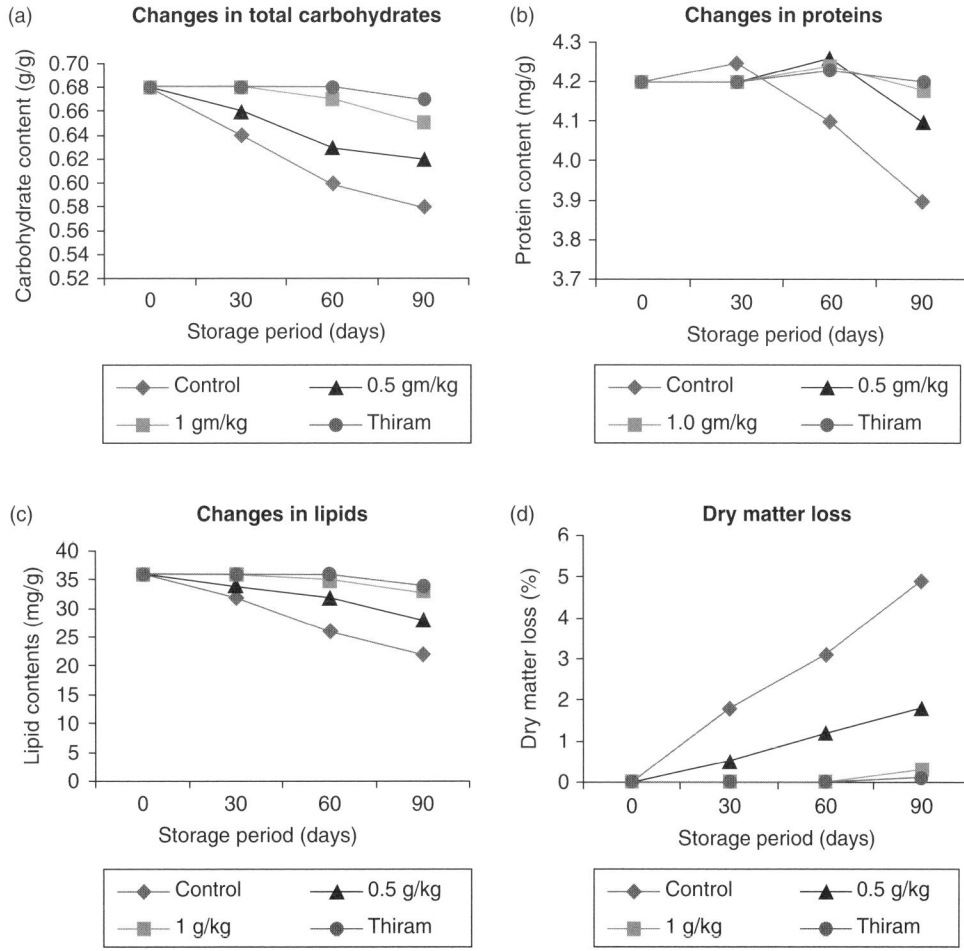

Fig. 4.8. (a–d) Comparative efficacy of the bioactive compound 2-hydroxy-4-methoxybenzaldehyde isolated from *D. hamiltonii* (0.5 g/kg and 1 g/kg) and thiram (2 g/kg) on fungi inducing nutritional losses in paddy grains stored up to 90 days.

prevent the biodeterioration of grains during storage. *D. hamiltonii* is an important edible plant and thus has an additional advantage as a herbal, eco-friendly remedy for the prevention of grain biodeterioration. Further investigations on the standardization of methods for high yield of the active compounds, toxicological aspects, development of formulations, treatment procedures and a package of practices to be adapted may pave the way for commercial exploitation.

4.5 Conclusion

Plants have been over exploited for human health needs while under exploited for plant health needs. One of the possible reasons for this may be the high-input cost that could be made good in human health care, while it may be difficult in plant health care. Considering the cost of environmental pollution and other related damages caused by the use of chemical pesticides in crop protection, the development of eco-friendly natural pesticides is a necessity. Screening plants for antimicrobial activity against phytopathogens in general and against biodeterioration-causing fungi in particular is the first step towards this goal. Subsequent to demonstrating the science behind the antimicrobial activity, research needs to be focused on developing an appropriate, cost-effective package of practices employing biotechnological approaches. Some of the possible approaches to decreasing the cost of production are: (i) identify the gene responsible for the production of the active component and considerably enhance the yield of the active component through bioengineering; (ii) isolate and identify the endophytes, if any, associated with the production of the active compound and develop fermentative methods for the large-scale production of the active component; (iii) develop a package of practice, preferably utilizing the edible plant part pieces or powder directly as an amendment in an appropriate quantity during grain storage; and (iv) the chemical synthesis of the active compound.

References

Aladesanmi, A.J. and Odediran, S.A. (2000) Antimicrobial activity of *Trichilia heudelotti* leaves. *Fitoterapia* 71, 179–182.

Al-Fatimi, M.A.A., Julich, W.D., Jansen, R. and Lindequist, U. (2006) Bioactive components of the traditionally used Mushrooms *Podaxis pistillaris. Evidence. Based Complementary and Alternative Medicine* 3, 87–92.

Bouamama, H., Noel, T., Villard, J., Benharref, A. and Jana, M. (2006) Antimicrobial activities of the leaf extracts of two Moroccan *Cistus* L. species. *Journal of Ethnopharmacology* 104, 104–107.

Carpano, S.M., Spegazzini, E.D., Rossi, J.S., Castro, M.T. and Debenedetti, S.L. (2003) Antifungal activity of *Terminalia australis. Fitoterapia* 74, 294–297.

Chowdhury, R., Hasan, C.M. and Rashid, M.A. (2003) Antimicrobial activity of *Toona ciliata* and *Amoora rohituka. Fitoterapia* 74, 155–158.

Cowan, M.M. (1999) Plant products as antimicrobial agents. *Clinical Microbiology Reviews* 12, 564–582.

Doughari, J.H. (2006) Antimicrobial activity of *Tamarindus indica* Linn. *Tropical Journal of Pharmaceutical Research* 5, 592–603.

Dubios, M., Gilles, K.A., Hamihon, J.R., Rebers, P.A. and Smith, F. (1956) Phenol sulphuric acid method for total carbohydrate. *Annual Chemistry* 26, 350.

Duke, S.O. (1990) Natural pesticides from plants In: Janick J. and Simon J.E. (eds) *Advances in New Crops*. Timber press, Portland, pp. 511–517.

Erasto, P., Grierson, D.S. and Afolayan, A.J. (2006) Bioactive sesquiterpene lactones from the leaves of *Vernonia amygdalina*. *Journal of Ethnopharmacology* 106, 117–120.

Fan, J.J. and Chen, J.H. (1999) Inhibition of Aflatoxin producing fungi by Welsh onion extract. *Journal of Food Protection* 62, 414–417.

Farooq, S. and Pathak, G.K. (1998) A comparative study of '*in vitro*' antimicrobial activity of total solvent extracts of some medicinal plants with Asava and Aarishta preparations. *Journal of Non-timber Forest Products* 5, 79–81.

Goun, E., Cunningham, G., Chu, D., Nguyen, C. and Miles, D. (2003) Antibacterial and antifungal activity of Indonesian ethnomedical plants. *Fitoterapia* 76, 592–596.

Gupta, M., Mazumder, U.K., Chaudhuri, H.A., Chaudhuri, R.K., Bose, P., Bhattacharya, S., Manikandan, L. and Patra, S. (2002) Antimicrobial activity of *Eupatorium ayapana*. *Fitoterapia* 73, 168–170.

Hurtado, R.D., Garrido, G.G., Gonzalez, S.P., Iznaga, Y., Gonzalez, L., Torres, J.M., Curini, M., Fpifano, F. and Marcotullio, M.C. (2003) Antifungal activity of some Cuban *Zanthoxylum* species. *Fitoterapia* 74, 384–386.

ISTA, (1999) Proceedings of the International seed testing association, International rules for seed testing. *Seed Science and Technology* 76, 481–484.

Janardhana, G.R., Raveesha, K.A. and Shetty, H.S. (1999) Mycotoxin contamination of Maize grains grown in Karnataka (India). *Food and Chemical Toxicology* 37, 863–868.

Joshi, B.C., Pandey, A., Chaurasia, L., Pal, M., Sharma, R.P. and Khare, A. (2003) Antifungal activity of stem bark of *Ailanthus excelsa*. *Fitoterapia* 74, 689–691.

Katyayani, B.M., Rao, P.M., Muralichand, G., Rao, D.S. and Satyanarayana, T. (2002)

Antimicrobial activity of bark of *Alangium salvifoloium* Linn. F. *Indian Journal of Microbiology* 42, 87–89.

Kiran, B. and Raveesha, K.A. (2004) Antifungal activity of seed extracts of *Psoralea corylifolia* L. *Plant Disease Research*. 20, 213–215.

Lis-Balchin, M., Patel, J., Hart, S. (1998) Studies on the mode of action of essential oils of scented-leaf *Pelargonium* (Geraniaceae). *Phytotherapy Research* 12, 215–217.

Lowry, O.H., Rosebrough, N.T., Farr, A.L. and Randall, R.J. (1951) Protein estimation by Lowrey's method. *Journal of Biological Chemistry* 193, 265.

Mahmoud, A.L.E. (1999) Inhibition of growth and aflatoxin biosynthesis of *Aspergillus flavus* by extracts of some Egyptian plants. *Letters in Applied Microbiology* 29, 334–336.

Mathekga, D.D.M., Meyer, J.J.M., Horn, M.M. and Drewes, S.E. (2000) An acylated phloroglucinol with antimicrobial properties from *Helichrysum caespititum*. *Phytochemistry* 53, 93–96.

Mohana, D.C., Raveesha, K.A. and Lokanath Rai, K.M. (2006) Herbal remedies for the management of seed-borne fungal pathogens by an edible plant *Decalepis hamiltonii* (Wight and Arn) *Archives of Phytopathology and Plant Protection* 1–12.

Mohana, D.C., Satish, S. and Raveesha, K.A. (2009) Antifungal avtivity of 2-Hydroxy-4-Methoxybenzaldehyde isolated from *Decalepis hamiltonii* (Wight and Arn) on seed borne fungi causing biodeterioration of Paddy. *Journal of Plant Protection Research* 49(3), 250–256.

Nagaraju, S., Jagan Mohan Rao, L. and Gurudatt, K.N. (2001) Chemical composition of the volatiles of *Decalepis hamiltonii* (Wight and Arn*). Flavour Fragrance Journal* 16, 27–29.

Neergaard, P. (1977) Seed Pathology. The MacMillan Press Ltd., London, Volumes I and II: 1–840.

Ngane, A.N., Biyiti, L., Bouchet, Ph. Nkengfack A. and Zollo, P.H.A. (2003) Antifungal activity of *Piper guineense* of Cameroon. *Fitoterapia* 74, 464–468.

Oros, G. and Ujvary, I. (1999) Botanical fungicides: natural and semi-synthetic ceveratrum alkaloids. *Pesticide Science* 55, 253–264.

Otshudi, A.L., Apers, L.S., Claeys, P.M., Pannecouque, C., Clercq, E.D., Zeebroeck, A.V., Lauwers, S., Fred, A. and Foriers, A. (2005) Biologically active bisbenzyliso-quinoline alkaloids from the root bark of *Epinetrum villosum*. *Journal of Ethnopharmacology* 102, 89–94.

Parekh, J. and Chanda, S. (2007) *In vitro* antimicrobial activity of *Trapa natans* L. fruit rind extracts in different solvents. *African Journal of Biotechnology* 6, 766–770.

Pinto, C.M.F., Maffia, L.A., Casali, V.W.D. and Cardoso, A.A. (1998) *In vitro* effect of plant leaf extracts on mycelial growth and sclerotial germination of *Sclerotium cepivorum*. *Journal of Phytopathology* 146, 421–425.

Sardari, S., Amin, G., Micetich, R.G. and Daneshtalab, M. (1998) Phytopharmaceuticals. Part I, Antifungal activity of selected Iranian and Canadian plants. *Pharmaceutical Biology* 36, 180–188.

Sharma, K., Shukla, S.D., Mehta, P. and Bhatnagar, M. (2002) Fungistatic activity of *Semecarpus anacardium* Linn.f. nut extract.

Indian journal of Experimental Biology 40, 314–318.

Sharma, P., Singh, S.D. and Rawal, P. (2003) Antifungal activity of some plant extracts and oils against seed borne pathogens of pea. *Plant Disease Research* 18, 16–20.

Singh, I. and Singh, V.P. (2000) Antifungal properties of aqueous and organic solution extracts of seed plants against *Aspergillus flavus* and *A. niger*. *Phytomorphology* 50, 151–157.

Sinha, P. and Saxena, S.K. (1999) Inhibition of fruit rot fungus and fruit fly by leaf extracts of onion (*Allium cepa*) and garlic (*Allium sativum*). *Indian Journal of Agricultural Sciences* 69, 651–653.

Stojanovic, G., Radulovi, N., Hashimoto, T. and Pali, R. (2005) *In vitro* antimicrobial activity of extracts of four *Achillea* species: The composition of *Achillea clavennae* L. (Asteraceae) extract. *Journal of Ethnopharmacology* 101, 185–190.

Yanez, C., Pezoa, J., Rodriguez, M., Nunez-Vergara, L.J. and Squella, J.A. (2005) Voltamerric behavior of a 4-nitroimidazole derivative Nitro radical anion formation and stability. *Journal of the Electrochemical Society* 152, J41–J56.

5

Some Natural Proteinaceous and Polyketide Compounds in Plant Protection and their Potential in Green Consumerization

L.A. SHCHERBAKOVA

Russian Research Institute of Phytopathology, Moscow, Russia

Abstract

Plant protection from diseases with the use of natural compounds, which are indigenous and biodegradable in the environment, fits the purpose of green consumerization. Natural compounds controlling plant pathogens belong to different chemical classes and are produced by a wide range of organisms. The structural and functional diversity of these compounds provides a great potential in crop protection technologies in correspondence with green consumerization objectives. Disease prevention by natural compounds results from either direct or plant-mediated influences on targeted pathogens. Some natural substances affecting causative agents directly do not produce a biocidal effect but specifically attack pathways related to their pathogenicity. Biogenic compounds influencing plants induce resistance or enhance tolerance to diseases. They elicit natural defence responses in plants and, unlike chemical pesticides, do not promote the occurrence of resistant forms in targeted pathogens. Two bacterial proteins, CspD and MF3, that have elicitor properties have been well documented to provide resistance against various phytopathogens and play a promising role in green consumerization. Fungal polyketides, statins, inhibiting a pathogenicity-related pathway in melanin-producing fungi are considered as prospective candidates for developing biopesticides and reducing the impact of plant diseases without killing pathogenic microorganisms.

5.1 Introduction

Plant diseases continue to cause considerable damage to global agriculture resulting in yield losses and deterioration of agricultural products, including their contamination with hazardous substances, e.g. mycotoxins, which induce toxicological problems in people and animals. Despite amazing progress achieved in crop protection due to breeding for resistance to diseases and availability of effective agrochemicals, a number of problems are faced in managing the diseases successfully. The resistant cultivars provide only partial plant protection from a disease because of the occurrence of new pathogenic races that are able to overcome the resistance. Sometimes, breeding for resistance is too complicated because of the diversity and high variability of plant pathogens and the variety of defence responses at different stages of pathogenesis and plant development.

Chemical protection gives a powerful tool for effective control of many disease-causing agents, but its implementation has a range of undesirable side impacts. Thus, the use of chemical pesticides is associated with environmental risks, such as pollution of soil, water and crop plants with xenobiotic pesticide residues, killing beneficial microorganisms or insects, thereby interfering with the natural management of pests and pathogens, a reduction in biological diversity, and so on. Sometimes, chemical pesticides are not always effective and many treatments with several fungicides are needed to control some fungal pathogens. As a result of continual pesticide application on plants and stored agricultural products, toxic residues of chemical pesticides may appear and accumulate in food. Lastly, perhaps the most disturbing consequence of chemical protection is the development of pesticide resistance in pathogens that makes a protective effect changeable and insufficiently effective. The health and ecological problems related to chemical pesticides are now receiving a great deal of attention. Recognition of these problems has created the phenomenon of green consumerization.

The biocontrol of phytopathogens as an additional or alternative plant protection method looks more natural and environmentally compatible than chemical treatments. The use of non-pathogenic or beneficial microorganisms and/or natural compounds, which are native and biodegradable in the environment, is the main objective of green consumerization. A few hundred biocontrol products based on numerous beneficial strains predominantly of *Trichoderma*, *Bacullis*, *Pseudomonas*, *Agrobacterium*, *Streptomyces* and *Gliocladium* species are produced, commercialized and used as biopesticides and/or biofertilizers.

Natural compounds effectively controlling plant pathogens and pests are represented by a diversity of substances belonging to different chemical classes and produced by a wide range of living organisms (Varma and Dubey, 1999). Currently, the list of these natural compounds is growing longer from year to year. Many natural products have already been used or recommended as biopesticides (e.g. various secondary plant metabolites such pyrethrins, glucosinolate-sinigrin-producing allyl isothiocyanate, carvone, azadirachtin, rotenone, ryania, nicotine, some essential oils and other volatile plant substances). The plant-mediated mode of action of some natural compounds, which does not establish contact with pathogens, decreases the risk of resistant pathogenic form selection. Because the search for natural substances fitting phytosanitary regulations and environmentally compatible cropping methods is a continuous active process, novel compounds are being discovered in plants, animals and microorganisms and are marketed in place of chemical pesticides in order to be adopted by so-called 'cottage industry' or governmentally regulated farming.

The disease-preventing effect of natural compounds that are used or can be used for crop protection is a result of a direct or plant-mediated influence on targeted pathogens. Affecting phytopathogens directly, the compounds interrupt pivotal metabolic pathways and/or disrupt vitally important structures that have as a consequence the death of the pathogens (biocidal effect). Some natural substances or their analogues do not produce lethal effects to plant pathogens but specifically influence pathways related to pathogenicity or/and toxigenesis, e.g. mycotoxin production (Kumar *et al.*, 2008; Shukla *et al.*, 2008). Such non-biocidal suppressors can impair the pathogenicity for plants as well as the toxic risk for people and animals. In the case of a plant-mediated mode of action, natural compounds elicit and activate defence responses in plants that result in induced resistance to diseases (e.g. Shcherbakova *et al.*, 2007; 2008; Odintsova *et al.*, 2009). Lastly, some of biogenic compounds influencing plants do not induce resistance but enhance tolerance to pathogens (sometimes by improving the physiological state of plants). The structural and functional diversity of natural compounds as well as their abundance provide a great potential for bringing plant protection technologies in line with green consumerization.

The main goal of this chapter is to analyse properties of a few chemically and functionally different natural compounds and demonstrate their prospective as promising candidates for new plant protection products with significant potential in relation to green consumerization. The chapter will focus on the characterization of two proteinaceous compounds of bacterial origin, the elicitor properties of which provide disease resistance, and also on fungal statins as natural compounds, which inhibit a pathogenicity-related pathway in melanin-producing fungi and can be used for developing biopesticides that reduce the harmful impact of crop diseases without killing pathogenic organisms. In addition, microbial and plant proteinaceous compounds that have been previously well documented to be promising for green consumerization will be briefly surveyed.

5.2 Proteinaceous Compounds of Microbial and Plant Origin in the Biocontrol of Plant Pathogens

Proteins are abundant and ubiquitous natural compounds strongly associated with all living organisms, hence it is possible to believe their involvement in plant protection will not conflict with green consumerization principles. Because proteinaceous compounds are important constituents of a plant immune system and take part in the natural plant defence against pathogens, scientists have emphasized their potential for plant protection for a long time. Basically, two general approaches to the practical realization of the protective potential of these compounds are the formulation of producers (biocontrol or non-pathogenic microorganisms) or the engineering of transgenic plants. The compounds synthesized by formulated biocontrol agents interact with pathogenic microflora or induce resistance in plants. Constitutive or inducible expression of protein-encoding genes transferred into plants confer them resistance to many economically important diseases.

Several groups of microbial and plant proteinaceous substances that have well-established potentialities in plant protection will be very briefly reviewed below. The antipathogenic properties of two proteins, which were isolated from biocontrol bacteria and relatively recently studied as putative elicitors of disease resistance in plants, will be described in more detail.

The biocontrol potential of proteinaceous compounds will be considered almost exclusively by using examples of fungal and bacterial pathogens. Their antipest properties will be not covered within the scope of this chapter. Nevertheless, it should be emphasized that many proteins and peptides discussed in this section (microbial peptides, defensins, proteinase inhibitors and lectins) are well known as biological insecticides or nematicides of great potentiality. Biopesticides based on microbial peptides or transgenic plants carrying genes of defensins, lectins and proteinase inhibitors are good tools for the fight against pests of plants.

Antimicrobial peptides produced by microorganisms and plants

The generation of antimicrobial peptides is one of the widespread natural defence mechanisms of innate immunity of living organisms, and these compounds are interesting targets for green consumerization. Plants, including agricultural crops, produce numerous defensive peptides. Microorganisms, including those used in plant protection, synthesize and excrete antimicrobial peptides that are often responsible for the biocontrol effect. In several authors' opinions, peptides have advantages over more elementary organic antimicrobial compounds when used for plant protection goals. Because the peptide molecules contain from 10 to 50 (sometimes up to 85) amino acid residues, they can more specifically interact with their protein targets in causative agents (Park *et al.*, 2009). Biopesticidal peptides can work against pathogens by inhibiting nucleic-acid and protein biosyntheses and enzyme activity or by interacting with the plasmalemma and destroying its integrity (Huang, 2000).

Antimicrobial peptides from biocontrol microorganisms

Peptides possessing antimicrobial effects have been reported to be produced by bacteria, fungi, plants, invertebrates and vertebrates (Garcia-Olmedo *et al.*, 1998; Zasloff, 2002; Degenkolb *et al.*, 2003; Bulet *et al.*, 2004; Nybroe and Sorensen, 2004; Carvalho and Gomes, 2009). According to the recent review focused on antimicrobial peptides related to plant-disease control (Montesinos, 2007), about 900 antimicrobial peptides are produced by living organisms via ribosomal or non-ribosomal synthesis. Biocontrol microorganisms have been reported to use both types of the peptide synthesis (Finking and Marahiel, 2004; Montesinos, 2007) and generate a wide range of these compounds. Antimicrobial peptides of microorganisms

(AMPM) have been classified into several groups of linear and cyclic peptides (fungal defensins, bacteriocins, peptaibols, cyclopeptides and pseudopeptides) based on their most essential structural characteristics, and have been inventoried according to name, composition and producer microorganisms (Montesinos, 2007).

The efficacy of the natural antimicrobial peptides produced by biocontrol microorganisms towards various plant pathogenic bacteria, fungi and oomycetes *in vitro* and *in vivo* has been well documented (Table 5.1).

Table 5.1. Instances of plant pathogens sensitive to antimicrobial peptides of microorganisms.

Targeted plant pathogens	Group of effective AMPM	Reference
Fungi		
Alternaria spp.	Pseudopeptides	Stein, 2005
Botrytis cinerea	Cyclopeptides	Lavermicocca *et al.*, 1997
	Fungal defensins	Ongena *et al.*, 2005
	Peptaibols	Vila *et al.*, 2001
	Pseudopeptides	Moreno *et al.*, 2003
		Moreno *et al.*, 2005
		Xiao-Yan *et al.*, 2006
		Stein, 2005
Bipolaris sorokiniana	Peptaibols	Xiao-Yan *et al.*, 2006
Colletotrichum spp.	Peptaibols	Xiao-Yan *et al.*, 2006
Erysiphe sp.	Cyclopeptides	Selim *et al.*, 2005
Fusarium avenaceum	Cyclopeptides	Selim *et al.*, 2005
Fusarium oxysporum	Peptaibols	Xiao-Yan *et al.*, 2006
		Ongena *et al.*, 2005
Fusarium solani	Cyclopeptides	Stein, 2005
Fusarium spp.	Fungal defensins	Vila *et al.*, 2001
		Moreno *et al.*, 2003
		Moreno *et al.*, 2005
Leptosphaeria maculans	Cyclopeptides	Pedras *et al.*, 2003
Magnaporthe grisea	Fungal defensins	Moreno *et al.*, 2003
	Pseudopeptides	Moreno *et al.*, 2005
		Stein, 2005
Monilina fructicola	Cyclopeptides	Gueldner *et al.*, 1988
Podosphaera fuca	Cyclopeptides	Romero *et al.*, 2007
Podosphaera sp.	Cyclopeptides	Stein, 2005
Rhizoctonia solani	Cyclopeptides	Asaka and Shoda, 1996
	Peptaibols	Bassarello *et al.*, 2004
	Pseudopeptides	Nielsen and Sorensen, 2003
		Ongena *et al.*, 2005
		Xiao-Yan *et al.*, 2006
		Stein, 2005
Sclerotinia sclerotium	Cyclopeptides	Pedras *et al.*, 2003
Sclerotium cepivarum	Peptaibols	Gouland *et al.*, 1995

Continued

Table 5.1. Continued.

Targeted plant pathogens	Group of effective AMPM	Reference
Sphaerotheca sp.	Pseudopeptides	Stein, 2005
Uncinula nector	Pseudopeptides	Stein, 2005
Ventura inaequalis	Cyclopeptides	Burr *et al.*, 1996
Phytophthora infestans	Cyclopeptides	De Bruijn *et al.*, 2007
Pythium intermedium	Cyclopeptides	De Souza *et al.*, 2003
Pythium ultimum	Cyclopeptides	Nielsen *et al.*, 2002
Bacteria		
Erwinia amylovora	Pseudopeptides	Brady *et al.*, 1999
		Jin *et al.*, 2003
Erwinia caratovora	Cyclopeptides	Selim, 2005
Clavibacter michiganesis	Peptaibols	Xiao-Yan *et al.*, 2006
Pseudomonas syringae	Cyclopeptides	Bais *et al.*, 2004
Rhodococcus fascians	Cyclopeptides	Bassarello *et al.*, 2004

Antimicrobial peptides from plants

A broad family comprising antimicrobial peptides produced by plants is called defensins. One of their functions in plants is the involvement in defence mechanisms against pathogens, pests and abiotic stresses (Terras *et al.*, 1995; Lay and Anderson, 2005). These are basic peptides structurally and function-ally related to the defensins of mammalia and insects. Molecular masses of plant defensins that are characterized at present range from 5 to 7 kDa. Their primary structure is formed with 45–55 amino acid residues. Two structural characteristics of the defensins, the presence of a pattern of conserved cysteine residues and a domain with extremely variable amino acid sequence, are typical components for mature peptide molecules of plant defensins. With rare exceptions, defensins found in plants have a similar globular spatial structure that is stabilized by a structural motif formed using disulfide bonds between eight cysteine residues. Possibly, defensins can aggregate *in vivo* into dimers or oligomers (Terras *et al.*, 1992). A certain correlation is observed between the primary structure and antimicrobial activity of some plant defensins. In general, the incorporation of basic amino acids (viz. arginine) that add positive charge to a peptide molecule resulted in a considerable increase in the antimicrobial activity (Terras *et al.*, 1992; De Samblanx *et al.*, 1997; Landon *et al.*, 2000).

In the overwhelming majority of cases, plant defensins possess inhibi-tory activity against fungi; however, the growth of Gram-negative bacteria is also arrested after exposure to these peptides (Terras *et al.*, 1993; Segura *et al*, 1998; Wong *et al*, 2006). An analysis of publications by Carvalho and Gomes (2009) showed many damaging plant pathogens of economically important crops can be inhibited *in vitro* with plant defensins. Various fungal species (*Alternaria brassicola*, *Alternaria solani*, *Botrytis cinerea*, *Cladosporium colocasiae*,

Cladosporium sphaerospermum, Colletotrichum lindemuthianum, Diploidia maydis, Magnaporthe grisea, Mycosphaerella arachidicola, Mycosphaerella fijinesis, Nectria haematococca, Penicillium digitatum, Penicillium expansum, Phaeoisariopsis personata, Physalospora piricola, Rhizoctonia solani, Septoria tritici, Verticilium albo-atrum, V. dahliae, and the toxigenic species *Fusarium culmorum, F. decemcellulare, F. graminearum, F. oxysporum, F. verticillioides* and *Aspergillus niger*); two species of oomycetes (*Phytophthora infestans* and *P. parasitica*) and also bacteria, viz. *C. michiganensis* and *Ralstonia solanacearum,* are among them. Effective concentrations of plant defensins differ depending on the tested peptides and the targeted pathogens (values of IC_{50}, a protein concentration that is required for 50% growth inhibition, vary from 1–100 µg/ml). The level of antimicrobial activity may be regulated with bivalent ions (Terras *et al.,* 1992; 1993; Osborn *et al.,* 1995; Segura *et al.,* 1998; Wong and Ng, 2005). Along with the growth inhibitory effect, some plant defensins cause morphological changes in fungal mycelia (Carvalho and Gomes, 2009).

Studies on the mode of action of antimicrobial plant defensins are in progress. There are confirmed hypotheses that an interaction with the cell membrane of microorganisms resulting in ion efflux and reactive oxygen species (ROS) generation significantly contributes to the mechanisms responsible for antifungal properties of plant defensins. Besides antimicrobial activity, plant defensins possess a range of biological functions (Lay and Anderson, 2005; Carvalho and Gomes, 2009).

The availability of antimicrobial peptides produced by plants or microorganisms through ribosomal synthesis for crop protection has been demonstrated by an increased disease resistance of transgenic plants expressing fungal and plant defensin genes (Montesinos, 2007; Carvalho and Gomes, 2009). Several agricultural crops, e.g. tobacco, tomato, rice, aubergine, papaya and canola, which are transformed with these genes and produce the corresponding peptides, have little or no disease development in laboratory, greenhouse or field experiments. A number of peptides produced by microorganisms are insecticidal or nematicidal. The ability of plant defensins to inhibit α-amylase and proteases can contribute to plant defence against pest insects (Lin *et al.,* 2007).

Enzymes, proteinase inhibitors, lectins and PR proteins

Lytic enzymes

To obtain nutrients, microorganisms synthesize various lytic enzymes that can attack polymeric compounds of different origin. Biocontrol agents can use these enzymatic activities on plant pathogens. Microbial chitinases, glucanases and proteases are lytic enzymes of most importance for the biocontrol of phytopathogens. These enzymes hydrolyse chitin, β-glucans and proteins, which can result in direct suppression of pathogen development or generate products that function as resistance inducers. For instance, biocontrol isolates of *Trichoderma harzianum* and *Trichoderma atroviride* produce endochitinase, β-1,3-glucanase and alkaline proteinase, which degrade plant pathogenic fungi *in vitro*, halt their growth *in planta* and play a role in mycoparasitism (Elad *et al.*, 1982; Benitez *et al.*, 1998; Lorito, 1998; Chernin and Chet, 2002). Oligosaccharides or chitosan derived from fungal cell walls exposed to microbial glucanases and chinases elicit a cascade of defence responses in plants: generation of ROS, induction of pathogenesis-related proteins (PR proteins) (including plant chitinases and glucanases), phytoalexins and lignification (Dyakov and Ozeratskovskaya, 2007).

The potential of lytic enzymes for plant-disease management was well demonstrated by studying chinolytic systems in the biocontrol bacteria *Bacillus cereus*, *Pantoea agglomerans*, *Pantoea dispersa*, and fungi, especially in the widely used biocontrol fungus *Trichoderma*. Chitinases produced by *Trichoderma* are effective on virtually all chitinous pathogens, non-toxic for plants and possess higher antifungal activity than such enzymes isolated from other sources. The antifungal activity of chitinases from *Trichoderma* can reach the level of some chemical pesticides (Lorito, 1998; Bonaterra *et al.*, 2003; Chang *et al.*, 2003; Gohel *et al.*, 2004).

Along with use of enzyme-producing biocontrol agents, there are several other application strategies for chinolytic enzymes. The most conventional approach consists of conferring resistance via engineering transgenic plants containing heterologous chitinase and glucanase genes. Overexpression of these genes in response to pathogen invasion can cause higher levels of the enzymes in the plant cells followed by a faster and effective neutralization of the pathogen. Indeed, transgenic broccoli, potato and tobacco plants expressing the *T. harzianum* endochitinase gene have been found to show resistance against *A. alternata, A. solani, B. cinerea* and *R. solani*. Transgenic tobacco and cabbage, carrying a bean chitinase gene were protected against *R. solani*. Transgenic cucumber, rice, grapevine, strawberry and wheat transformed with chitinase genes from rice (*Oryza sativa*) were resistant to *B. cinerea, R. solani, M. grisea, Sphaerotheca humuli* and *F. graminearum*, respectively (Gohel *et al.*, 2006). Expression of exochitinase genes in transgenic apple trees confers resistance to apple scab (*Venturia inaequalis*), a pathogen which is controlled by multiple applications of chemical fungicides during the growing season (Bolar *et al.*, 2000). These results show the broad potential for the microbial chitinase transgenesis into plants for controlling fungal phytopathogens. The additional strategies are related to fermentation and different ways of improving the enzyme producers (Gohel *et al.*, 2006).

Other cases suggesting feasibility of crop protection with enzymes can be illustrated by the examples of constructing transgenic potato carrying glucose oxidase gene from *A. niger* or apple, potato and tobacco plants expressing the bacteriophage T4 lysozyme gene (Wu *et al.*, 1997). Glucose oxidase is an enzyme involved in generating plant ROS. Expression of the glucose oxidase gene led to accumulation of peroxide ions in plant tissues that increased resistance to fungal diseases, e.g. to late blight (*P. infestans*), wilt (*Verticillium dahliae)* and early blight *(A. solani)*. Lysozymes are widespread enzymes that hydrolyse peptidoglycan of bacterial cell walls. Apple plants with the T4L gene showed significant resistance to the fire blight agent *E. amylovora* (Ko *et al.*, 2000), while potato and tobacco was resistant to

E. carotovora subsp. *carotovora* (During *et al.*, 1993). However, lysozyme excretion can have adverse effect on soil microbiota. Thus, the growth of *B. subtilis* has been observed to be suppressed in rhizosphere transgenic T4 lysozyme-producing potato plants (Ahrenholtz *et al.*, 2000).

Lectins

Plant lectins are a heterogeneous collective of proteins that specifically bind carbohydrates in a reversible way and take part in phytopathogen recognition. Interacting with like components, lectins can attach to the cell surface. The structure and functions of these compounds are discussed in detail (e.g. Chrispeels and Raikhel, 1991). Lectins are well known naturally occurring insecticides of widespread effect. Some free lectins strongly affect microbe growth in plants and probably contribute to inhibiting pathogenesis. There are chitin-specific lectins synthesized in the phloem and translocated via vessels. These findings suggest that lectins are potential antifungal agents.

Proteinase (protease) inhibitors of plant origin

Plant inhibitors of proteinases are a large group of peptides or small proteins able to bind proteolytic enzymes of different organisms with competitive inhibition of their activity. In plants, they are abundant in seeds and storage organs, where their content can be up to 10% of water-soluble proteins. These compounds are considered as reserve proteins and regulators of protein status or enzyme activity in plants. The proteinase inhibitors differ in substrate specificity, have various isoforms, and their oligomers can combine or dissociate with an influence on the inhibitor properties.

A defensive function of proteinase inhibitors towards insects was initially revealed when insects, after feeding, became inactive as a result of trypsin inactivation. Protease inhibitors of plant origin were also shown to be active against plant pathogenic nematodes. Since many phytopathogenic fungi and bacteria secrete extracellular proteolytic enzymes, which play an important role in pathogenesis (Valuyeva and Mosolov, 2004), plants use inhibition of such enzymes as a defence strategy towards these microorganisms (Ryan, 1990; Habib and Khalid, 2007).

There are now ample data on protease inhibitors effective against phytopathogens *in vitro* and *in vivo*. For instance, inhibitors from potato inactivate proteinases secreted by *F. solani* or *F. sambucinum* into cultural liquid. Inhibitors from buckwheat and pearl millet suppress spore germination and the growth of many fungi including *Aspergillus flavus*, *Aspergillus parasiticus*, *F. moniliforme*, *F. oxysporum*, *A. alternata* and *Trichoderma reesei*. Inhibitors of proteinases are accumulated in response to pathogen invasion and prevent disease development (e.g. in tomato inoculated with *P. infestans*). In some cases, correlation between disease resistance and the constitutive inhibitors is found (e.g. between wheat resistance to smut, or lupine and soybean to fusarial wilt). Cells of potato tubers treated with elicitors, such as salicylic or arachidonic acids, are able to excrete potatin and three chymotrypsin inhibitors (Habib and Khalid, 2007).

At least 14 genes encoding different protease inhibitors alone or in combination with other heterologous genes have been reported to be transferred into cultured plants, which showed increased resistance predominantly to insects (Valuyeva and Mosolov, 2004). Plant protease inhibitors have also been used to engineer resistance against viruses in transgenic plants. For example, expression of the gene encoding the cysteine proteinase inhibitor from rice, oryzacystatin, by transgenic tobacco was found to confer resistance against tobacco etch virus and potato Y virus, replication of which depends on cysteine proteinase activity (Valuyeva and Mosolov, 2004; Habib and Khalid, 2007).

In the green consumerization context, it is important that some of the available proteinase inhibitors are inactive for non-pathogenic microorganisms and do not inhibit activity of proteinases of animal origin. Thus, chestnut cystatin, which strongly inhibits the protease activity and the growth of pathogenic *B. cinerea*, *Colletotrichum graminicola*, and *Stagonospora nodorum*, has no effect on the protease activity and the growth of the saprophyte *Trichoderma viride* (Pernas *et al.*, 1999), and a proteinase inhibitor extracted from bean seeds specifically suppresses serine proteinase of pathogenic *C. lindemuthianum* but does not influence animal trypsin and chymotrypsin activity (Valuyeva and Mosolov, 2004). Non-specific antipathogenic activity of protease inhibitors suggests that transgenic crops producing inhibitors of insecticidal or nematicidal proteinases may be incorporated into integrated systems of plant protection against pests and pathogens. A further advantage of this approach is the possibility that inhibitory activity against proteinases could be combined with the insecticidal activity of lectins, resulting in a synergistic antipathogenic effect.

PR proteins

Plants have numerous defence mechanisms that are activated in response to pathogen attacks, abiotic stresses and chemicals that mimic pathogen challenge. Production of proteins related to pathogenesis (PR proteins) is one such inducible mechanism (Van Loon *et al.*, 1994; Van Loon *et al.*, 2006). These proteins are not detectable at all or are present only at basal concentrations in healthy plant tissues. Since the term PR proteins based on the above characteristics was often used for all constitutive plant proteins for which content or increasing activity was induced by microorganisms, the new term 'inducible defence-related proteins' was recently introduced (Van Loon *et al.*, 2006).

PR proteins consist of a large variety of families with members that differ in occurrence, expression and biological activities; they are divided into 17 classes (Sels *et al.*, 2008). A range of the above-mentioned plant proteins and peptides such as chitinases, β-1,3-glucanases, peroxidase, defensins and proteinase inhibitors are PR proteins and vice versa; some typical PR proteins are antimicrobial and inhibit pathogen growth *in vitro*. For instance, tobacco PR-1a is antifungal, and tomato proteins of the PR-1 family inhibit zoospore germination and reduce pathogenicity of *P. infestans*. The ability to disrupt fungal membranes has been shown for toumatin-like proteins (permatins) of the PR-5 class. Overexpression of PR protein genes in plants renders disease resistance. For example, the high level expression of PR-1 in transgenic tobacco plants promotes control of *Perenospora tabacina* and *Phytophthora nicotiana* (Dyakov and Ozeratskovskaya, 2007).

Biochemical functions of PR proteins, their role in defence mechanisms, engineering of transgenic plants with enhanced resistance to plant pathogens and characteristics of individual PR proteins have been comprehensively surveyed by many researchers (Van Loon, 1985; Loon and Van Strien, 1999; Van Punja, 2001; De Lucca *et al.*, 2005; Edreva, 2005; Van Loon *et al.*, 2006; Sels *et al.*, 2008).

Proteinic inducers of plant resistance as a promising strategy for green consumerization

The protective effect of some natural compounds against plant diseases results from the induction of plant resistance to pathogens rather than from biocidal activity towards the causative agents. This strategy is worthy of special consideration because it avoids a direct effect on a pathogen and involves a plant-mediated mode of action. This activates the natural defence responses of plants, minimizing the probability of the targeted pathogens developing resistance.

Active defence mechanisms are initiated in plants upon recognition of structural and chemical characteristics particular to a pathogen, collectively referred to as pathogen-associated molecular pattern (PAMPs). The PAMP components represented by compounds of different chemical origin were named as general elicitors (or general inducers). These elicitors initiate a conserved set of plant defence responses such as ROS production, deposition of callose, protein phosphorylation, and transcriptional activation of early response genes, resulting in PAMP-triggered immunity or induced resistance. Natural compounds conferring disease resistance on plants in the manner of general elicitors is of interest because of the non-specific character of the induced resistance – lots of pathogens can be controlled with one active compound. The success of applying elicitor compounds is directly dependent on understanding their properties and mechanisms of action. As well as all other biogenic or abiogenic, general or specific inducers, proteinaceous elicitors produced by plant pathogens (e.g. glicoproteins, flagellins, elongation factor Tu, elicitins and transglutaminases from *Phytophthora* spp., proteins and peptides from *Cladosporium flavum*, monolicollin, cold shock proteins, harpins etc.) are the subject of genetic and biochemical research of signalling pathways and molecular mechanisms underlying plant resistance to diseases. Non-protein small compounds (plant signalling molecules and hormones) participating in signalling as well as chemical compounds that have been found to mimic elicitors are also being studied in this way. Transgenic plants with genes encoding microbial elicitors (e.g. elicitins) are created for research purposes using biotechnological methods.

Regarding the application, general elicitors to plant protection, chitosan, arachidonic, salicylic and β-aminobutyric acids may be referred to as examples of active ingredients that have been formulated and introduced into agricultural practice as commercial biogenic non-protein inducers of plant resistance. For instance, a supplement of tomato growth substratum with chitosan suppressed the root rot caused by *F. oxysporum* f. sp. *radicis-lycopersici* (Lafontaine and Benhamou, 1996). Moreover, several chitosan formulations collectively known as 'chitozars', which are effective against root rots, late blight and powdery mildew on cereals, potato and legumes, have been included in biological protection systems of these crops in Russia. Among protein elicitors, bacterial harpins are the most important. Harpins (hypersensitivity response and pathogenicity) are components of a transport system used by bacteria to transfer proteins through bacterial and plant membranes. In the bacterial genome, harpins are products a *hrp*-gene cluster. One of the first harpins studied was HrpN, the product of *hprN* gene expressed in *E. amylovora*. HrpN has been shown to be an acidic glycine-rich thermostable protein with a molecular mass of 44 kDa that induces resistance in *Arabidopsis* (Wei and Beer, 1996). Harpin has been commercialized and marketed as a product that enhances resistance of some field, ornamental and vegetable crops to many diseases, and that stimulates plant growth and flowering.

In the coming years, we can expect the agricultural use of other elicitor proteins for increasing plant resistance to diseases both by engineering of plants containing elicitor protein genes and by exposure of plants to protein elicitor-based products. Two bacterial proteins with plant-protecting activity are considered below as examples of very promising compounds for developing both resistant transgenic crops and products for plant treatment.

Cold shock protein CspD as an elicitor of disease resistance in plants

Cold shock protein D (CspD) isolated from *Bacillus thuringiensis* culture broth is a thermostable low molecular weight protein (molecular mass 7.2 kDa). The *B. thuringiensis* gene encoding CspD was cloned and sequenced (*CspD*, # AY272058 in GenBank). Nucleotide and amino acid sequences of CspD show high homology with other bacterial CSPs (Dzhavakhiya *et al.*, 2000, Kromina and Dzhavakhiya, 2004).

CSPs are highly conserved proteins produced by various bacteria constitutively or in response to cold shock. They have also been studied in plants, e.g. in *Arabidopsis*, tobacco and wheat. CSPs are involved in cell growth and adaptation to low temperatures. They bind nucleonic acids and are considered to be putative translation anti-terminators. Bacterial CPSs contain a specific conserved cold shock domain (CSD), which occurs in CSPs from other organisms. For instance, a fragment of high homology to bacterial CSD was found in a family of plant CSPs referred to as glycine-rich proteins (GRPs). Expression of *grp*-genes is dependent on the level of plant hormones (indolyl acetic, salicylic, abscisic acids), illumination and the plant development phase, and is upregulated by cold and wounding. The synthesis of GRPs can also be enhanced upon pathogen attack both in compatible and incompatible plant–pathogen combinations. Alien CSPs are recognized by plants as components of PAMP and elicit resistance to some pathogens. Thus, CSPs from *Micrococcus lysodeikticus* non-specifically induced defence responses in plants. It was found that a peptide consisting of 15 amino acid residues (csp15) that represented a consensus sequence of RNA-binding PNP-1 and RNP-2 motifs was responsible for eliciting activity towards some *Solanaceae* plants (Felix and Boller, 2003). The peptide csp15 induced an 'oxidative burst' in tobacco (*Nicotiana tabacum*, cv. Havanna 425) and potato (*Solanum tuberosum*), but was ineffective towards rice and cucumber cells.

CspD from *B. thuringiensis* has been reported to induce resistance both in monocotyledonous and dicotyledonous crops against a wide range of pathogens: from filamentous microorganisms (fungi and oomycetes) to viruses (Dzhavakhiya et al., 2000). For instance, CspD applied by dropping onto wheat leaves or by spraying potato and rice seedlings with water- or bovine serum albumin (BSA)-stabilized CspD solutions produced resistance against *S. nodorum*, *P. infestans* and *M. grisea*, respectively. Effective crop protection was observed in both greenhouse and field experiments. Exposure of tobacco plants to CspD induced resistance to tobacco mosaic virus (TMV) and potato X-virus. Peptide csp15 of CspD protein (VKWFNAEKGFGFITP) also showed resistance-inducing activity on plants and in model plant–pathogen systems. Treatments of tobacco leaf halves with 1–10 µmol csp15 in 0.1% BSA a day before inoculation with TMV resulted in a drastic reduction of lesion spot number on the treated halves as compared to 0.1% BSA-treated (control) halves or whole control leaves of the same plant (Kromina and Dzhavakhiya, 2004).

Several tests were carried out to identify which host defence responses to the pathogen challenge are activated by CspD and csp15. Applying csp15 to the surface of discs cut from potato tubers (cv. Istrinskiy, R1) increased the hypersensitive response of potato cells to the incompatible *P. infestans* race r4 with a decreased number of dead cells and induced accumulation of salicylic acid in the tuber tissues. Both CspD and csp15 have no fungitoxicity. Addition of the peptide to the suspension of cultured tobacco cells activated the H^+ pump and caused a reversible change in the extracellular pH. The resistance induced with csp15 or CspD in plants has a systemic character (Kromina and Dzhavakhiya, 2004).

To transfer the CspD gene into tobacco plants, the pBilt7 plasmid, containing the *CspD* expression cassette (*P35S/CspD/pACaMV*), on pBin19 vector background was constructed and transformed into *A. tumefaciens*. As a result, seven lines of cv. Xanthi (NN) and five lines of cv. Samsung (nn) were produced. Expression of *CspD* in these lines was confirmed by real-time PCR. The transgenic tobacco plants had the same habitus as control plants and produced fertile projeny. The transgenic lines showed increased resistance to *Alternaria longipes* and TMV that coincided with the range of antipathogenic activity observed for the elicitor protein per se. Importantly, the level of *CspD* expression coincided with the resistance level to the both pathogens among all tested lines as well as inside any one line. These findings lead to the assumption that disease resistance of transgenic tobacco is due to defence responses that are elicited with CspD synthesized inside the plant (Kromina and Dzhavakhiya, 2006).

The above results can be used in breeding programmes for production of crop cultivars with resistance against a wide range of phytopathogens or CspD-based biopesticides providing complex protective effects against fungi, oomycetes and viruses.

MF3 protein, bacterial peptidyl-prolyl cis/trans isomerase, conferring plant resistance to pathogens

The thermostable protein MF3 (16.9 kDa) is produced by biocontrol *Pseudomonas fluorescens* isolate 197. After cloning and sequencing the MF3-encoding gene from *P. fluorescens* 197, a full primary structure of the protein was determined, and high homology between amino acid sequences of MF3 and peptidyl-prolyl *cis/trans* isomerases of the FK506-binding protein (FKBP) type was found (Dzhavakhiya *et al.*, 2005; Shumilina *et al.*, 2006).

Peptidyl-prolyl *cis/trans* isomerases (PPIases) are enzymes that accelerate the slow *cis/trans* isomerization of the peptide bond between proline residues and the adjacent amino acid in protein. Initially, *de novo* synthesized polypeptides have the *trans* conformation, which stabilizes the polypeptide chain, but to form the native protein structure, about 7% of the peptide bonds formed with proline residues need to be isomerized into the *cis* conformation. PPIases take part in the renaturation of denatured proteins, in protein synthesis *de novo* and in the formation of biologically active conformations of some polypeptides. PPIases are highly conservative enzymes produced by prokaryotes, eukaryotes, viruses and phages. They have been classified into three families, referred to as cyclophilins, FKBPs, which regulate signal systems of eukaryotic cells, and parvulins.

In plants, FKBPs were initially detected in *Vicia fava*, *Triticum aestivum* and *Arabidopsis thaliana*. It was found that *A. thaliana* contains 23 FKBPs and FKBP-like low- or high-molecular weight proteins. Plant PPIases of the FKBP type control plant fertility, cell division and differentiation as well as the perception of plant hormones. Plant FKBPs are expressed both constitutively and in response to various stresses such as wounding, salinity or heat shock. PPIases produced by phytopathogens can contribute significantly to the pathogenic process. For instance, the plant pathogenic fungi *M. grisea* and *B. cinerea* produce cyclophilins. Mutations in the cyclophylin genes impair virulence of these fungi. The *M. grisea* mutants form low-turgor appressoria unable to penetrate through leaf cuticle. Deletion of the cyclophylin gene does not influence *M. grisea* growth *in vitro* but affects formation of fungal conidia (Viaud *et al.*, 2002).

It is likely that PPIases are involved in plant–pathogen relationships. On the one hand, some pathogens are in need of host plant PPIases for establishing the infection process. On the other hand, interaction of plant PPIase with pathogen proteins can result in induction of disease resistance. Thus cysteine proteinase AvrRpt2, that is transported by *Pseudomonas syringae* to

Arabidopsis cells during pathogen invasion must be folded with the plant cyclophylin ROC1 to form an active conformation able to cleave plant proteins. Along with other proteins, the activated bacterial protease cleaves protein RIN4, which is combined in *Arabidopsis* in an inactive complex with a signal protein RPS2. Proteolysis of RIN4 releases RPS2 that triggers defence responses (Mackey *et al.*, 2003). PPIases of one organism have been reported to inhibit the growth of another organism by means of competitive binding with a receptor, and probably some PPIases can suppress growth of cultured plant pathogens. For instance, cyclophylin C-CyP isolated from Chinese cabbage (*Brassica campestris L. ssp. pekinensis*) inhibits *in vitro* growth of several fungi including *Rhizoctonia solani*, *B. cinerea*, *F. solani* and *F. oxysporum* (Lee *et al.*, 2007).

Various experiments have shown that MF3 induced resistance in various monocotyledonous and dicotyledonous plants against several fungal and viral pathogens. MF3 was found to protect tobacco against TMV, potato Y virus and *A. longipes*, and barley against *Bipolaris sorokiniana*. At the same time, this protein did not influence TMV infectivity, was not fungitoxic and had no phytotoxicity towards tobacco plants and cereals. Moreover, treatment of barley and wheat seeds, naturally infected by *B. sorokiniana* and *F. culmorum*, with MF3 promoted formation of a well-developed root system in the diseased barley and wheat seedlings, that conferred plant tolerance to root rots and was consistent with growth-stimulatory functions for PPIases in plants (Dzhavakhiya *et al.*, 2005; Shumilina *et al.*, 2006).

Comparative analysis of MF3 with 45 proteins of different homology levels revealed two conserved sequences in the MF3 polypeptide chain. Proteolysis with trypsin inside one of these conserved sequences produced fragments that did not induce resistance to TMV in tobacco leaves. These data allowed the assumption that the analysed sequence contained a motif responsible for the resistance-inducing activity of MF3. Such a motif, IIPGLEKALE GKAVGDDLEVAVEPEDAYG, was detected and named MF3-29 because it was found to consist of 29 amino acid residues. This fragment was necessary and sufficient for induction of tobacco resistance to TMV. Biological tests on isolated tobacco leaves showed that treatments of tobacco leaves with chemically synthesized oligopeptide MF3-29 at concentrations 0.5, 5 and 50 nM were as effective against the virus as the whole protein at the same concentrations.

In order for products based on protein elicitors to effectively control plant pathogens, they have to access plant receptors recognizing PAMPs. The large size of the molecules or hydrophobic barriers on the plant surface (such as cuticle) can impede physical contact or chemically mediated recognition of elicitor proteins and the subsequent induction of defence responses in plants. To solve this problem, special molecular carriers facilitating elicitor transport to plant cells should be developed. Various polycationic molecules, especially chitosan, which are used now for the delivery of large biological molecules (DNA or proteins) to their outer or intracellular receptors, look promising as putative carriers of proteinaceous elicitors. Experiments with the wheat leaf spot agent *S. nodorum* and turnip mosaic virus (TuMV) demonstrated that MF3 integration with chitosan, a non-toxic and biodegradable natural polymer with elicitor properties, enhanced resistance-inducing activity of MF3, and enlarged the scope of the antipathogenic action of this protein. No inhibitory effect on fungal growth *in vitro* was observed if MF3, chitosan or the MF3–chitosan complex was added to culture media. Neither the protein nor chitosan applied separately on wheat leaves protected plants against *S. nodorum*. However, leaf treatment with the MF3–chitosan complex before plant inoculation with the pathogen significantly reduced disease severity. Spraying cabbage seedlings (cv. Krautman) with the MF3– chitosan-complex resulted in at least a week's delay in TuMV accumulation, whereas virus spread was not delayed in seedlings sprayed with MF3 alone or chitosan alone (Shumilina *et al.*, 2005).

Rape lines (cv. Westar) with an *mf3* gene insertion showed increased resistance to fungal pathogens *Plasmodiophora brassicae* and TuMV. Most of the transgenic plants carried more than one gene insertion. The bacterial protein concentrations in some lines amounted to 37 pg/mg of fresh weight. An increased level of rape resistance to TuMV was kept independently of MF3 variation in plant tissues. Identically high resistance was observed at eight different protein concentrations in transgenic lines. This may suggest that tiny amounts of MF3 are sufficient for conferring plant resistance to the virus. It was shown that five transgenic lines of the six that were resistant to TuMV also expressed an enhanced endurance to *P. brassicae* for 50 days after inoculation. The disease symptoms were significantly less severe on roots of transgenic plants as compared to symptoms developing on non-transgenic rape roots or roots of transgenic lines that had lost the *mf3* insertion. It is believed that the best lines obtained may be included in breeding programmes for creating rape cultivars with resistance against both pathogens (Dzhavakhiya *et al.*, 2005; Shumilina *et al.*, 2006).

Thus, the wide range of elicitor activity of CspD and MF3, as well as the possibility to use general approaches of biocontrol, gene engineering technology and plant treatments in CspD- and MF3-based plant protection demonstrate the great potential of proteinaceous elicitors of natural defence mechanisms for the environmentally safe control of phytopathogens. Toxicology studies and in-house laboratory tests demonstrated that the proteins are non-toxic to animals and plants. Both of these compounds are patented as new proteins with plant protecting properties (EP0868431A1 and PCT WO2005/061533 A1). Due to the systemic character of the resistance induced with CspD and MF3, the resistance spreads to the whole plant even if only the seeds or leaves are treated. After a single treatment with the protein solutions, induced resistance remained effective for not less than 3 weeks. In many cases this can cover the period of highest risk infection. The large-scale introduction of transgenic cultivars resistant to a certain pathogen is sometimes limited because different pathogens dominate when the climatic and soil conditions change. Moreover, there is the problem of unpredictable seasonal alteration of pathogens, which decreases the effect of pre-sowing treatments. Crop engineering or/and developing new products based on general elicitors such as CpsD or MF3 inducing 'universal' resistance would promote overcoming these limitations. Searching for novel proteins and peptides would promote progress in this field. Finding elicitor activity in MF3 protein belonging to PPIases that were never meant to be designated as disease resistance inducers shows desired properties may be discovered during screening of natural biologically active compounds which were previously studied and characterized as non-hazardous for people and the environment. This presents additional prospects for developing new biocontrol products complying with green consumerization requests.

5.3 Statins as New Promising Candidates for Biopesticides

One of the effective plant disease management strategies is based on the control of metabolic relationships in a plant–pathogen system. This can be reached by using compounds that alter or partially block specific pathways of biosynthesis in plants or microorganisms, resulting in derangements in the trophic relationships of pathogens with host plants, or in the production of microbial metabolites related to pathogenicity. Such imbalances may not cause the rapid death of pathogens but can lead to a reduction of their vital capacity or/and impair their pathogenic properties. Similar metabolic effects may be caused by plant inhibitors of fungal and bacterial proteases as described above, as well as by the inhibitors of insect and nematode enzymes. Statins, secondary metabolites of microbial origin belonging to the polyketide group, represent another class of natural compounds that have recently been discovered to modulate metabolic plant–pathogen interactions. The metabolic pathway targeted by statins is sterol biosynthesis.

Sterols are well known to be ubiquitous and important components of outer membrane and intracellular membranes of eukaryotic organisms and to play essential roles in their physiology. These compounds are required for growth, development and reproduction of plant pathogens. A number of organisms including insects, nematodes and oomycetes are not able to synthesize sterols or even their precursors. Sterol-dependent phytopathogens obtain free sterols or intermediates from host plants and may be controlled by compounds inducing alterations in content, availability or composition of plant sterols.

Currently, several chemical fungicides are available that inhibit the synthesis of ergosterol in fungi. They prevent cellular membrane formation, stop fungal growth and may suppress sporogenesis in established infection agents. However, statins are natural inhibitors of sterol biosynthesis that presumably can cause similar effects. Statins were found to be produced by different microorganisms, predominantly by filamentous fungi. They inhibit the enzyme β-hydroxy-β-methylglutaryl-CoA reductase (HMG-CoA reductase) and prevent sterol biosynthesis at the level of conversion of HMG-CoA to mevalonic acid. Statins are widely applied in medicine as drugs against atherosclerogenic diseases because they effectively decrease the blood cholesterol level. Thus, the biological compatibility and safety of statins are well proven.

Recently, studies were made to investigate the potential of two statins, lovastatin and compactin, in plant protection against diseases (Dzhavakhiya and Petelina, 2008; Ukraintsteva, 2008). In these studies, both statins were obtained by means of microbial synthesis using the 'superproducer' strains *Aspergillus terreus* 45-50 (lovastatin) and *Penicillium citrinum* 18-12 (compactin); these statins were examined in greenhouse and field experiments with plant treatments and *in vitro* tests on several pathogenic fungi. As statins are insoluble in water, the authors used aqueous solutions of the statin sodium salts that are referred to below as compactin and lovastatin.

The studied statins were found to possess fungicidal activity *in vitro* against a number of plant pathogenic fungi. The addition of lovastatin at concentrations from 0.001 to 0.1% to agar media inhibited growth of *M. grisea*, *S. nodorum* and *Coletotrichum atramentarium*. Compactin arrested *in vitro* growth of *Cladosporium cucumerinum*, *S. nodorum* and *M. grisea* at the same concentration range. In addition, 0.001% compactin solution significantly reduced germination of *S. nodorum* spores while the 0.01% solution completely suppressed spore germination in this fungus. Among tested plant pathogens, the fungus *M. grisea*, rice blast agent, showed the highest sensitivity to both statins (for lovastatin IC_{50} was about 0.003%), while *S. nodorum* was most resistant to lovastatin ($IC_{50} = 0.02\%$) and *C. cucumerinum* tended to compactin resistance. Besides a growth-inhibitory effect, the presence of the statins in nutrition media resulted in a discolouration of fungal mycelia in all fungi, especially in *M. grisea* and *C. atramentarium* exposed to lovastatin. This observation suggested suppression of fungal melaninogenesis. In all experiments, dicolouration-inducing concentrations of the statins were at least tenfold lower than growth-inhibiting concentrations.

The studied statins were revealed to possess disease-preventing properties. They delay and decrease disease development on treated plants when applied simultaneously or prior to inoculation with pathogens. Lovastatin and compactin protected wheat against *S. nodorum* and tobacco from *A. longipes*. In addition, lovastatin showed protective activity against *M. grisea* on rice, whereas compactin was effective against *Puccinia graminis* on wheat and *P. infestans* on potato. Both fungicidal and protective effects were strongly dose-rate dependent, but fungitoxic concentrations also made a toxic impact on plants, whereas protection from diseases was provided by far lower concentrations that were non-phytotoxic. Thus, almost total or 27–36% *S. nodorum* growth inhibition *in vitro* was observed at 0.1% and 0.01% lovastatin concentrations in nutrition media, respectively. Both doses showed phytotoxicity on wheat leaves, but only the higher concentration caused retardant effect on growing plants. When 0.1% lovastatin solution was used for wheat seed soaking for 1.5–2 h, reduction of seedling length averaged 50%, and 0.01% concentration caused no inhibitory influence on plant growth. A considerable reduction of the disease index was found after application of only 0.0005% lovastatin solution on isolated wheat leaves. The applied concentration did not suppress *S. nodorum* germination and *in vitro* growth but prevented disease development with 94 and 72% protection efficacy at 3 and 7 days after inoculation, respectively. This suggests that a mechanism other than fungicidal activity contributes to the protective effect of statins (Dzhavakhiya and Petelina, 2008).

The resistance of fungi to extreme environmental conditions and biotic stresses is largely determined by their ability to produce protective high-molecular-weight pigments. One of the common fungal pigments in cell walls is melanin. Melanin is a coloured polymer produced by many plant pathogenic fungi, including *M. grisea, S. nodorum, C. lagenarium* and *C. atramentarium*, through the pentaketide pathway and depends on the availability of acetyl-CoA. Melanin plays a significant role in the infectivity of these fungi. For instance, the ability of the rice pathogen *M. grisea* to penetrate into tissues of the host plant is directly associated with the presence of melanin in the fungus. Strains of *M. grisea* and *C. lagenarium* defective in melanin formation lose pathogenicity and are incapable of forming mycelial overgrowth in host plants. Revertants restoring wild-colour type regain pathogenicity (Dzhavakhiya *et al.*, 1990).

Interestingly, antimelanogenic compounds, aminoalkylphosphinates, a family of phospho-analogues of natural amino acids, suppress biosynthesis of melanin and some toxic metabolites in the polyketide pathway of *M. grisea*, and are also fungicidal. The aminoalkylphosphinates serve as analogues of alanine, a precursor to pyruvic acid that is required for melanin biosynthesis. Exposure of fungi to the phospho-analogues of amino acids inactivates pyruvate dehydrogenase, thus inhibiting synthesis of acetyl-CoA and melanin (Zhukov *et al.*, 2004) as well as aflatoxins (Khomutov, Khurs, Shcherbakova, Mikityuk, Dzavakhiya and Zhemchuzhina; unpublished data). The discovery that non-fungicidal lovastatin and compactin concentrations can induce mycelium de-pigmentation and decrease disease severity on plants suggests that their protective effect may be associated with impairing pathogenicity due to an effect on melanin biosynthesis in causative agents. Statins and fungal melanin are both polyketides. Thus, the metabolism of the two compounds is interrelated, and it is not improbable that statins can negatively mediate a stage of the polyketide pathway involved in the melanization of plant pathogenic fungi.

Although the mode of protective action of statin is required for understanding and further research, first small-plot field trials suggest they may be of certain interest from a practical point of view. For example, one pre-planting treatment of potato tubers by soaking in 0.1% compactin solution for half an hour resulted in a 1-month delay of late blight (*P. infestans*) emergence on plants and a slower course of disease. Only extremely high statin concentrations of 0.5% undesirably influenced plant physiological characteristics. By the end of the growing season, an insignificant reduction of potato late blight was observed on plants arising from the treated tubers, but these plants produced fewer diseased tubers. There is reason to suppose that lovastatin and compactin possess anti-phytoviral activity (Ukraintseva, 2008).

Further research of the mechanisms responsible for protective activity could help statins take a fitting place among the biopesticides of tomorrow. Lovastatin and compactin per se might serve as base molecules for biochemical engineering of active analogues, and the approaches used to studying their plant-protecting properties might be implicated in screening of other pathogen-controlling compounds within the statin group and their derivatives. Such screening might lead to the detection of new natural polyketides with high protective activity that would not have an adverse effect on plants. Since the agricultural application of statin-based formulations does not demand as high a level of statin purification as that in the pharmacological industry, using a relatively simple isolation procedure (Dzhavakhiya, 2008) and the 'superproducer' strains suggests that statin-based plant-protection technology would be a more economical and ecologically safe strategy than any method using chemical pesticides.

5.4 Conclusion

The disease-preventing effect of natural compounds is a result of a direct or plant-mediated influence on targeted pathogens. The compounds can affect phytopathogens directly, interrupting pivotal metabolic pathways and results in the death of the pathogens (biocidal effect). Some natural substances or their analogues specifically influence pathways related to pathogenicity and toxigenesis. In the case of a plant-mediated mode of action, natural compounds elicit and activate defence responses in plants that result in induced resistance to diseases. In addition, some biogenic compounds influencing plants do not induce resistance but enhance tolerance to pathogens by improving the physiological state of plants. The structural and functional diversity of natural compounds as well as their abundance provide a great potential for bringing plant-protection technologies in line with green consumerization.

References

Ahrenholtz, I., Harms, K., De Vries, J. and Wackernagel, W. (2000) Increased killing of *Bacillus subtilis* on the hair roots of transgenic T4 lysozyme-producing potatoes. *Applied and Environmental Microbiology* 66, 1862–1865.

Asaka, O. and Shoda, M. (1996) Biocontrol of *Rhizoctonia solani* damping-off of tomato with *Bacillus subtilis* RB14. *Applied and Environmental Microbiology* 62, 4081–4085.

Bais, H.P., Fall, R. and Vivanco, J.M. (2004) Biocontrol of *Bacillus subtilis* against infection of Arabidopsis roots by *Pseudomonas syringae* is facilitated by biofilm formation and surfactin production. *Plant Physiology*, 143, 1–13.

Bassarello, C., Lazzaroni, S., Bifulco, G., Lo Cantore, P., Iacobellis, N.S., Riccio, R.,

Gomez-Paloma, L. and Evidente, A. (2004) Tolaasins A–E, five new lipodepsipeptides produced by *Pseudomonas tolaasii*. *Journal of Natural Products* 67, 811–816.

Benitez, T., Limon, C., Delgado-Jarana, J. and Rey, M. (1998) Glucanolytic and other enzymes and their genes. In: Harman, G.E., Kubicek, C.P. (eds) *Trichoderma and Gliocladium: Enzymes, biological control and commercial application*, Taylor & Francis, London, 2, 101–128.

Bolar, J.P., Norelli, J.L., Wong, K-W., Hayes, C.K., Harman, G.E. and Aldwinckle, H.S. (2000) Expression of endochitinase from *Trichoderma harzianum* in transgenic apple increases resistance to apple scab and reduces vigor. *Phytopathology* 90, 72–77.

Bonaterra, A., Mari, M., Casalini, L. and Montesinos, E. (2003) Biological control of *Monilinia laxa* and *Rhizopus stolonifer* in postharvest of stone fruit by *Pantoea agglomerans* EPS125 and putative mechanisms of antagonism. *International Journal of Food Microbiology* 84, 93–104.

Brady, S.F., Wright, S.A., Lee, J.C., Sutton, A.E., Zumoff, C.H., Wodzinski, R.S., Beer, S.V. and Clardy, J. (1999) Pantocin B, an antibiotic from *Erwinia herbicola* discovered by heterologous expression of cloned genes. *Journal of American Chemical Society* 121, 11912–11913.

Bulet, P., Stocklin, R. and Menin, L. (2004) Antimicrobial peptides: from invertebrates to vertebrates. *Immunolgical Reviews* 198, 169–184.

Burr T.J., Matteson M.C., Smith C.A., Corral-Garcia M.R. and Huang T.Z. (1996) Effectiveness of bacteria and yeasts from apple orchards as biological control agents of apple scab. *Biological Control* 6, 151–157.

Carvalho, A.O. and Gomes, V.M. (2009) Plant defensins – Prospects for the biological functions and biotechnological properties. *Peptides* 30, 1007–1020.

Chang, W.T., Chen, C.S. and Wang, S.L. (2003) An antifungal chitinase produced by *Bacillus cereus* with shrimp and crab shell powder as a carbon source. *Current Microbiology* 47, 102–108.

Chernin, L. and Chet, I. (2002) Microbial enzymes in the biocontrol of plant pathogens and pests. In: Burns, R.G. and Dick, R.P. (eds) *Enzymes in the Environment: Activity, Ecology and Applications.* Marcel Dekker, New York, pp. 171–226.

Chrispeels, M.J. and Raikhel, N.V. (1991) Lectins, lectin genes, and their role in plant defense. *Plant Cell* 3, 1–9.

De Bruijn, I., De Kock, M.J.D., Yang, M., de Waard, P., van Beek, T.A. and Raaijmakers, J.M. (2007). Genome-based discovery, structure prediction and functional analysis of cyclic lipopeptide antibiotics in *Pseudomonas* species. *Molecular Microbiology* 63, 417–428.

De Lucca, A.J., Cleveland, T.E. and Wedge, D.E. (2005) Plant-derived antifungal proteins and peptides. *Canadian Journal of Microbiology* 51, 1001–1014.

De Samblanx, G.W., Goderis, I.J., Thevissen, K., Raemaekers, R., Fant, F., Borremans, F., Acland D. P., Osborn R. W., Patel S. and Broekaert W.F. (1997) Mutational analysis of a plant defensin from radish (*Raphanus sativus* L.) reveals two adjacent sites important for antifungal activity. *Journal of Biological Chemistry* 272, 1171–1179.

De Souza J.T., de Boer M., deWaard P., Van Beek T.A. and Raaijmakers J.M. (2003) Biochemical, genetic and zoosporicidal properties of cyclic lipopeptide surfactants produced by *Pseudomonas fluorescens. Applied and Environmental Microbiology* 69, 7161–7172.

Degenkolb, T., Berg, A., Gams, W., Schlegel, B. and Grafe, U. (2003) The occurrence of peptaibols and structurally related peptaibiotics in fungi and their mass spectrophotometric identification via diagnostic fragment ions. *Journal of Peptide Science* 9, 666–678.

During, K., Porsch, P., Fladung, M. and Lorz, H. (1993) Transgenic potato plants resistant to the phytopathogenic bacterium *Erwinia carotovora. The Plant Journal* 3, 587–598.

Dyakov, Yu.T. and Ozeratskovskaya, O.L. (2007) Vertical pathosystem: Resistance genes and their products. Immune response. In: Dyakov, Yu.T., Dzhavakhiya, V.G. and Korpela, T. (eds) *Comprehensive and Molecular Phytopathology.* Elsevier, the Netherlands, pp. 181–314.

Dzhavakhiya, V., Aver'yanov, A.A., Minayev, V.I., Ermolinsky, B.S., Voinova, T.M., Lapikova V.P., Petelina, G.G. and Vavilova, N.A. (1990) Structure and functions of melanin in cell wall of rice blast agent, micromicete *Pyricularia oryza* cav. *Zhurnal Obshchei Biologii. Journal of General Biology* 51, 528–535.

Dzhavakhiya, V., Filipov, A., Skryabin, K., Voinova, T., Kouznetsova, M., Shulga, O., Shumilina, D., Kromina K., Pridannikov, M., Battchikova, N. and Korpela, T. (2005) Proteins inducing multiple resistance of plants to phytopathogens and pests. *Patent* PCT WO2005/061533 A1, 2005-07-07.

Dzhavakhiya, V.G., Nikolaev, O.N., Voinova, T.M., Battchikova, N.V., Korpela T. and Khomutov, R.M. (2000) DNA sequence of gene and amino acid sequence of protein from *Bacillus thuringiensis*, which induces non-specific resistance of plants to viral and fungal diseases. *Journal of Russian Phytopathological Society* 1, 75–81.

Dzhavakhiya, V.V. (2008). Studying the potentiality of lovastatin using for plant protection from diseases and development of a statin production technology based on new producer strains. Ph.D. thesis. Russian Research Institute of Phytopathology, Russia.

Dzhavakhiya, V.V. and Petelina, G.G. (2008) Lovastatin influence on phytopathogenic fungi. *Agro XXI* 4–6, 33–35.

Edreva, A. (2005) Pathogenesis-related proteins: research progress in the last 15 years. *General and Applied Plant Physiology* 31, 105–124.

Elad, Y., Chet, I. and Henis, Y. (1982) Degradation of plant pathogenic fungi by *Trichoderma harzianum*. *Canadian Journal of Microbiology* 28, 719–725.

Felix, G. and Boller, T. (2003) The highly conserved RNA-binding motif RNP-1 of bacterial cold shock proteins is recognized as an elicitor signal in tobacco. *Journal of Biological Chemistry* 278, 6201–6208.

Finking, R. and Marahiel, M.A. (2004) Biosynthesis of nonribosomal peptides. *Annual Review of Microbiology* 58, 453–488.

Garcia -Olmedo, F., Molina, A., Alamillo, J.M. and Rodriguez-Palenzuela, P. (1998) Plant defense peptides. *Biopolymers* 47, 479–491.

Gohel, V., Megha, C., Vays, P. and Chhatpar, H.S. (2004) Strain improvement of chitinolytic enzyme producing isolate *Pantoea dispersa* for enhancing its biocontrol potential against fungal plant pathogens. *Annals of Microbiology* 54, 503–515.

Gohel, V., Singh, A., Vimal, M., Ashwini, P. and Chhatpar, H.S. (2006) Bioprospecting and antifungal potential of chitinolytic microorganisms. *African Journal of Biotechnology* 5, 54–72.

Gouland, C., Hlimi, S., Rebuffat, S. and Bodo, B. (1995) Trichorzins HA and MA, antibiotic peptides from *Trichoderma harzianum*. *Journal of Antibiotics* 48, 1248–1253.

Gueldner, R.C., Reilly, C.C., Pusey, P.L., Costello, C.F., Arrendale, R.F., Cox, R.H., Himmelsbach, D.S., Crumley, F.G. and Cutler, H.G. (1988) Isolation and identification of iturins as antifungal peptides in biological control of peach brown rot with *Bacillus subtilis*. *Agricultural Food Chemistry* 36, 366–370.

Habib, H. and Khalid, M.F. (2007) Plant protease inhibitors: a defense strategy in plants. *Biotechnology and Molecular Biology Review* 2, 068–085.

Huang, H.W. (2000) Action of antimicrobial peptides: Two-state model. *Biochemistry* 39, 8347–8352.

Jin, M., Wright, S., Beer, S. and Clardy, J. (2003) The biosynthetic gene cluster of pantocin A provides insights into biosynthesis and a tool for screening. *Angewandte Chemie International Edition* 42, 2902–2905.

Ko, K., Norelli, J.L., Reynoird, J-P., Boresjza-Wysocka, E., Brown, S.K. and Aldwinckle, H.S. (2000) Effect of untranslated leader sequence of AMV RNA 4 and signal peptide of pathogenesis-related protein 1b on attacin gene expression, and resistance to fire blight in transgenic apple. *Biotechnology Letters* 22, 373–381.

Kromina, K.A. and Dzhavakhiya, V.G. (2004) Peptide csp15, which represents consensus sequence of RNA-binding motif RNP-1 of bacterial cold shock proteins, induces non-specific resistance of plants to viral and fungal pathogens. *Abstractbook of the International Joint Workshop on PR-Proteins and Induced Resistance*. RISO & University of Fribourg, Denmark, pp. 144.

Kromina, K.A. and Dzhavakhiya, V.G. (2006) Expression of bacterial gene *CspD* in tobacco plants results in enhanced resistance to fungal and viral phytopathogens. *Molecular Genetics, Virology and Microbiology* (Russia) 1, 31–34.

Kumar, A., Shukla, R., Singh, P., Prasad, C.S. and Dubey, N.K. (2008) Assessment of *Thymus vulgaris* L. essential oil as a safe botanical preservative against post harvest fungal infestation of food commodities. *Innovative Food Science and Emerging Technologies* 9, 575–580.

Lafontaine, P.J. and Benhamon, N. (1996) Chitosan treatment: an emerging strategy for enhancing resistance of greenhouse tomato to infection by *Fusarium oxysporum* f. sp. *radicilycopersici. Biocontrol Science and Technology* 6, 111–124.

Landon, C., Pajon, A., Vovelle, F. and Sodano, P. (2000) The active site of drosomycin, a small insect antifungal protein, delineated by comparison with the modeled structure of Rs-AFP2, a plant antifungal protein. *The Journal of Peptide Research* 56, 231–238.

Lavermicocca P., Iacobellis N.S., Simmaco M. and Graniti A. (1997) Biological properties and spectrum of activity of *Pseudomonas syringae* pv. *syringae* toxins. *Physiological Molecular Plant Pathology* 50, 129–140.

Lay, F.T. and Anderson, M.A. (2005) Defensins – Components of the Innate Immune System in Plants. *Current Protein and Peptide Science* 6, 85–101.

Lee, J.R., Park, S.C., Kim, J.Y., Lee, S.S., Park, Y., Cheong, G.W., Hahm, K.S. and Lee, S.Y. Molecular and functional characterization of a cyclophilin with antifungal activity from Chinese cabbage. (2007) *Biochemical and Biophysical Research Communications* 353, 672–678.

Lin, K.F., Lee, T.R., Tsai, P.H., Hsu, M.P., Chen, C.S. and Lyu, P.C. (2007) Structure-based protein engineering for a-amylase inhibitory activity of plant defensin. *Proteins* 68, 530–540.

Lorito, M. (1998) Chitinolytic enzymes and their genes. In: Harman, G.E. and Kubicek, C.P. (eds) *Trichoderma and Gliocladium: Enzymes, Biological Control and Commercial Application.* Taylor & Francis, London, volume 2, 73–100.

Mackey, D., Belkhadir, Y., Alonso, J.M., Ecker, J.R. and Dangl, J.L. (2003) *Arabidopsis* RIN4 Is a Target of the Type III Virulence Effector AvrRpt2 and Modulates RPS2-Mediated Resistance. *Cell* 112, 379–389.

Montesinos, E. (2007) Antimicrobial peptides and plant disease control. *FEMS Microbiology Letters* 270, 1–11.

Moreno, A.B., Martiinez, A., Borja, M. and SanSegundo, B. (2003) Activity of the antifungal protein from *Aspergillus giganteus* against *Botrytis cinerea. Molecular Plant-Microbe Interactions.* 93, 1344–1353.

Moreno, A.B., Penas, G., Rufat, M., Bravo, J. M., Estopa, M., Messeguer, J. and SanSegundo, B. (2005) Pathogen-induced production of the antifungal AFP protein from *Aspergillus giganteus* confers resistance to the blast fungus *Magnaporthe grisea* in transgenic rice. *Molecular Plant-Microbe Interactions.* 18, 960–972.

Nielsen, T.H. and Sorensen, J. (2003) Production of cyclic lipopeptides by *Pseudomonas fluorescens* strains in bulk soil and in the sugar beet rhizosphere. *Applied and Environmental Microbiology* 68, 3416–3423.

Nielsen, T.H., Sorensen, D., Tobiasen, C., Andersen, T.R. and Christophersen, C. (2002) Antibiotic and biosurfactant properties of cyclic lipopeptides. *Applied and Environmental Microbiology* 68, 3416–3423.

Nybroe, O. and Sorensen, J. (2004) Production of cyclic lipopeptides by fluorescent pseudomonads. In: Ramos, J-L. (ed.) *Pseudomonas, Biosynthesis of Macromolecules and Molecular Metabolism.* Kluwer Academic/Plenum Publishers, New York, USA, 147–172.

Odintsova, T, Shcherbakova, L., Fravel, D., Egorov, T. and Suprunova, T. (2009) Discovery of a novel protein, a putative elicitor from a biocontrol *Fusarium oxysporum*, inducing resistance to Fusarium wilt in tomato. *Abstracts of XIV International Congress on Molecular Plant-Microbe Interactions.* Quebec City, Canada, 22pp.

Ongena, M., Jacques, P., Toure, Y., Destain, J., Jabrane, A. and Thonart, P. (2005) Involvement of fengycin-type lipopeptides in the multifaceted biocontrol potential of *Bacillus subtilis. Applied Microbiology and Biotechnology* 69, 29–38.

Osborn, R.W., De Samblanx, G.W., Thevissen, K., Goderis, I., Torrekens, S., Van Leuven, F., Attenbrough S., Rees S.B. and Broekaert W. F. (1995) Isolation and characterisation of plant defensins from seeds of *Asteraceae, Fabaceae, Hippocastanaceae* and *Saxifragaceae. FEBS Letters* 368, 257–262.

Park, H-Y., Park, H-C. and Yoon, M-Y. (2009) Screening for peptides binding on *Phytophthora capsici* extracts by phage display. *Journal of Microbiological Methods* 78, 54–58.

Pedras, M.S., Ismail N., Quail J.W. and Boyetchko S.M. (2003) Structure, chemistry, and biological activity of pseudophomins A and B, new cyclic lipodepsipeptides isolated from the biocontrol bacterium *Pseudomonas fluorescens*. *Phytochemistry* 62, 1105–1114.

Pernas, M., Lopez-Solanilla, E., Sanchez-Monge, R., Salcedo, G. and Rodriguez-Palenzuela, P. (1999) Antifungal Activity of a Plant Cystatin. *MPMI* 12, 624–627.

Punja, Z.K. (2001) Genetic engineering of plants to enhance resistance to fungal pathogens—a review of progress and future prospects. *Canadian Journal of Plant Pathology* 23, 216–235.

Romero D., de Vicente A., Rakotoaly R.H., Dufour S.E., Veening J.W., Arrebola E., Cazorla F., Kuipers O.P., Paquot M. and Perez-Garcia A. (2007) The iturin and fengycin families of lipopeptides are key factors in antagonism of *Bacillus subtilis* towards *Podosphaera fusca*. *Molecular Plant-Microbe Interactions* 20, 430–440.

Ryan, C.A. (1990) Proteinase inhibitors in plants: genes for improving defenses against insects and pathogens. *Annual Review of Phytopathology* 28, 425–449.

Segura, A., Moreno, M., Molina, A. and García-Olmedo, F. (1998) Novel defensin subfamily from spinach (*Spinacia oleracea*). *FEBS Leterst* 435, 159–162.

Selim, S., Negrel J., Govaerts G., Gianinazzi S. and van Tuinen D. (2005) Isolation and partial characterization of antagonistic peptides produced by *Paenibacillus* sp. strain B2 isolated from the sorghum mycorrhizosphere.*Applied and Environmental Microbiology* 71, 6501–6507.

Sels, J., Mathys, J., De Coninck, B.M.A., Cammue, B.P.A. and De Bolle, M.F.C. (2008) Plant pathogenesis-related (PR) proteins: A focus on PR peptides. *Plant Physiology and Biochemistry* 46, 941–950.

Shcherbakova L.A., Dorofeeva L.L., Devyatkina, G.A., Sokolova, G.D. and

Fravel, D.R. (2008) Control of wheat root rots under field conditions with compounds produced by *Fusarium sambucinum* strain FS-94. *Journal of Plant Pathology*, 90, 338.

Shcherbakova, L., Fravel, D. Kromina, K. and Shumilina, D. (2007) Plant-mediated interactions of two biocontrol Fusaria with host and non-host plant pathogens. *Proceedings of XIII International Congress on Molecular Plant-Microbe Interactions*, IS-MPMI, Italy, Sorrento, 395–396.

Shukla, R., Kumar, A., Prasad, C.S., Srivastava, B. and Dubey, N.K. (2008) Antimycotic and antiaflatoxigenic potency of *Adenocalymma alliaceum* Miers. on fungi causing biodeterioration of food commodities and raw herbal drugs. *International Biodeterioration and Biodegradation* 62, 348–351.

Shumilina, D.V., Il'ina, A.V., Kulikov, S.N. and Dzhavakhiya, V.G. (2005) Elicitor activity of MF-3 protein from *Pseudomonas fluorescens* and combination of MF3-protein with chitosan in different host-pathogen pairs. *Advances in Chitin Sciences* 8, 275–278.

Shumilina, D., Krämer, R., Klocke, E. and Dzhavakhiya, V. (2006) MF3 (peptidyl-prolyl *cis-trans* isomerase of FKBP type from *Pseudomonas fluorescens*) – an elicitor of non-specific plant resistance against pathogens. *Phytopathologia Polonica* 41, 39–49.

Stein, T. (2005) *Bacillus subtilis* antibiotics: structures, syntheses and specific functions. *Molecular Microbiology* 56, 845–857.

Terras, F.R.G., Eggermont, K., Kovaleva, V., Raikhel, N.V., Osborn, R.W., Kester, A., Rees, S.B., Torrekens, S., Leuven, F.V., Vanderleyden, J., Cammue, B.P. and Broekaert, W.F. (1995) Small cysteine-rich antifungal proteins from radish: their role in host defense. *Plant Cell* 7, 573–588.

Terras, F.R.G., Schoofs, H.M.E., De Bolle, M.F.C., Van Leuven, F., Rees, S.B., Vanderleyden, J., Cammue, B.P. and Broekaert, W.F. (1992) Analysis of two novel classes of plant antifungal proteins from radish (*Raphanus sativus* L.) seeds. *Journal of Biological Chemistry* 267, 15301–15309.

Terras, F.R.G., Torrekens, S., Van Leuven, F., Osborn, R.W., Vanderleyden, J., Cammue, B.P. and Broekaert, W.F. (1993) A new family of basic cysteine-rich plant antifungal proteins from Brassicaceae species. *FEBS* 316, 233–240.

Ukraintseva, S.N. (2008) Potentiality of compactin using as the protection frame of plants against pathogenic organisms. In: *Anniversary proceeding digest 'Fifty years on guard of home food safty'*, RAAS, ARRIP, B. Vyazyomy, Russia, 488–496.

Valueva, T.A. and Mosolov, V.V. (2004) Role of inhibitors of proteolytic enzymes in plant defence against phytopathogenic microorganisms. *Biochemistry* (Moscow*)* 69, 1305–1309.

Van Loon, L.C. (1985) Pathogenesis-related proteins. *Plant Molecular Biology* 116, 111–116.

Van Loon, L.C., Pierpont, W.S., Boller, T. and Conejero, V. (1994) Recommendations for naming plant pathogenesis-related proteins. *Plant Molecular Biology Reporter* 12, 245–264.

Van Loon, L.C., Rep, M. and Pieterse, C.M. (2006) Significance of inducible defense-related proteins in infected plants. *Annual Review of Phytopathology* 44, 135–162.

Van Loon, L.C. and Van Strien, E.A. (1999) The families of pathogenesis-related proteins, their activities, and comparative analysis of PR-1 type proteins. *Physiological and Molecular Plant* Pathology 55, 85–97.

Varma, J. and Dubey, N.K. (1999) Perspective of botanical and microbial products as pesticides of tomorrow. *Current Science* 76, 172–179.

Viaud, M.C., Balhadere, P.V. and Talbot, N.J.A (2002) *Magnaporthe grisea* cyclophilin acts as a virulence determinant during plant infection. *The Plant Cell* 14, 917–930.

Vila L., Lacadena V., Fontanet P., Martinez A. and SanSegundo B. (2001) A protein from the mold *Aspergillus giganteus* is a potent inhibitor of fungal plant pathogens. *Molecular Plant-Microbe Interactions* 14, 1327–1331.

Wei, Z.M. and Beer, S.V. (1996) Harpin from *Erwinia amylovora* induces plant resistance. *Acta Horticulturae* 411, 223–225.

Wong, J.H. and Ng, T.B. (2005) Sesquin, a potent defensin-like antimicrobial peptide from ground beans with inhibitory activities toward tumor cells and HIV-1 reverse transcriptase. *Peptides* 26, 1120–1126.

Wong, J.H., Zhang, X.Q., Wang, H.X. and Ng, T.B. (2006) A mitogenic defensin from white cloud beans (*Phaseolus vulgaris*). *Peptides* 27, 2075–2081.

Wu, G.S., Shortt, B.J., Lawrence, E.B. , Leon, J., Fitzsimmons, K.C., Levine, E.B., Raskin, I. and Shah, D.M. (1997) Activation of host defense mechanisms by elevated production of H_2O_2 in transgenic plants. *Plant Physiology* 115, 427–435.

Xiao-Yan S.S, Qing-Tao S., Shu-Tao X., Xiu-Lan C., Cai-Yun S. and Yu-Zhong Z. (2006) Broad-spectrum antimicrobial activity and high stability of trichokonins from *Trichoderma koningii* SMF2 against plant pathogens. *FEMS Microbiological Letters* 260, 119–125.

Zasloff, M. (2002) Antimicrobial peptides of multicellular organisms. *Nature* 415, 389–395.

Zhukov, Yu.N., Vavilova, N.A., Osipova, T.I., Khurs, E.N., Dzavakhiya V.G. and Khomutov, R.M. (2004) Fungicidal activity of phosphinic analogues of amino acids involved in methionine metabolism. *Doklady Biochemistry and Biophysics* 397, 210–212.

6

Natural Products as Allelochemicals in Pest Management

ROMAN PAVELA

Crop Research Institute, Prague, Czech Republic

Abstract

Allelochemicals are one of the most plentiful groups of substances in the vegetable kingdom. The significance and use of these chemicals in nature are as varied as allelochemicals themselves. Allelochemicals also include a group of substances called allomones. These substances are created by plants as a defence against phytophagous insects and comprise repellents, anti-ovipositants and antifeedants. This group of plant metabolites can be of practical use in many areas of human activity. This chapter deals with possibilities of using plant extracts containing allomone in the protection of plants against pests. Substances with antifeedant effects, in particular, appear to be highly promising for the development of new, environmentally safe insecticides. In this chapter, we have therefore focused mostly on this group of substances and critically considered the perspective of using antifeedant substances in plant protection.

6.1 Introduction

Plants have developed alongside insects since the very beginning of their existence. Thanks to their common history, very fragile mutual relationships have formed between individual plant and insect species, which we are trying to clarify today using scientific approaches. On the one hand, plants utilize insects, especially as their pollinators, and on the other, insects use plants as the source of their food. Therefore, for balanced mutual interactions between plants and insects, communication and mutual influence must exist between them to prevent uncontrolled excessive reproduction of any species in order to maintain equilibrium in the ecosystem with maximum possible biodiversity. Plants thus created many strategies in the course of their co-evolution to protect themselves efficiently against insect pests, and the insects try to circumvent such strategies. Understanding their mutual relationships helps us not only to understand the world around us but also provides information that may lead to a practical use.

As mentioned above, there are many types of plant–insect interactions; however, the interaction between plants as a source of food and phytophagous insects as their pests is one of the most interesting and important ones for practical applications in agriculture. In this case, it is a relationship between 'food' on one hand and 'consumer' on the other; the plants were forced to develop numerous defensive mechanisms to prevent uncontrolled destructive pest attacks. The most important defensive plant mechanisms include the synthesis of biologically active compounds, the so-called secondary metabolites. First, such substances may provide direct insecticide effects, causing mortality of phytophagous insects, and/or second, they may exert indirect insecticide effects, only influencing insect behaviour in some manner. Knowledge of such relationships leads not only to useful information necessary for cultivating plants with resistance, but also to the direct utilization of extracts in cultural plant protection using the so-called botanical insecticides.

Although the first group of compounds, i.e. those with direct insecticidal activity, has been used by humans as extracts for millennia, both in fighting phytophagous pests as well as against parasites or storage pests, the other group of compounds, generally called allelochemicals, has become the subject of more profound interest only in recent decades (Pavela, 2007a).

The term allelochemicals (from Greek allelon: 'one another') is used to describe the chemicals involved in interspecific interactions. It is defined as a chemical significant to organisms of a species different from its source. Allelochemicals are divided into four subgroups, depending on whether the emitter, the receiver, or both benefit in the interaction.

- An allomone (from Greek allos – 'another'; horman – 'to stimulate') is defined as a chemical substance, produced or acquired by an organism, which evokes in the receiver a reaction adaptively favourable to the emitter, e.g. a plant emits allomones to deter herbivores.
- A kairomone (from Greek kairos – 'opportunistic') is defined as a chemical substance, produced or acquired by an organism, which evokes in the receiver a reaction adaptively favourable to the receiver but not to the emitter, e.g. secondary plant compounds help herbivores in finding plants to feed on.
- A synomone (from Greek syn – 'with or jointly') is defined as a chemical substance, produced or acquired by an organism, which evokes in the receiver a reaction adaptively favourable to both the emitter and the receiver. This group of allelochemicals includes floral scents and nectars that attract insects and other pollinators and substances that play an important role in symbiotic relationships.
- An apneumone (from Greek a-pneum – 'breathless or lifeless') is defined as a substance, emitted by a non-living material, which evokes a reaction adaptively favourable to the receiving organism, but detrimental to another organism that may be found in or on the non-living material. For example, parasites or predators are attracted to non-living substances in which they may find another organism, their host or prey, by apneumones released from the non-living substance.

In many cases not a single semiochemical has an effect on its own but different groups of chemicals in a precisely defined mixture act in an effectively combined manner. In general, it can be said that the whole group of allelochemicals may find its application in plant protection. Nevertheless, metabolites falling in the allomone group represent the most studied group with the most potential at present.

6.2 Allomones – A Prospective Group of Substances for Alternative Plant Protection

As explained above, this group of substances includes plant metabolites that have some type of negative effect on insect behaviour providing benefit to the plants. The chemicals can be divided as follows according to their mode of action:

1. Repellents
2. Anti-ovipositants
3. Antifeedants

Repellents are substances that directly deter insects from settling on the plant. They include an entire group of simple aromatic hydrocarbons, which may be released into the environment, thus having a direct effect on insect chemoreceptors (Koul, 2005). Repellency often tends to be connected with anti-oviposition, as such substances deter females from settling on nutritive plants and prevent oviposition at the same time. However, anti-oviposition need not always be connected with repellency. It may be connected with anti-feedancy very often – when upon settling on the plant, the female finds that the plant cannot provide food of good quality or acceptable for its descendants, and thus she flies off to seek a more suitable plant. The group of substances with an anti-oviposition effect includes a whole range of chemicals from simple aromatic terpenes, phenols or alkaloids to molecules falling in the group of polyphenols or limonoids (Koul, 2005).

The last group of substances – antifeedants – deter phytophagous insects against food consumption. This group of substances has been studied on a large scale in recent times and the use of such substances in plant protection against pests is connected with their significant potential. This group of chemicals also range from simple aromatic terpenes, phenols or alkaloids to molecules falling in the group of polyphenols or limonoids. At present, there are high expectations for this group of substances due to their practical use in plant protection (Isman, 1994; Koul, 2005; Pavela, 2007b).

The good prospects for the practical use of antifeedants are based on several factors. Most of the repellent substances fall into the group of aromatic hydrocarbons, which are volatile in the environment and their efficiency time is therefore reduced depending on period of application and dosage used. Numerous substances belonging to the polyphenols and higher terpenes are included in the category of antifeedants, which may have a much longer persistence time, thereby extending the efficiency time of products (Pavela and Herda, 2007a,b). Moreover, their antifeedant efficiency is often connected with further biological activity such as growth inhibition and cumulative mortality.

Allomones with antifeedant activity include a high percentage of substances that also meet the following requirements for development of new botanical insecticides:

- They are natural plant metabolites, safe for health and the environment;
- Such substances provide specific effects on individual pest species – they thus show high selectivity;
- Mixtures of substances with synergistic effects can be obtained, which prevent the development of pest resistance;
- Antifeedant substances usually show other biological effects such as growth inhibition, mortality and reduced fertility of surviving individuals, which increase the practical efficiency of the products (Isman, 2002; Koul, 2005).

All these characteristics increase the prospects for the use of antifeedant substances in plant protection. However, products based on allomones should be applied as a preventative measure, for example, at the time of pest invasion or its initial occurrence or at the time of larval incubation. Although a large number of biologically active substances and extracts from plants with antifeedant effects have been tested, only a limited number of products have been subjected to a detailed field application and in the development of plant protection products (Isman, 1994).

6.3 Current State of Antifeedant Research

The history of the research of substances that show antifeedant effects dates back to the 1930s. As early as 1932, Metzger and Grant tested about 500 plant extracts against *Popilla japonica*, although results were not substantially encouraging. Later, Pradhan *et al.* (1962) evaluated extracts of the Indian neem tree, *Azadirachta indica*, that prevented feeding by the desert locusts. Although terrestrial plants produce a diverse array of secondary metabolites, probably more than 100,000 unique compounds (Isman, 2002), today, about 900 compounds have been identified to possess feeding deterrence against insects (Koul, 2005).

In addition to various compounds isolated from plants or synthesized as insect antifeedants, several studies demonstrate the antifeedant efficacy in metabolite mixtures of plant essential oils or total extracts against a variety of insect species.

In recent years studies have revealed the antifeedant potential of plant essential oils against postharvest pests, aphids, thrips, lepidopterans, termites and mite pests (Hori, 1999; Hou-HouaMin *et al.*, 2002; Koul, 2005; Isman, 2006). Similarly, during the past few years much emphasis has been placed on demonstrating the antifeedant efficacy in total plant extracts (Mancebo *et al.*, 2000a,b; Wang *et al.*, 2000; Pavela, 2004a,b; Zhang *et al.*, 2004;

Debrowski and Seredynska, 2007) as they seem to exhibit the activity as multicomponent systems. However, it is also well known that antifeedants show interspecific variability (Isman, 1993). The existence of such inter-specific differences, as shown for many insect species, is encouraging with a view to searching selectively for specific feeding deterrents.

As already indicated, present research provides very important information on plant substances and their antifeedant efficiency. Nevertheless, it must be noted that the antifeedant efficiency of a significant number of known antifeedant compounds is dependent on the exposure period and their concentration. After a long exposure period, some of these compounds lose their antifeedant efficiency due to the development of resistance in the insects consuming the contaminated food (Koul, 2005). Nevertheless, such compounds may have other biological effects, at the same time, such as cumulative mortality or perhaps larval growth inhibition.

However, it should be emphasized that most of the antifeedant research is in the preliminary trial stage, although the activity of more than 900 compounds and several hundreds of plant extracts are known (Koul, 2005). However, it is expected that in the near future some novel efficacious plant-based compounds will be formulated as antifeedants, from the huge biodiversity provided by nature. Attention is currently being paid to promote the indigenous antifeedant plants by extolling their practical application to the farmers.

6.4 Mechanism of Antifeedant Action

Food selection among insect herbivores is a highly specialized phenomenon. While olfactory and physical aspects of plants or their organs can be important in insect host finding and acceptance (Miller and Strickler, 1984), the choice of food is based primarily upon contact chemoreception of various allelochemicals (Frazier, 1986; Stadler, 1992). In particular, dietary experience has influenced the ability of insects to taste plant chemicals that may have served as signals of suitability or unsuitability. Certain dietary constituents appeared to suppress the development of taste sensitivity to deterrents in an insect (Renwick, 2001). Avoidance of allelochemicals, when looked at from a behavioural point of view, is the outcome of interactions with chemoreceptors characterized by broad sensitivity to a spectrum of deterrents (Mullin *et al.*, 1994).

According to Schoonhoven *et al.* (1992), there are four basic reasons why the chemosensory perception of feeding deterrents by phytophagous insects warrants special attention:

- Feeding deterrents are apparently more important in host-plant recognition than phagostimulants.
- A huge number of feeding deterrents exist, with variable molecular structures adding to their diversity.
- There are fewer deterrent receptors.
- Different deterrents may elicit different behavioural reactions, indicating the presence of a differential sensory coding system.

Studies of the chemosensoric insect system are only at the initial point of research, and knowledge of the mode of action of the substances is superficial and should be studied in detail. Such a lack of knowledge is particularly caused by the fact that suitable technologies for chemosensoric investigations have been available only relatively recently, making it possible to perform reliable measurements of response to substance at insect chemoreceptors.

Although research of chemoreceptors is important for a general understanding of efficiency of individual antifeedant substances, from the practical point of view, experiments based on simple biological tests are those mostly used in antifeedancy studies.

Bioassays against insects have been used for decades as a means of elucidating the activity of many chemical components or extracts. The major goals achieved by employing bioassay techniques are to determine the roles of naturally occurring chemicals, identify the mechanism of resistance in crop plants and to find various insect control agents. The basic design to study deterrents is to present to an insect a substrate with the candidate chemical and to measure the response of the insect. Therefore, substrate choice and presentation are important factors for a successful bioassay. Both natural and artificial substrates are used, depending upon the goal of the experiment. On one hand one may emphasize that artificial substrates offer uniformity, but at the same time studies have shown that thresholds for the same deterrent may vary as much as 1000 times between natural and artificial substrates (Schoonhoven, 1982), perhaps due to differences in porosity or uptake rates by the insect. For sucking insects, the principal artificial substrate used has been a chemically defined liquid presented between natural or artificial membranes (Koul, 2005).

However, whatever the substrate may be, it is important that no textural differences should occur between the control and test substrates. Colour differences may also influence insects during testing. Care is needed to ensure the least hindrance with the presentation to the insect chemoreceptors, which should be in the usual way. Natural substrates could be whole plant, leaves, leaf discs, or specialized substrates such as twigs, blocks of wood, board, and paper towel discs. Artificial substrates usually include agar-based artificial diets, simple liquid-based artificial diets, styropors, or discs of foamed polystyrene, or polyurethane, and glass fibre discs (Koul, 2005).

Leaf discs are commonly used in preference or consumption bioassays with chewing insects. These assays are important in estimating the biological potential of the antifeedant effect of plant extracts in screening studies, and they correspond as much as possible to the conditions of the practical application. However, it must be emphasized that these assays are short term. For the purpose of practical use, further biological assays must be performed in extracts or substances showing the best biological activity, which will be of a long-term nature and will provide evidence on the practical applicability of the substances or extracts in plant protection. Nevertheless, such assays are irreplaceable as screening assays.

Individual types of biological assays used for evaluating antifeedant efficiency are discussed by Koul (2005). However, in general, such assays can be divided into two groups according to the mode of the experiment: a choice assay or a no-choice assay.

The principle is that insects can choose either control or treated discs (choice) or insects may be exposed to the test substance only (no choice). The no-choice situation often is more representative of our agricultural system, especially for monophagous species, but at the same time it is very sensitive (Fig. 6.1).

The general procedure adopted in this test is that measured leaf discs are punched out from substrates and treated either on one side or both sides with a known quantity of test material in a carrier solvent. It is preferable to use emulsified solutions in water in order to avoid interference with leaf disc texture due to solvents (Isman, 2002). A method has been described by which leaf surfaces can be covered with a uniform amount of a test chemical for bio-assay with leaf-feeding insects. Chemicals are dissolved in gelatine solutions, which can be sprayed evenly and which will adhere well to many leaf surfaces. Upon accurate application, the dosage per leaf area can be determined, which is an important practical viewpoint for the application itself.

After application, the leaf discs are dried at room temperature and then fed to candidate insects. Usually the arenas used are Petri dishes of variable sizes in which one treated and one control disc is placed (choice), or both the leaf discs are treated (no choice). In certain experiments five to ten treated and untreated leaf discs are used and placed alternately in the Petri dishes in a choice situation. The number of larvae introduced into each arena is variable depending upon the size and stage of the larvae used. There is also considerable variation in the duration of experiments, both long term and short

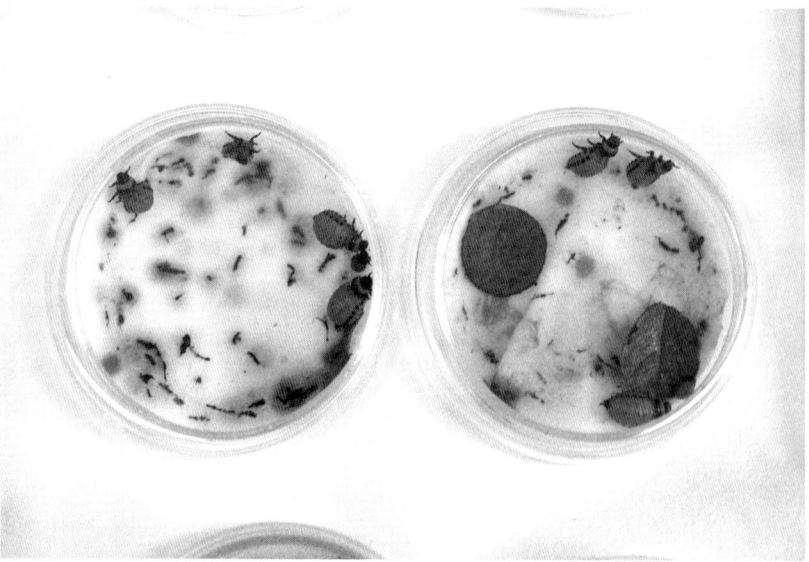

Fig. 6.1.   No-choice test with extract obtained from *Leuzea carthamoides* against larvae *Leptinotarsa decemlineata*, 48 h after application of 1% extract.

term. The consumption in each experiment is measured using various digitizing leaf area meters.

The bioassay is conducted for a very short duration (2 to 8 h) or until 50% of either disc is consumed. Several formulas and ways of denomination are presented in the literature for calculating the biological effect (Koul, 2005). Nevertheless, the following formula to calculate feeding deterrence provides the highest accuracy:

Feeding deterrence (%) = (C – T)/(C + T) × 100, where C and T are the consumption of control and treated discs, respectively (Koul, 2005; Pavela *et al.*, 2008).

As mentioned above, no-choice assays are most important from the practical point of view, because they are closest in nature to practical application. It is thus advisable to favour such assays.

When 90–100% feeding deterrence is obtained in the assays, efficient concentration (EC_{50} and EC_{95}) can be determined in subsequent assays, which is another important parameter for the mutual comparison of extract or substance efficiency. Moreover, it is also of equal importance to determine the time during which the given insect does not consume food treated with EC_{95}, as insects may become habituated to antifeedant substances or the biologically active substances may be degraded due to the action of the environment (Koul, 2005). The insects can thus overcome the initial resistance against food treated with antifeedant substances, which may cause the products to lose their primary efficiency. However, it must be emphasized that in this respect, information on the period during which the insect does not consume food treated with antifeedant substances is very sporadic, and therefore the experiments must be completed with such information.

6.5 Current Practical Use of Antifeedant Substances

A long way still remains ahead in order to arrive at the full application potential offered by plant allomones in plant protection. No commercial products based purely on antifeedant efficiency are yet available. Nevertheless, some products that primarily contain substances with an insecticide effect, and at the same time also show an antifeedant and anti-oviposition effect, are used at present.

Perhaps the most widely known application of antifeedant effects relates to extracts made of the Indian plant *Azadirachta indica* A. Juss (syn. *Melia azadirachta*). *A. indica* has been well known in India and neighbouring countries for more than 2000 years as one of the most versatile medicinal plants having a wide spectrum of biological activity. *A. indica* A. Juss and *M. azedarach* are two closely related species of Meliaceae.

Extracts made of seeds of this plant contain numerous biologically active substances with insecticidal, fungicidal and bactericidal effects, used in many industries, ranging from medicine to agriculture. Many works have been published concerning the plant itself, its useful substances, and biological efficiency including antifeedant effects (Jacobson, 1989; Schmutterer, 1990; Ascher, 1993).

In the case of *A. indica*, substances falling in the group of limonoids (azadirachtin, salanin, nimbin etc.) are responsible for antifeedant activity. Both primary and secondary antifeedant effects have been observed in the case of azadirachtin (Ascher, 1993). Primary effects include the process of chemoreception by the organism (e.g. sensory organs on mouthparts which stimulate the organism to begin feeding), whereas secondary processes are effects such as gut motility disorders due to topical application only (Schmutterer, 1990; Ascher, 1993). Inhibition of feeding behaviour by azadirachtin results from the blockage of input receptors for phagostimulants or by the stimulation of deterrent receptor cells or both (Mordue and Blackwell, 1993). In a recent study by Yoshida and Toscano (1994), the relative consumption rate of *Heliothis virescens* larvae treated with azadirachtin was 25% of the control, equivalent to the lowest assimilation efficiency of all natural insecticides tested. In another study, larvae of *Heliothis virescens* consumed less food, gained less weight, and were less efficient at converting ingested and digested food into biomass (Barnby and Klocke, 1987). Sensitivity between species to the antifeedant effects of azadirachtin is profound. Order Lepidoptera appear most sensitive to azadirachtin's antifeedant effects, with Coleoptera, Hemiptera and Homoptera being less sensitive (Mordue and Blackwell, 1993).

A whole range of commercial products based on azadirachtin are sold at present; however, they utilize another significant effect of azadirachtin type tetranotriterpenoids, namely the growth inhibition effect (Schmutterer, 1990).

Extracts from the Indian tree of the *Pongamia* genus are another example of the commercial application of products based on antifeedant and anti-oviposition effects. This genus has one species only, that is *Pongamia pinnata* L. (syn. *P. glabra* Vent.; *Derris indica* Lamk.) which belongs to family Leguminosae; subfamily Papilionaceae (Kumar and Kalindhar, 2003).

P. pinnata is a rich source of flavonoids, the B-ring is either linked to a furan or pyran ring. Some of these flavonoids are known to have biological activity. Antifeedant activities of various extracts of *P. pinnata* were observed against many insect pests of different crops. Under laboratory conditions, 0.1% water emulsion of pongam oil (so-called karanj oil) showed antifeedant activity against *Amsacta moorei* Butler (Verma and Singh, 1985) or *Spodoptera litura* F. (Kumar and Kalindhar, 2003). The pongam oil is known to possess strong repellent activity for egg-laying against many insect pests (Khaire *et al.*, 1993; Kumar and Kalindhar, 2003). The aqueous extracts of seeds and plants of this species are known to possess ovicidal action against *Phthorimaea operculella* Zell. and *Helopeltis theivora* Waterh. (Deka *et al.*, 1998).

The repellent activities, including host deterrence and anti-oviposition, of pongam oil against the adults of the common greenhouse whitefly *Trialeurodes vaporariorum* Westwood were tested in greenhouses (Pavela and Herda, 2007a,b). Chrysanthemum plants treated with different concentrations (0.5–2.0%) of water-suspended pongam oil showed relatively long-lasting host deterrent and anti-oviposition effects on the adults of greenhouse whitefly. Although the repellent effect declined in time and concentration, strong effects on the reduction of oviposition were found, while this effect lasts at least 12 days after application, dependent on concentration.

Another example of commercially produced extracts with antifeedant effects is represented by aromatic plants. For aromatic plants, substances responsible for the smell and those that may be isolated using distillation or supercritical extraction have been studied the most. This group of substances includes the mono- and di- sesquiterpenes, phenols and some other hydrocarbons similar in structure (Pavela, 2008a).

Many monoterpenes from plant sources have been evaluated as feeding deterrents against insects (Koul, 1982). However, capillin, capillarin, methyl eugenol and ar-curcumene isolated from *Artemisia capillaris* show promise as antifeedant compounds against cabbage butterfly larvae, *Pieris rapae* crucivora. The relative strong antifeedant activity of capillin and capillarin suggest that the C=O carbonyl group instead of CH_2 methylene group, a C≡C in a side chain and a lactone ring are some of the many factors that contribute to the biological activity (Yano, 1987). Various derivatives of these base compounds such as methyl eugenol reveal that the 3,4-dimethyl group and 1-substituent of 3,4-dimethoxy-1-substituted benzenes contribute to the antifeedant activity (Yano and Kamimura, 1993).

Aromatic hydrocarbons show a significant direct insecticide activity, so consequently they have been best studied (Pavela, 2006b, 2008a,b). Nevertheless, antifeedant efficiency is connected not only with aromatic hydrocarbons in aromatic plants, but also with other polyphenolic substances synthesized by such plants as part of their defence strategy against diseases and pests.

These polyphenolic substances are currently being studied and because of their huge biochemical variety, it is highly probable that an antifeedant activity will be discovered in the near future that would lead to practical application.

Recent results of biological activity studies of substances obtained from Eurasian region plants, too, provide evidence that primary research is necessary and may lead to commercial application. Very strong, long-lasting antifeedant activity of seed extracts and root extracts of *Leuzea carthamoides* plants against the Colorado potato beetle (*Leptinotarsa decemlineata*) was found (Pavela, 2004a, 2006a) (Fig. 6.2. a,b). This antifeedant efficiency against the above mentioned pest led to the formulation of a new product based on antifeedant activity. The research on this aspect, which has been performed both in our laboratory and in departments around the world, is thus providing its first results.

6.6 Prospects for Products Based on Antifeedant Substances

Many feeding inhibitors from plant sources have so far given excellent results in laboratory conditions. In field situations only a few of them are satisfactory alternatives to traditional pest management. The chemical control is usually

(a)

(b)

Fig. 6.2. Efficiency of extract obtained from *L. carthamoides* against larvae *Leptinotarsa decemlineata*. (a) Control; (b) 15 days after application.

with broad-spectrum insecticides, and they have to be broad spectrum by necessity. They have to sell in amounts large enough to accommodate financial development, research, and marketing.

Nevertheless, the use of antifeedants in pest-management programmes has enormous appeal. They satisfy the need to protect specific crops while avoiding damage to non-target organisms so the potential value is great. In fact, insect damage to plants results from feeding or from transmission of pathogens during feeding; therefore, the chemicals that reduce pest injury by rendering plants unattractive or unpalatable can be considered as potential substitutes for conventional insecticides. The host choice of generalists and to some extent specialists may be modified when feeding inhibitors are used. The range of insect species targeted may be chosen by either the chemical structure of the inhibitor or the composition of a mixture of inhibitors, if different inhibitors are active against different species within the range. Therefore, a multi-component defence strategy of plants themselves could be used, as shown in number of recent studies with non-azadirachtin types of limonoid inhibitors (Koul, 2005) where non-azadirachtin limonoids have two different modes of action, such as feeding deterrence and physiological toxicity, which play a significant role in the potentiation effect. Moreover some other plant extracts have high potential in commercial application (Pavela, 2007b).

For the research to be successful, further assays must be performed and suitable, new substances with high antifeedant efficiency must be sought. Important research will also concern determining synergism of the effects of biologically active substances as such synergism seems to be very significant in increasing the biological activity of substances that have insecticide effects (Pavela, 2008b).

Most feeding inhibitors are less stable chemicals than traditional insecticides and act with lower residual activity and environmental impact. Natural predators and parasitoids remain unharmed by feeding deterrents targeting the herbaceous host insects. As the target sites of antifeedants are different, pesticide-resistant insect populations will still be affected by feeding inhibitors. Multi-component tactics will also slow down the resistance development to these new compounds. In fact, lack of resistance is very useful for the practical application of antifeedants as it is unlikely that oligophagous insects could develop general resistance to such deterrents, because this would result in a rapid change of their host-plant range, which is determined mainly by the occurrence of such chemicals in the non host plants. Different molecular structures of possible antifeedant compounds could be another advantage. The blend of active constituents might diffuse the selection process, mitigating the development of resistance compared to that expected with a single active ingredient. This also supports the earlier mentioned contention that combination mixtures of antifeedants could be more effective than individual compounds.

The huge variety of defensive mechanisms of plants, including the synthesis of allelochemicals, thus provides a research focus in seeking new environmentally safe products to provide plant protection against phytophagous insects. However, such research should be intensified and needs full cooperation between basic or applied research, manufacturers and the users. Such cooperation may lead subsequently to an important reduction in the dependence of agriculture on chemical industries synthesizing toxic insecticides, as well as increasing the quantity of safe foodstuffs and improving the health of this planet's inhabitants.

Acknowledgement

This study was supported by grants of the Czech Republic Ministry of Education, Youth and Sports (ME09079).

References

Ascher, K.R.S. (1993) Nonconventional insecticidal effects of pesticides available from the Neem tree, Azadirachta indica. *Archives of Insect Biochemistry and Physiology* 22, 433–449.

Barnby, M.A. and Klovme, J. A. (1987) Effects of azadirachtin on the nutrition and development of the Tobacco Budworm, *Heliothis virescens* (Fabr) (Lepidoptera: Noctuidae). *Journal of Insect Physiology* 33, 69–75.

Debrowski, Z.T. and Seredynska, U. (2007) Characterization of the two-spotted spider mite (*Tetranychus urticae* Koch, Acari: Tetranychidae) response to aqueous extracts from selected plant species. *Journal of Plant Protection Research* 47, 113–124.

Deka, M.K., Singly, K. and Handique, R. (1998) Antieedant and repellent effect of pongam (*Pongamia pinnata*) and wild sage (*Lantana camarata*) on tea mosquito bug (*Helopeltis theivora*). *Indian Journal of Agricultural Sciences* 68, 274–276.

Frazier, J.L. (1986) The perception of plant allelochemicals that inhibit feeding. In: Brattsten, L.B. and Ahmad, S. (eds) *Molecular Aspects of Insect Plant Associations.* Plenum Press, New York, pp. 1–42.

Hori, M. (1999) Antifeeding, settling inhibitory and toxic activities of Labiatae essential oils against the green peach aphid, Myzus persicae (Sulzer) (Homoptera:Aphididae). *Applied Entomology and Zoology* 34, 113–118.

Hou, HouaMin, Xing, Zhang, Hou, H.M. and Zhang, X. (2002) Effect of essential oil of plants on three lepidopterous insects: antifeeding and growth inhibition. *Acta Phytophylacica Sinica* 29, 223–228.

Isman, M.B. (1993) Growth inhibition and antifedant effects of azadirachtin on six noctuids of regional economic importance. *Pesticide Science* 38, 57–63.

Isman, M.B. (1994) Botanical insecticides and antifeedants: New sources and perspectives. *Pesticide Research Journal* 6, 11–19.

Isman, M.B. (2002) Insect antifeedants. *Pesticide Outlook* 13, 152–157.

Isman, M.B. (2006) Botanical insecticides, deterrents, and repellents in modern agriculture and an increasingly regulated world. *Annual Review of Entomology* 51, 45–66.

Jacobson, M. (1989) *Focus on phytochemical pesticides, vol. I. The neem tree 1989.* CRC Boca Raton, FL/USA.

Khaire, V.M., Kachare, B.V. and Mote, U.N. (1993) Effect of vegetable oils on mortality of pulse beetle in pingeon pea seeds. *Seed Research* 21, 78–81.

Koul, O. (1982) Insect feeding deterrents in plants. *Indian Review Life Sciences* 2, 97–125.

Koul, O. (2005) *Insect Antifeedants.* CRC Press, Bota Racon, USA.

Kumar, S.M.B. and Kalidhar, S.B. (2003) A review of the chemistry and biological activity of *Pongamia pinnata.* *Journal of Medicinal and Aromatic Plant Science* 25, 441–465.

Mancebo, F., Hilje, L., Mora, G.A. and Salazar, R. (2000a) Antifeedant activity of plant extracts on Hypsipyla grandella larvae. *Revista Forestal Centroamericana* 31, 11–15.

Mancebo, F., Hilje, L., Mora, G.A. and Salazar, R. (2000b) Antifeedant activity of *Quassia amara* (Simaroubaceae) extracts on *Hypsipyla grandella* (Lepidoptera: Pyralidae) larvae. *Crop Protection* 19, 301–305.

Metzger, F.W. and Grant, D.H. (1932) Repellency of the Japanese beetle of extracts made from plants immune to attack. *USDA Technical Bulletins* 299, 21.

Miller, J.R. and Strickler, K.L. (1984) Finding and accepting host plants. In: Bell, W.J. and Carde, R. (eds) *Chemical Ecology of Insects*. Sinauer Associates, Sunderland, USA, pp. 127–157.

Mordue, A.J. and Blackwell, A. (1993) Azadirachtin: an update. *Journal of Insect Physiology* 39, 903-924.

Mullin, C.A., Chyb, S., Eichenseer, H., Hollister, B. and Frazier, J.L. (1994) Neuroreceptor mechanisms in insect gustation: A pharmacological approach. *Journal of Insect Physiology* 40, 913–931.

Pavela, R. (2004a) The effect of ethanol extracts from plants of the family Lamiaceae on Colorado Potato Beetle adults (*Leptinotarsa decemlineata* SAY). *National Academy Science Letters – India* 27(5-6), 195–203.

Pavela, R. (2004b) Insecticidal activity of certain medicinal plants. *Fitoterapia* 75 (7-8), 745–749.

Pavela, R. (2006a) The antifeedant effect of extracts from Leuzea carthamoides (Willd.) DC. on Leptinotarsa decemlineata Say. In: Govil, J.N., Singh, V.K., Arunachalam, C. (eds.), *Recent Progress in Medicinal Plants* Vol. 14 - Biopharmaceuticals, Studium Press, Houston, USA, pp. 305–314.

Pavela, R. (2006b) Insecticidal activity of essential oils against cabbage aphid *Brevicoryne brassicae*. *Journal of Essential Oil-Bearing Plants* 9(2), 99–106.

Pavela, R. and Herda, G. (2007a) Repellent effects of pongam oil on settlement and oviposition of the common greenhouse whitefly *Trialeurodes vaporariorum* on chrysanthemum. *Insect Science* 14, 219–224.

Pavela, R. and Herda, G. (2007b) Effect of pongam oil on adults of the greenhouse whitefly *Trialeurodes vaporariorum* (Homoptera: Trialeurodidae). *Entomolia Generalis* 30(3), 193–201.

Pavela, R. (2007a) The feeding effect of Polyphenolic compounds on the Colorado Potato Beetle (*Leptinotarsa decemlineata* Say). *Pest Technology* 1(1), 81–84.

Pavela, R. (2007b) Possibilities of botanical insecticide exploitation in plant protection. *Pest Technology* 1, 47–52.

Pavela, R. (2008a) Insecticidal properties of several essential oils on the house fly (*Musca domestica* L.). *Phytotherapy Research* 22(2), 274–278.

Pavela, R. (2008b) Acute and synergistic effects of some monoterpenoid essential oil compounds on the House Fly (*Musca domestica* L.). *Journal of Essential Oil-Bearing Plants* 11(5), 451–459.

Pavela, R., Vrchotová, N. and Šerá, B. (2008) Growth inhibitory effect of extracts from *Reynoutria* sp. plants against *Spodoptera littoralis* larva. *Agrociencia* 42, 573–584.

Pradhan, S., Jotwani, M.S. and Rai, B.K. (1962) The neem seed deterrent to locusts. *Indian Farming* 12, 7–71.

Renwick, J.A.A. (2001) Variable diets and changing taste in plant-insect relationships. *Journal of Chemical Ecology* 27, 1063–1076.

Schmutterer, H. (1990) Properties and potential of natural pesticides from the Neem tree, *Azadirachta indica*. *Annual Review of Entomology* 35, 271–297.

Schoonhoven, L.M. (1982) Biological aspects of antifeedants. *Entomologia Experimentalis et Applicata* 31, 57–69.

Schoonhoven, L.M., Blaney, W.M. and Simmonds, M.S.J. (1992) Sensory coding of feeding deterrents in phytophagous insects. In: Bernays, E. A. (ed.) *Insect-Plant Interactions, Vol 4* CRC Press, Boca Raton, FL/USA, pp. 59–99.

Stadler, E. (1992) Behavioral responses of insects to plnat secondary compounds. In: Rosenthal, G.A. and Berenbaum, M.R. (eds) *Herbivores: Their Interaction with Secondary Plant Metabolites; Evolutionary and Ecological Processes*. Academic Press, San Diego, USA, pp. 44–88.

Verma, S.K. and Singh, M.P. (1985) Antifeedant effects of some plant extracts on *Amsacta moorei* Butler. *Indian Journal of Agricultural Sciences* 55, 298–299.

Wang, S.F., Liu, A.Y., Ridsdill-Smith, T.J. and Chisalberti, E.L. (2000) Role of alkaloids in resistance of yellow lupin to red legged earth mite *Halotydeus destructor*. *Journal Chemical Ecology* 26, 429–441.

Yano, K. (1987) Minor components from growing buds of Artemisia capillaris that act as insect antifeedants. *Journal of Agricultural and Food Chemistry* 35, 889–891.

Yano, K. and Kamimura, H. (1993) Antifeedant activity toward larvae of *Pieris rapae* crucivora of phenol ethers related to methyleugenol isolated from *Artemisia capillaris*. *Bioscience Biotechnology and Biochemistry* 57, 129–130.

Yoshida, H.A. and Toscano, N.C. (1994) Comparative effects of selected natural insecticides on *Heliothis virescens* (Lepidoptera: Noctuidae) larvae. *Journal of Economic Entomology* 87, 305–310.

Zhang, W., McAuslane, H.J. and Schuster, D.J. (2004) Repellency of ginger oil to *Bemisia argentifolia* (Homoptera: Aleyrodidae) on tomato. *Journal of Economic Entomology* 97, 1310–1318.

7 Potency of Plant Products in Control of Virus Diseases of Plants

H.N. Verma[1] and V.K. Baranwal[2]

[1]Jaipur National University, Jaipur, India; [2]Advanced Centre of Plant Virology, New Delhi, India

Abstract

The exploitation of the inherent resistance phenomenon and manipulation of inducible defence in plants is currently receiving much attention by researchers to control virus infection. Recent advances in the molecular biology of resistance to virus infection have presented new approaches for making susceptible crops resistant against virus infection. These approaches include pathogen-derived resistance to viruses (coat-protein-mediated resistance, movement-protein-mediated resistance, replicase- and protease-mediated resistance) and virus resistance through transgenic expression of antiviral proteins of non-viral origin. Endogenously occurring substances in a few higher plants have also been reported to induce systemic resistance in susceptible hosts against virus infections. Ribosome-inactivating proteins (RIPs) may also play an important role in the prevention of virus infection.

7.1 Introduction

Viruses prove to be a menace to humans and the environment because of their disease-causing nature. They harm crops and cause economic losses. The recent outbreak of cotton leaf curl virus disease in cotton in the northern cotton-growing region of India has led to a huge yield loss of cotton fibre. Approximately 12,000 hectares of cotton were affected by leaf curl virus disease during 1996 in Rajasthan alone. An annual loss of US$300 million is caused by mungbean yellow mosaic virus (MYMV) by reducing the yield of black gram, mungbean and soyabean (Varma *et al.*, 1992). The worldwide losses caused by viral diseases are estimated at about US$60 billion per year.

To reduce losses, scientists have explored several strategies to control virus infection. However, it is the exploitation of the inherent resistance phenomenon and manipulation of inducible defence in plants that are receiving much attention from researchers. The common approach for introducing resistance against a virus in crops has been achieved through conventional plant breeding. Limited success has been achieved through this method. Recent advances in the molecular biology of resistance to virus infection have provided new approaches to making susceptible crops resistant against virus infection. These approaches include pathogen-derived resistance to viruses (coat-protein-mediated resistance, movement-protein-mediated resistance, replicase- and protease-mediated resistance) and virus resistance through the transgenic expression of antiviral proteins of non-viral origin (Baulcombe, 1994).

Resistance of plants to virus diseases may be broadly categorized into two groups: (i) constitutive; and (ii) induced. Constitutive resistance is heritable and occurs in cultivars, which have gene(s) conferring resistance to viral infection, whereas induced resistance has to be conferred afresh upon a susceptible plant and is normally not heritable. Induced resistance operates through the activation of natural defence mechanisms of the host plant. The two forms of induced resistance are systemic acquired resistance (SAR) and systemic induced resistance (SIR). In both SAR and SIR, plant defences are preconditioned by prior infection or treatment that results in resistance (or tolerance) against subsequent challenge by a pathogen or parasite. Great strides have been made over the past 20 years in understanding the physiological and biochemical basis of SAR and SIR. Much of this knowledge is due to the identification of a number of chemical and biological elicitors, some of which are commercially available for use in conventional agriculture. However, the effectiveness of these elicitors to induce SAR and SIR as a practical means to control various plant diseases is just being realized.

The infection of plants by necrotizing pathogens, including fungi, bacteria and viruses, induces systemic resistance to subsequent attack by the pathogens. This resistance is called SAR (Kessman *et al.*, 1994; Ryals *et al.*, 1994). It can also be activated in numerous plants by pre-inoculation with biotic inducers including pathogens (Sticher *et al.*, 1997). Endogenously occurring substances in a few higher plants have been reported to induce systemic resistance in susceptible hosts against virus infections. Such plant extracts have been used for protecting economically important crops against virus infections (Verma and Baranwal, 1989). Endogenously occurring virus inhibitors may also be ribosome-inactivating proteins (RIPs) (Barbieri and Stirpe, 1982; Mansouri *et al.*, 2006; Zhang *et al.*, 2007). Virus infection is prevented if a mixture of RIP and virus is exogenously applied on the leaf surface of a susceptible host. RIPs presumably inhibit virus infection by entering the cytoplasm along with the virus particle and inhibiting protein synthesis on host ribosomes, thus preventing early virus replication (Reddy *et al.*, 1986).

On the basis of their antiviral activity, virus inhibitors from plants can be grouped into two categories:

1. Plant products that inhibit virus infection by inducing an antiviral state either at the site of application (local resistance) and/or at a remote site (systemic resistance) when applied a few minutes or hours prior to virus challenge (Verma and Mukerji, 1975; Verma and Awasthi, 1979, 1980; Verma *et al.*, 1984; Verma *et al.*, 1995a; Prasad *et al.*, 1995; Verma *et al.*, 1996). Such plant products have been called systemic resistance inducers (SRIs).

2. Basic proteins from plants that function by inactivating ribosomes of the host and have been called RIPs (Barbieri *et al.*, 1993, 2003; Van Damme *et al.*, 2001).

7.2 Systemic Resistance Inducers (SRIs)

Induction of systemic resistance by plant extracts has been reviewed from time to time (Verma, 1985; Verma and Prasad, 1992; Verma *et al.*, 1995a; Verma *et al.*, 1998). Physico-chemical characteristics of systemic resistance inducers from plants such as *Boerhaavia diffusa* (Verma and Awasthi, 1979, 1980; Srivastava, 1995); Mirabilis jalapa (Verma and Kumar, 1980); *Cuscuta reflexa* (Awasthi, 1981); *Clerodendrum aculeatum* (Verma *et al.*, 1984; Verma *et al.*, 1996; Kumar *et al.*, 1997; Srivastava *et al.*, 2008); *Bougainvillea spectabilis* (Verma *et al.*, 1985; Verma and Dwivedi, 1984; Srivastava, 1995); *Pseuderanthemum atropurpureum* (Verma *et al.*, 1985) have been studied in some detail (Table 7.1).

Induced resistance operates through the activation of natural defence mechanisms of the host plant. Extracts from brinjal (Verma and Mukherjee, 1975) and a few other higher plants such as *Boerhaavia diffusa* (Verma *et al.*, 1979), *Bougainvillea* (Verma and Dwivedi, 1984), *Clerodendrum* (Verma *et al.*, 1996; Kumar *et al.*, 1997) and *Datura* (Verma *et al.*, 1982) induce systemic resistance to viral multiplication in plants. The active products present in these

Table 7.1. Characteristics of systemic resistance inducers obtained from some higher plants.

Characteristics	SRI-yielding plants					
	BD	CA/CI	BS	PA	CR	MJ
Source	Root	Leaf	Leaf	Leaf	Leaf	Leaf
Thermal inactivation points (°C)	80	90	80	0	70	90
Nature	Glyco-protein	Basic Protein	Protein	0	Protein	Protein
Molecular weight (kDa)	30	34/29 and 34	28	0	14–18	24
Active against	TMV	TMV	TMV	TMV	TMV	TMV
	SHRV	SHRV	SHRV	SHRV	SHRV	SHRV
	GMV	GMV	CGMMV	CGMMV	GMV	PVY
					PLRV	TmYMV

BD = *Boerhaavia diffusa*; CA = *Clerodendrum aculeatum*; CI = *C. inerme*; BS = *Bougainvillea spectabilis*; PA = *Pseuderanthemum atropurpureum*; CR = *Cuscuta reflexa*; MJ = *Mirabilis jalapa*. TMV = Tobacco mosaic virus; SHRV = Sunnhemp rosette virus; GMV = Gomphrena mosaic virus; TRSV = Tobacco Ringspot virus; TmYMV = Tomato Yellow Mosaic virus; CGMMV = Cucumber green mottle mosaic virus; PVY = Potato virus Y; PLRV = Papaya leaf reduction virus.

extracts have no direct effect on viruses; their antiviral activity is mediated by host cells in which they induce the antiviral state. All groups of plants respond to treatment with extracts of these plants. The active products in plant extracts inducing resistance are mostly small molecular weight proteins which sometimes may be glycosylated. These proteins are highly thermostable and can withstand prolonged treatment at alkaline/acidic pH. Plant products from different species vary in molecular weight and may differ in other characteristics too. Incubation of viruses with these antiviral agents has no effect on viral infectivity, but they exert their antiviral effect by rendering host cells incapable of supporting viral replication. Resistance development is inhibited by actinomycin D and cyclohexamide. This implies that resistance is dependent on DNA-coded information of the cell and the antiviral activity of the antiviral agent is indirect. The antiviral agents showed no host specificity and were active against a wide range of viruses (Faccioli and Capponi, 1983).

Systemic resistance inducers obtained from plants have been shown to be effective against a wide range of viruses (Verma and Awasthi, 1979). Leaf extract of *Clerodendrum aculeatum* is active against TMV, SHRV, GMV, TmYMV, (Verma *et al.*, 1984), and an extract of *Mirabilis jalapa* is effective against TmYMV, PLRV, CMV, CGMMV (Verma and Kumar, 1980). The extract-treated plants become resistant to attack by diverse pathogens such as fungi, bacteria and viruses (McIntyre *et al.*, 1981).

Mechanism of systemic induced resistance by botanicals

Systemic induced resistance against virus infection by botanicals is not yet fully understood. The botanical resistance inducers themselves do not act on the virus directly. Verma and Awasthi (1980) demonstrated the *de novo* synthesis of a virus inhibitory agent (VIA) in untreated leaves of *Nicotinia glutinosa*, whose basal leaves were treated with root extract of *B. diffusa*. These induced substances inhibited almost completely tobacco mosaic virus in *N. glutinosa*, *Datura stramonium* and *D. metel* but inhibition of tobacco ring spot virus or Gomphrena mosaic virus in *Chenopodium amaranticolor* was less pronounced. In another study Verma and Dwivedi (1984) found that VIA induced in *C. tetragonoloba* inhibited completely the infection of tobamoviruses in all the seven hypersenstive hosts tested. Yet in another study by Khan and Verma (1990), it was observed that VIA produced in *C. tetragonoloba* following treatment with extract of *Pseuderanthemum bicolor* inhibited completely SHRV, TMV, cucumber green mottle mosaic virus and PVX in their respective hypersensitive hosts, namely *C. tetragonoloba*, *D. stramonium*, *C. amaranticolor* and *G. globosa*. However, the VIA produced by *B. spectabilis* in *N. tabacum* cv. Samsun NN, *N. glutinosa*, and *D. stramonium* was less effective against TMV (Verma and Dwivedi, 1984). Thus, it appears that induction of systemic resistance by botanicals is non-specific and is effective against a broad spectrum of viruses.

The production of VIA is maximum after 24 h of application of *B. diffusa* root extract in *N. glutinosa*. However, *B. spectabilis* induced maximum VIA activity after 48 h of its application in *N. glutinosa*. Thus, VIA production in a host is time specific but a general phenomenon. Properties of VIA produced in *C. tetragonoloba* after application of *B. spectabilis* leaf extract has been studied by Verma and Dwivedi (1984). The VIA could be precipitated by ammonium sulfate and hydrolysed by trypsin but not by ribonulease indicating that it was a protein rather than a nucleic acid. The VIA induced by *Pseuderanthemum bicolor* in *C. tetragonoloba* was also proteinaceous with a molecular weight of 15 kDa (Khan and Verma *et al.*, 1990). The production of VIA was sensitive to actinomycin D (Verma and Dwivedi, 1984). The mobile signal produced VIA in the entire plant system. It appears that the putative messenger becomes active soon after induction and starts producing VIA, reaches a maximal concentration and then starts to decline.

Commonly associated with systemic acquired resistance induced by pathogens is the systemic synthesis of several families of serologically distinct low molecular weight pathogenesis-related (PR) proteins. The localization and timing of some PR proteins suggested their possible involvement in acquired resistance against viruses. However, definite proof that the induction of PR proteins causes the acquired resistance has not been given. Plobner and Leiser (1990) did not find the production of PR proteins during systemic resistance induced by carnation extract in Xanthi-nc tobacco plants. Absence of PR proteins during induction of SAR by botanicals suggests that another biochemical chain of reaction, other than one operating in pathogen / chemical induced SAR, might be operating during botanical induced systemic resistance against virus infection in plants.

Durrant *et al.* (2004) reported that SAR is a mechanism of induced defence that confers long-lasting protection against a broad spectrum of microorganisms. SAR requires the signal molecule salicylic acid (SA) and is associated with accumulation of PR proteins, which are thought to contribute to resistance.

Hansen (1989), while reviewing antiviral chemicals for plant disease control, gave the following characteristics of an ideal antiviral compound that will serve all purposes for virus disease management in crops:

- Soluble in water or non-phytotoxic solvents.
- Effective against at least some agriculturally important viruses at non-phytotoxic concentrations.
- Easily taken up by plants and distributed throughout the system.
- Non-toxic by itself and in its catabolic forms to humans, plants and wildlife.

Botanical resistance inducers can be classified as ideal virus-suppressing agents, as they have all the characteristics of an ideal antiviral compound. The resistance-inducing proteins from *Boerhaavia diffusa* and *Clerodendrum aculeatum* can be applied directly by spraying on systemic hosts, for management of some commonly occurring virus diseases under greenhouse conditions or field conditions. The agricultural role of endogenous antiviral substances of plant origin has been reviewed by Verma *et al.* (1995).

The induced antiviral state in *N. glutinosa* by SRI from *C. aculeatum* decreased considerably after 3 days (Verma and Varsha, 1994). However, the resistance inducing ability of *C. aculeatum* SRI could be enhanced up to 6 days by priming it with certain proteinaceous additives (Verma and Versha, 1994). The activity is probably enhanced by modification or enhanced stability of proteinaceous inducers by these additives. Thus, one unexploited approach to engineering virus resistance is the manipulation of inducible defences in plants. The production of systemic resistance by the use of botanicals will be effective against a broad spectrum of viruses and will not break down when plants are exposed to high temperatures.

Suppression of disease symptoms by true inhibitors may be accomplished either by acting on the first stage of the infection process, which is the adsorption of the virus into the host cell, or by blocking or competing with the virus receptor sites on the leaf surface (Ragetli, 1957; Ragetli and Weintraub, 1962) or by affecting the susceptibility of the host by altering host cell metabolism (Verma and Awasthi, 1979). Bozarth and Ross (1964) suggested the phenomenon of SAR as a result of the initial infection by which a signal was generated at the site of application and transported throughout the plant to respond more effectively to the subsequent infection.

Induction of systemic resistance by resistance inducers obtained from plant extracts was first detailed by Verma and Mukerjee (1975). Extracts from a few plants induce an antiviral state by acting through an actinomycin D (AMD) sensitive mechanism (Verma and Awasthi, 1979; Verma *et al.*, 1984). AMD is an inhibitor of protein synthesis at the transcription level. Concomitant application of AMD with SRI reversed the induction of resistance in susceptible plants. However, induction of resistance remains unaffected when AMD is applied 12 h post-treatment. This gives an indication that plant-extract-induced resistance is a host-mediated response (Verma *et al.*, 1979).

The activity at a distance from the point of application might be explained by the supposition that the SRI present in the plant extract selectively attaches at the surface, and a type of chain reaction starts that elicits the transcription of defence-related genes, leading to the production of a new VIA (Verma and Awasthi, 1979).

Ribosome inactivating proteins (RIPs)

The basic proteins that function by inactivating the ribosome of the host have been called RIPs. The RIPs have been shown to be N-glycosidases, which remove a specific adenine base in a conserved loop of the 28s rRNA of eukaryotic organisms (Endo and Tsurugi, 1987; Endo *et al.*, 1987) or the 23s rRNA of prokaryotes (Hartely *et al.*, 1991). Such damaged ribosomes can no longer bind the elongation factor-2 (Gessner and Irvin, 1980; Rodes and Irvin, 1981). The RIPs damage the ribosome, arrest protein synthesis and cause cell death. RIPs show antiviral activity against both animal and plant viruses (Barbieri and Stirpe, 1982) and have been classified into two types (Stirpe *et al.*, 1992; Barbieri *et al.*, 1993). Type-I RIPs consist of a single polypeptide chain that is enzymatically active. These are scarcely toxic to animals and inhibit protein synthesis in cell-free systems, but have little or no effect on whole cells. The three well known antiviral proteins PAPs, dianthins and MAP belong to this category. Type-II RIPs contain two types of polypeptide chains. Chain A is linked to chain B through a disulfide bond. Chain B binds the toxin to the cell surface and chain A enzymatically inactivates the ribosomes (Olsnes and Pihl, 1982). These are toxic to cells and inhibit protein synthesis in intact cells and in cell-free systems. Several similarities exist among the type-I RIPs. They are all basic proteins with a molecular weight in the range 26–32 kDa. They have an alkaline isoelectric point and are usually stable, being resistant to denaturing agents and protease. A majority of the RIPs are glycoproteins. Type-I RIPs are strongly immunogenic. Strocchi *et al.* (1992) established that cross reactivity between RIPs obtained from unrelated plants was either very weak or absent.

Recently cloning and expression of antiviral/ribosome-inactivating protein from *Bougainvillea xbuttiana* was reported by Choudhary *et al.* (2008). They reported that full-length cDNA encoding ribosome inactivating / antiviral protein (RIP/AVP) consisted of 1364 nucleotides with an open reading frame (ORF) of 960 nucleotides encoding a 35.49 kDa protein of 319 amino acids.

The three well known antiviral proteins, namely PAPs (from *Phytolacca americana*), dianthins (from *Dianthus caryophyllus*) and MAPs (from *Mirabilis jalapa*) belong to the category of type-I RIPs.

Pokeweed antiviral proteins (PAPs)

The antiviral protein present in the leaves of *Phytolacca americana* was purified to homogeneity and its molecular weight determined as 29 kDa (Irvin, 1975). This protein, called pokeweed antiviral protein (PAP), had a PI of 8.1 (Irvin, 1983). The antiviral effect of PAP was most pronounced when it was co-inoculated with the virus. PAP also inhibited virus infection when applied prior to virus challenge. Local lesion formation by TMV on *N. tabacum* cv. Xanthi-nc was inhibited by nearly 70% even after 48 h of treatment. The virus inhibitory effect of PAP increased with the decrease in the time lapse between treatment and challenge inoculation. PAP was less effective in preventing virus infection when applied a short time after virus inoculation. No inhibition was observed when PAP was applied 50 min after virus infection (Chen *et al.*, 1991). PAP reduced infectivity of several mechanically transmitted RNA and DNA viruses when the purified virus or sap from virus-infected plants was mixed with an equal volume of PAP solution and the mixture rubbed on the leaves of the local-lesion hosts (*N. glutinosa*/TMV; *Chenopodium quinoa*/ CMV; *C. amaranticolor*/TMV, CMV, alfalfa mosaic virus, PVY; *Gomphrena globosa*/PVX) or systemic hosts (*Brassica campestris*/cauliflower mosaic virus; *N. benthamiana*/African cassava mosaic virus). PAP, thus, appears to be a general inhibitor of virus infection (Tomlinson *et al.*, 1974; Stevens *et al.*, 1981;

Chen *et al.*, 1991; Picard *et al.*, 2005). PAP also shows antiviral activity against several animal viruses. It is toxic to cells infected with poliovirus (Ussery *et al.*, 1977) and influenza virus (Tomlinson *et al.*, 1974). It inhibits multiplication of herpes simplex virus type 1 (Arnon and Irvin, 1980) and human immunodeficiency virus (HIV; Zarling *et al.*, 1990). *P. americana* is now known to contain three proteins (PAP I, II and III) with similar biological properties. A new insight into the antiviral mechanism of PAP is that PAP depurination of Brome mosaic virus RNA impedes both RNA replication and subgenomic RNA transcription (Picard *et al.*, 2005).

Rajmohan *et al.* (1999) reported that PAP isoforms PAP-I, PAP-II and PAP-III depurinate RNA of HIV-I. A non-toxic PAP mutant inhibiting pathogen infection via a novel SA-independent pathway was reported by Zoubenko *et al.* (2000). PAP inhibits translation by depurinating the conserved sarcin/ricin loop of the large ribosomal RNA. Depurinating ribosomes are unable to bind elongation factor 2, and, thus, the translocation step of the elongation cycle is inhibited. Ribosomal conformation is required for depurination that leads to subsequent translation inhibition (Mansouri *et al.*, 2006).

Carnation antiviral proteins (Dianthins)

Sap from carnation leaves shows virus inhibitory activity (Van Kammen *et al.*, 1961; Ragetli *et al.*, 1962). Dianthin 30 and 32 were isolated from the leaves of *Dianthus caryophyllus* (Ragetli *et al.*, 1962). Local lesion production by TMV on *N. glutinosa* was inhibited by 100% when the inhibitor was co-inoculated with the virus (Stevens *et al.*, 1981). The molecular weights as determined by SDS–PAGE are 29.5 and 31.7 kDa, respectively (Stirpe *et al.*, 1981). Immunoelectrophoresis revealed that dianthin 32 is distributed in the growing shoots and in the young and old leaves of *D. caryophyllus* and dianthin 30 is distributed throughout the plant (Reisbig and Bruland, 1983). The two are glycoproteins containing mannose and show a weak cross reaction. The nucleotide sequence of cDNA encoding dianthin 30 has been determined (Legname *et al.*, 1991). The carnation proteins are also inducers of systemic resistance (Plobner and Leiser, 1990). Cho *et al.* (2000) performed isolation and characterization of cDNA encoding ribosome inactivating protein from *Dianthus sinensis* L.

Mirabilis antiviral protein (MAP)

The roots, leaves and stem of *Mirabilis jalapa* show high inhibitory activity against plant viruses. The *Mirabilis jalapa* leaf extract, when used as a foliar spray 24 h prior to virus inoculation, suppressed disease symptoms on a few systemic hosts (tomato/tomato yellow mottle virus; *Cucumis melo* var. momordica/CMV; *Cucumis sativa*/cucumber green mottle mosaic virus; tomato/tomato yellow mosaic virus; urd/yellow mosaic of urd) (Verma and

Kumar, 1980). A 50–60% reduction of the virus content in the treated plants was observed in the infectivity assays. *M. jalapa* extract was able to check the population of aphids and whiteflies and, thereby, control the natural spread of a few viruses on the systemic hosts (Verma and Kumar, 1980).

MAP isolated from root inhibits mechanical transmission of TMV, PVY, cucumber green mottle mosaic virus, and turnip mosaic virus on local-lesion and systemic hosts and can induce systemic resistance of a low order when applied to basal leaves (Kubo *et al.*, 1990). The purified protein consists of a single polypeptide without a sugar moiety and has a molecular weight of 24.2 kDa. It is a basic protein rich in lysine content, with a PI of 9.8 (Kubo *et al.*, 1990). The complete amino acid sequence of MAP has been determined. It consists of 250 amino acids and its molecular weight as determined from the sequence is 27,833 kDa. The native MAP was resistant to protease digestion (Habuka *et al.*, 1989).

MAP produced by *M. jalapa* cells in suspension culture showed comparable biological activity with that of the roots and leaves and also reacted positively with anti-MAP serum (Ikeda *et al.*, 1987). Several nutritional and hormonal factors also affect the formation of MAP by *M. jalapa* cells in suspension culture (Ikeda *et al.*, 1987). Bolognesi *et al.* (2002) reported ribosome-inactivating and adenine polynucleotide glycosylase activities in *M. jalapa* L. tissues.

Ricinus (RICIN A) antiviral protein

The ricinus antiviral protein is a type II RIP isolated from *Ricinus communis* that has a molecular weight of 65 kDa. Its is a heterodimer consisting of chain A similar to the type-I RIP and chain B linked together by a disulfide bond. Chain B has great similarity to mammalian lectins A. Once inside the cell chain A acts as a type-I RIP and inactivates protein translation machinery. The cDNA for ricin has been cloned (Lamb *et al.*, 1985). Both A and B chains are encoded by a single gene which has no intron. Detection of ricin and other ribosome-inactivating proteins by an immuno-polymerase chain reaction assay was reported by Lubelli *et al.* (2006).

Role of RIPs

The above examples of RIPs show their important role against both plant viruses and also RNA- and DNA-containing animal viruses (Barbieri *et al.* 1993). The mechanism of antiviral activity suggested for both plant and animal system involves increased permeability of and easier entry of RIPs into virus-infected cells, blocking of protein synthesis and reduced virus multiplication (Barbieri *et al.*, 1993).

Virus resistance previously observed in transgenic plants expressing coat protein genes, and so on, has been specific for the virus from which the genes are derived or for closely related viruses (Beachy *et al.*, 1990) but transgenic tobacco and tomato plants expressing the pokeweed antiviral gene are found to be resistant to a broad spectrum of plant viruses (Lodge *et al.*, 1993).

A problem often encountered in using type-I RIPs is the fact that they cannot inhibit protein synthesis in intact cells. However, when coupled to type-II RIP, they can be used effectively. The toxicity of plant materials containing type-II RIPs have long been known and has great medicinal potential. Recently, apoptosis was described in both lymphoid tissue and in the intestine of abrin- and ricin-poisoned rats. Apoptosis was also observed in tissue culture of cancer cells treated with ricin.

The antiviral activity of the RIP could be due to inactivation of ribosome of the infected plant cell. As compared to the usefulness in the plants, the role of RIPs in human and animal systems has been much more widely documented. In spite of all these efforts, the biological significance of RIPs in nature is not yet known and moreover, their antiviral action does not seem to depend upon the inhibition of host ribosomes (Chen *et al.*, 1993).

Great potential exists today in elucidating the possible significance of RIPs and their exact antiviral role which may not always depend upon inhibition of host ribosomes. Their co-action with another antiviral mechanism also needs to be explored, especially their place in the cascade of events following viral infection up to establishment of resistance, both systemic or localized.

7.3 Pathogen-induced Systemic Acquired Resistance (SAR)

The infection of plants by necrotizing pathogens, including fungi, bacteria and viruses, induces systemic resistance to subsequent attack by the pathogens. This resistance is called systemic acquired resistance (SAR) (Kessman *et al.*, 1994; Ryals *et al.*, 1994; Prasad *et al.*, 2001). Also, it can be activated in numerous plants by pre-inoculation with biotic inducers including pathogens (Sticher *et al.*, 1997). One of the prominent features of SAR is that resistance is expressed against pathogen which can be widely different from the initial infecting pathogens.

In a range of plant species, the development of necrotic lesions in response to pathogen infection leads to induction of generalized disease resistance in uninfected tissue. Thus, TMV inoculated hypersensitive tobacco cultivar develops systemic resistance against TMV (virus), *Phytophthora parasitica* var. *nicotinae* (fungi) and *Pseudomonas tabaci* (bacteria). TMV also induced resistance against *Peronospora tabacina* and reduced reproduction of the aphid *Myzus persicae* (McIntyre *et al.*, 1981). Thus a single viral agent induced resistance in tobacco against diverse challenges.

The first report of virus induced SAR came in 1952. Primary inoculation of the lower leaves of *D. barbatus* with carnation mosaic virus (CarMV) resulted in the development of fewer lesions on the upper leaves upon challenge inoculation with CarMV (Gilpatrick and Weintraub, 1952). Virus-induced resistance was further substantiated by Ross (1961a, 1961b) and Loebenstein (1963). Cucumber plants infected with tobacco necrosis virus

(TNV) protected the plant systemically against disease caused by the fungus *Colletotrichum lagenarium* (Jenns and Kuc, 1977). SAR expessed in *Vigna* plants following inoculation with TNV against challenge by TNV was not expressed against challenge by a CMV which infects the host systemically (Pennazzio and Roggero, 1991). Associated with SAR was the stimulation of ethylene-forming enzyme activity. Infection of ecotype Dijon of *Arabidopsis thaliana* with turnip crinkle virus (TCV) leads to the resistance against further infection by TCV or *Pseudomonas syringae* (Uknes *et al.*, 1993). SAR in cucumber plants against powdery mildew disease, caused by *Sphaerotheca fuliginea* (Schlechtend Fr.) Pollacci, was induced by localized infection in cucumber cotyledons with TNV (Farrag *et al.*, 2007).

The SAR phenomenon is observed both in dicotyledonous and mono-cotyledonous plants; it provides the third and final line of defence against pathogens. The first line of defence consists of genetically inherited resistance mechanisms that make plants constitutively resistant to the majority of pathogens present in the environment. The second line of defence is activated in the immediate vicinity of the infected or wounded site in an attempt to prevent the spread of pathogens throughout the plant. The local resistance response develops more rapidly than SAR and involves cell-wall and cuticle strengthening, synthesis of toxins, antifeedants and the production of defence-related proteins including the PR proteins. In addition to long-distance signal molecules, local resistance may be partially mediated through relatively immobile endogenous elicitors, which include oligogalacturonide fragments of the plant cell wall (Lamb and Dixon, 1990). Several lines of evidence suggest that endogenous SA is a signal molecule in SAR. Involvement of SA in SAR came from the discovery that endogenous SA increases by at least 20-fold in the virus-inoculated leaves of tobacco (Malamy *et al.*, 1990). The increase coincides with the appearance of hypersensitive response (HR) lesions on the inoculated leaves. Accumulation of SA increased with the intensity of HR and was proportional to the dose of virus inoculum (Yalpani *et al.*, 1991). Tissue accumulation of SA in TMV inoculated xanthi-nc tobacco paralleled or preceded detectable increase in the levels of PR-1 mRNA in both inoculated and uninoculated leaves (Malamy *et al.*, 1990).

7.4 SAR and the Role of PR Proteins

SAR strongly correlates with the coordinate expression of at least nine families of genes (*sar* genes), of which several encode the PR proteins (Ward *et al.*, 1991). Generally, tissues in which a significant amount of PR protein has been induced are more resistant to infection by pathogen than those that lack PR proteins. PR proteins were first discovered in 1970 in tobacco plants reacting hypersensitively to TMV infection (Gianinazzi *et al.*, 1970; Van Loon and Kammen, 1970). PR proteins are produced in plants in response to infection by viruses, bacteria (Metraux and Boller, 1986), fungi (Gianinazi *et al.*, 1980) and viroids (Conejero *et al.*, 1979) and are also synthesized in response to chemical treatments and specific physiological stresses (Stintzi *et al.*, 1993).

Several classes of the PR proteins either possess direct antimicrobial activity or are closely related to classes of antimicrobial proteins. These include β-1,3-glucanase, chitinase, cysteine-rich proteins related to thaumatin and PR-1 proteins. An *in vivo* role in disease resistance has not been demonstrated for any of the PR proteins. Alexander *et al.* (1993) demonstrated that constitutive high level expression of PR 1-a in transgenic tobacco results in tolerance to infection by *Peronospora tabacina* and *Phytophthora parasitica* var. *nicotinae*. On the other hand, transgenic tobacco plants expressing PR 1-b gene exhibited no reduction in the severity of TMV symptoms (Cutt *et al.*, 1989). Quite often SAR was not correlated with the induction of PR proteins (Kopp *et al.*, 1989; Ye *et al.*, 1989; Cohen *et al.*, 1993; Kessman *et al.*, 1994). Fraser found a poor correlation between the levels of PR 1-a protein and also gave evidence for the occurrence of PR proteins in leaves of healthy tobacco plants during flowering (Fraser, 1981; 1982). In conclusion, most of the evidence shows that the PR proteins are closely associated with, and not necessarily responsible for, induced resistance.

7.5 Induced Resistance and the Role of Induced Antiviral Proteins

Virus-induced new antiviral protein components

Antiviral substances are formed in plants responding hypersensitively to virus infection (Verma and Prasad, 1992) and have been recognized as phosphorylated glycoproteins (Faccioli and Capponi, 1983), glycoproteins (Wieringa and Dekker, 1987), RNA (Kimmins, 1969), traumatic acid (Kato and Misawa, 1976), and protein-like substances (Nienhaus and Babovic, 1978). Chadha and MacNeill (1969) found the formation of an antiviral principle in tomato plants systemically infected with TMV. These antiviral compounds are not generally specific to the plants. Thus, antiviral substances produced in capsicum plants could reduce PVX infection on *Gomphrena globosa* as well as *Solanum tuberosum* (Nagaich and Singh, 1970).

The presence of an antiviral factor (AVF) was established in virus-infected plants which could decrease the number of local lesions produced by TMV and PVY (Sela and Applebaum, 1962). Partially purified AVF from TMV infected *N. glutinosa* plants was found to contain both protein and RNA (Sela *et al.*, 1964). It was sensitive to ribonuclease and was resistant to proteolytic enzymes (Sela *et al.*, 1966).

Mozes *et al.* (1978) established that the purified AVF is a phosphorylated glycoprotein of molecular weight 22 kDa on SDS gels. It is sensitive to pronase under conditions suitable for proteolysis of glycoproteins. It remains active after treatments with SDS and is stable at pH 2.0. It resembles interferon in many of its properties (Mozes *et al.*, 1978).

Previously it was believed that TMV infection was necessary for AVF production. But Edelbaum *et al.* (1983) found that TMV infection could be substituted by treatment with a mixture of Poly (I), Poly (C), cAMP and cGMP for the induction of active AVF in leaves and callus cultures of *N. glutinosa*.

Inhibitor of virus replication

Loebenstein and Gera (1981) reported, for the first time, an inhibitor of virus replication (IVR). IVR is released into the medium from TMV-infected cells of Samsun NN plants and inhibits replication in protoplasts from local-lesion-responding Samsun plants. IVR is detected as soon as 24 h after inoculation of protoplasts and it is effective when applied up to 18 h after inoculation. IVR is neither host nor virus specific. It inhibits TMV, CMV and also PVX (Gera and Loebenstein, 1983). Gera and Loebenstein (1983) reported that IVR also inhibited TMV replication in intact leaves when applied to cut stems or when used as a spray.

Plant-extract-induced virus inhibitory agent

It would be interesting to study the entire cascade of events and properties of the virus inhibitory agent (VIA) induced by antiviral agents. Treatment of lower / upper leaves of hypersensitive or systemic hosts of virus with antiviral agents results in the development of resistance throughout the plant a few hours later. This is detectable by challenge inoculation with viruses producing local lesions or systemic symptoms. The lesions are either reduced or totally absent and systemic symptoms are either milder or totally suppressed. Plant extracts or the semi-purified proteins from these plants stimulate the hosts to produce VIAs that spread to surrounding tissues and other plant parts (Verma *et al.*, 1996). The VIAs have been isolated from leaves of plants treated with phytoproteins and they have been shown to inactivate the viruses *in vitro* (Verma and Awasthi, 1980; Verma and Dwivedi, 1984; Verma *et al.*, 1996). The production of VIA leading to resistance seems to be an activation of a pre-existing system and hence is easily stimulated. VIA is able to move from one leaf to another through the vascular system of the plant.

Therefore, the antiviral agents from plants can be broadly grouped into two categories based on their mode of action: (i) those affecting virus *in vitro*; and (ii) those affecting via host plants.

Amongst the latter, we can distinguish (1) those that act by affecting host susceptibility and which need to be applied in leaf tissue before or at the time of virus inoculation, e.g. protein inhibitors from *Phytolacca* spp., *Dianthus* spp., *Chenopodium* spp. and so on, and (2) those acting by inducing the host resistance mechanism, which can truly be called antiviral agents, and act by obstructing the establishment of virus. They exert their effect when applied a few hours before virus inoculation. Their effect is visible even on non-treated parts of susceptible plants (Verma and Mukherjee, 1975; Verma and Awasthi, 1979; Verma *et al.*, 1979, 1980, 1982, 1985, 1995, 1996; Kumar *et al.*, 1997).

Verma and Awasthi (1980) reported that the synthesis of VIA is inhibited if AMD is applied soon after extract treatment. The VIA synthesized is neither virus specific nor host specific. Extracts containing VIA when incubated with the virus reduced their infectivity. VIAs from a few hosts have been characterized. The VIA synthesized in the leaves of *N. glutinosa*, treated with *B. diffusa* root extract reduced infectivity of TMV on *N. glutinosa*, *Datura stramonium* and *D. metel* (Verma and Awasthi, 1980). It was, however, less effective in inhibiting TRSV and GMV on *C. amaranticolor*. The VIA production was maximum after 24 h treatment with the extract. The VIA has a proteinaceous nature (Verma and Awasthi, 1980). VIA synthesized in *C. tetragonoloba* plants following treatment with *B. spectabilis* leaf extract prevented infection of tobamoviruses in seven hypersensitive hosts (Verma and Dwivedi, 1984) and its production was maximum after 48 h treatment. The VIA showed characteristics of a protein (Verma and Dwivedi, 1984). VIA produced in the same host upon treatment with *Pseuderanthemum bicolor* extract completely prevented infection of SHRV, TMV, CGMMV, and PVX on *C. tetragonoloba*, *D. stramonium*, *C. amaranticolor* and *G. globosa*, respectively (Khan and Verma, 1990).

VIA and the antiviral state are induced concomitantly with the systemic resistance induction in host plants, following treatment with certain plant extracts. VIA is an inducible gene product like AVF, IVR and PR proteins. The phytoproteins occurring in the root / leaf extracts of *Boerhaavia diffusa*, *Clerodendrum aculeatum*, *Clerodendrum inerme* and *B. spectabilis* etc. possess strong systemic-resistance-inducing properties (Verma and Prasad, 1992).

Clerodendrum aculeatum and *Boerhaavia diffusa*, the two potent antiviral plants

Clerodendrum and *Boerhaavia* are economically important plants because many of their species possess medicinal properties. Almost every part of these plants is credited with some medicinal properties and is employed in traditional Indian system of natural therapy.

Virus inhibitors from *Boerhaavia diffusa* and *Clerodendrum aculeatum* inhibit several plant viruses. These inhibitors modify the susceptibility of host plants towards virus infection. The roots of *B. diffusa* are a rich source of a basic protein, which has been used for inducing systemic resistance in many susceptible crops against commonly occurring viruses (Verma and Awasthi, 1979, 1980; Verma *et al.*, 1979; Awasthi *et al.*, 1984, 1985, 1989; Verma *et al.*, 1995, 1999). This protein or antiviral agent was active against spherical and tubular viruses in hypersensitive hosts such as *Datura metel*, *Nicotiana tabacum* var Ky-58, *N. glutinosa*/TMV, *Cyamopsis tetragonoloba*/SHRV, *Vigna sinensis*/SHRV, and systemic hosts such as *Nicotiana tabacum* c.v NP-31 / TMV, *Crotolaria juncea*/SHRV, *N.glutinosa*/TRSV, when applied a few hours (17–24 h) before virus inoculation or when tested after mixing with virus inoculum (Verma and Awasthi, 1979; Awasthi *et al.*, 1984). The inhibitor is a basic glycoprotein (70–80% protein, 8–13% carbohydrate) with a molecular weight of

16–20 kDa as determined by gel-filtration chromatography (Verma *et al.*, 1979). The protein has a PI of around 9 and has a molecular weight of 30 kDa (Srivastava, 1995). The RIP was found to be extremely thermostable (Verma and Awasthi, 1979). Following treatment with the systemic RIP, the host produces a VIA. The VIA shows characteristics of a protein and reduces infectivity of the viruses both *in vitro* and *in vivo* (Verma and Awasthi, 1980). Upon gel filteration on Sephadex G-25, two active fractions exhibiting protein characteristics were recovered (Verma and Awasthi, 1980).

The protein occurring in *B. diffusa* functions as a signal molecule and is of great interest as it has a role in stimulating the defence system of plants against viruses (Singh, 2006; Verma *et al.*, 2006). The VIA is present in both treated and untreated leaves (Verma and Awasthi, 1980). Micropropagation of *B. diffusa* has been carried out for producing the systemic RIP for viral disease management (Gupta, 1999; Gupta *et al.*, 2004).

Susceptible healthy hosts upon treatment with *C. aculeatum* extract develop complete resistance within 4–6 h. The SIR is reversed by the simultaneous application of AMD (Verma *et al.*, 1984). Treatment with SRI from *C. aculeatum* leaves, triggers accumulation of a new defensive VIA in treated and non-treated leaves of healthy host plants (Verma *et al.*, 1984; Verma *et al.*, 1996, 1999; Srivastava 1999). The SRI in crude sap is resistant to denaturation by organic solvents and extremely thermostable (Verma *et al.*, 1984). The resistance-inducing activity of *C. aculeatum* SRI (CA-SRI) is not affected by exogenous application of proteases (Verma *et al.*, 1996). Leaf extract of *C. aculeatum* is most effective for controlling virus diseases in crop plants (Verma *et al.*, 1995a; Srivastava 1999; Srivastava *et al.*, 2004). Leaf extract, when sprayed on susceptible host plants, prevents infection of mechanically and white fly transmitted viruses in several hosts, e.g. tomato yellow mosaic virus, tobacco mosaic virus, sunnhemp rosette virus and tobacco leaf curl virus (Verma *et al.*, 1984; Verma *et al.*, 1995a) (Table 7.2).

Table 7.2. Plants containing virus inhibitory activity.

Name	Family	Reference
Amaranthus albus	Amaranthaceae	Smookler, 1971
Alternanthera brasiliana	Amaranthaceae	Noronha *et al.*, 1983
Acacia arabica	Leguminosae (Fabaceae)	Gupta and Raychauduri, 1971a
Acer insulera	Aceraceae	Yoshi *et al.*, 1954
Agava americana	Agavaceae	Simon, 1963
Ailanthes excelsa	Simaroubaceae	Patel and Patel, 1979
Alternanthere ficoidea	Amaranthaceae	Noronha *et al.*, 1983
Amaranthus caudatus	Amaranthaceae	Smooker, 1971
Argemone mexicana	Papaveraceae	Patel and Patel, 1979
Beta vulgaris	Chenopodiaceae	Paliwal and Narinai, 1965

Continued

Table 7.2. Continued.

Name	Family	Reference
Boerhaavia diffusa	Nyctaginaceae	Verma and Baranwal, 1988; Mehrotra et al., 2002; Sukhdev, 2006; Singh, 2006
Bougainvillea spectabilis	Nyctaginaceae	Verma and Dwivedi, 1983
Brassica oleracea	Brassicaceae	Verma, 1973
Callistemon lanceolatus	Myrtaceae	Gupta and Raychaudhuri, 1971a,b;
Carissa edulis	Apocynaceae	Tolo et al., 2006
Celosia plumose	Amaranthaceae	Patil, 1973
Chenopodium amaranticolor	Chenopodiaceae	Smookler, 1971 Alberghina, 1976
Chenopodium ambrisioides	Chenopodiaceae	Verma and Barranwal, 1983
Cinchona ledgeriana	Rubiaceae	Gupta and Raychauduri, 1971b
Citrus medicatimonum	Rutaceae	Ray, et al., 1979
Clerodendrum aculeatum	Verbenaceae	Verma et al., 1984; Srivastava 1999; Srivastava et al., 2004; Singh, 2006
Clerodendrum fragrans	Verbenaeceae	Verma et al., 1984
Cocos nucifera	Arecaceae	Gendron, 1950
Cucurbita maxima	Cucurbitaceae	Weintrab and Willison, 1983
Cuscuta reflexa	Cucurbitaceae	Awasthi, 1982
Datura metel	Solanaceae	Verma and Awasthi, 1975
Datura stramonium	Solanaceae	Paliwal and Narinai, 1965
Dianthus caryophyllus	Caryophylaceae	Van Kammen et al., 1961
Eucalyptus tereticornis	Myrtaceae	Ray et al., 1979
Eugenia jambolana	Myrtaceae	Verma et al., 1969
Euphorbia hitra	Euphorbiaceae	Weeraratne, 1961
Gomphrena globosa	Amaranthaceae	Grasso and Shephard, 1978
Gyandropsis pentaphylla	Asteraceae	Paliwal and Nariani, 1965
Helianthus annus	Asteraceae	Johari, et al., 1983
Hordeum vulgare	Gramineae (Poaceae)	Leah, et al., 1991
Mirabilis jalapa	Nyctaginaceae	Kataoka et al., 1991
Opuntia robusta	Cactaceae	Simons et al., 1963
Petunia hybrida	Solanaceae	Singh, 1972
Phyllanthus frafernus	Euphorbiaceae	Saigopal et al., 1986
Phytolacca dodecandra	Phytolaccaceae	Reddy et al., 1984
Pseudoranthemum atropurpureum	Acanthaceae	Verma et al., 1985
Rumex hastatus	Polygonaceae	Singh, et al., 1977
Solanum nigrum	Solanaceae	Vasudeva and Nariani, 1952
Terminala chebula	Combretaceae	Gupta and Raychauduri, 1971b
Tetragonia expansa	Tetragoniaceae	Benda, 1956
Zingiber officinale	Zingiberaceae	Ray et al., 1979

The systemic resistance inducer from *Clerodendrum aculeatum* and *Clerodendrum inerme* have been purified and characterized. A 34-kDa basic protein was isolated from the leaves of *C. aculeatum*: 64 µg/ml of protein provided complete protection of untreated leaves against TMV infection in

N. tabacum Samsun NN (Verma *et al.*, 1996). Two basic proteins of 29 and 34 kDa (CIP-29 and CIP-34) were isolated from the leaves of *C. inerme*. A resistance-inducing protein of 31 kDa was isolated from the same plant (Parveen *et al.*, 2001). The minimum amount of purified proteins required to induce systemic resistance varied from 16 µg/ml for CIP-29 to 800 µg/ml for CIP-34 (Prasad *et al.*, 1995). CAP-34, a protein from *C. aculeatum*, when applied to lower leaves of *Carica papaya*, suppressed the disease caused by papaya ring spot virus (PRSV). In the control papaya plants the typical disease symptoms of PRSV, that is, mosaic to filiformy, appeared in 95% of the plants between 30 and 60 days after virus inoculation. In the CAP-34-treated papaya plants the symptoms appeared in only 10% of the plants during the same period. The presence of PRSV was determined by ELISA and RT–PCR (Srivastava *et al.*, 2009). It appears that protein from *C. aculeatum* is a good candidate for utilization in management of virus diseases. The systemic virus inhibitory activity of these proteins is due to the host-mediated phenomenon of formation / accumulation of a new virus inhibitory protein in the treated plants showing systemic resistance. When TMV is mixed with the induced virus inhibitory protein, the virus is completely inhibited. The fact that CA-SRI also behaves as an RIP, as it inhibited *in vitro* protein synthesis in rabbit reticulocyte lysate and wheat germ lysate (Kumar *et al.*, 1997), strongly confirms this belief that antiviral proteins generally have the properties of RIPs. The sequence of the CA-SRI protein showed varying homology (11–54%) with the RIPs from other plant species (Kumar *et al.*, 1997). However, the absence of hybridization between the CA-SRI gene and the DNA / RNA of *Mirabilis, Bougainvillea*, rice, pea and tobacco shows that the virus inhibitory genes do not react with leaf proteins from *Mirabilis, Bougainvillea, Boerhaavia*, rice, pea and tobacco. These only indicate that antiviral proteins from different plants may behave differently and can be used specifically in particular host–virus systems.

The pre-inoculation spray of SRIs from *C. aculeatum, C. inerme, Boerhaavia diffusa, B. spectabilis* and *Pseuderanthemum bicolor* modified the susceptibility of several host plants such as tomato, tobacco, mungbean, urdbean, bhendi and sunnhemp against subsequent infection by viruses. The treatment helps to protect the susceptible hosts during the vulnerable early stages of development. Induced resistance appears to be a universal process in all susceptible hosts and can be used to advantage in plant protection. Pre-inoculation sprays (four sprays) of *B. spectabilis* leaf extract protected plants of *Cucumis melo* against cucumber green mottle mosaic virus (CGMMV), *Crotalaria juncea* against sunnhemp rosette virus (SRV) and *Lycopersicon esculentum* against TMV for 6 days (Verma and Dwivedi, 1983). Since the duration of resistance conferred by SRIs was up to only 6 days, it was realized that the durability of resistance needed to be prolonged under field conditions to achieve better protection against virus infection. It has been shown that proteinaceous modifiers, such as papain, enhanced the activity of CA-SRI and also prolonged the durability of induced resistance against sunnhemp rosette tobamovirus in *Crotalaria juncea* up to 12 days (Verma and Varsha, 1995). In another study, five weekly sprays of CA-SRI in potted plants of tomato in open fields protected the plants from natural virus infection for more than 2 months (unpublished results). However, in another study it was shown that weekly sprays of leaf extract of *C. aculeatum* delayed the symptom appearance of leaf curl virus in tomato and promoted the growth of the plants (Baranwal and Ahmad, 1997). A natural plant compound called NS-83 was shown to reduce and delay the disease symptoms by TMV, PVX and PVY in tobacco and tomato plants under field conditions and the fruit yield in tomato was increased by 23.4% (Xin-Yun *et al.*, 1988). Plant extracts from *C. inerme* and *Ocimum sanctum* provided a high degree of resistance against tobacco chlorotic mottle virus in cowpea, possibly by induction of systemic induced resistance (Mistry *et al.*, 2003). Sumia *et al.* (2005) demonstrated inhibition of local lesions and systemic infection induced by Tobacco streak virus in cowpea and French bean by pre-treatment of a proteinaceous substance from seeds of *Celosia cristata.* Foliar sprays with aqueous leaf extract of *C. aculeatum* plant since the sprouting stage at fortnightly intervals in the field could significantly protect *Amorphophallus campanulatus* against infection by virus. Maximum reduction in disease incidence and symptom severity was exhibited by plants which received six sprays. Maximum plant growth along with considerable increases in corm weight was also observed in such plants (Khan and Awasthi, 2006).

It appears that antiviral phytoproteins trigger the host defence mechanism in a specific manner either by signal transduction, as demonstrated in the case of *Phytolacca* antiviral protein (PAP), and/or by increased synthesis of antiviral proteins in host plants treated by systemic resistance inducers as for *C. aculeatum.* While success has been achieved in developing transgenics that have genes for mutant PAP without RIP activity (Smirnov *et al.*, 1997), it remains to be seen how these transgenics can be utilized under field conditions. On the other hand, SRIs such as CA-SRI show a potential for use under field conditions. However, a larger quantity of SRIs would be required for their wider application. Cloning of the genes for CA-SRI and their expression has been achieved in an *Escherichia coli* expression vector. The expressed protein has been shown to inhibit protein synthesis (Kumar *et al.*, 1997). A detailed study is still required to determine whether or not a transgenic with native and mutant CA-SRI genes would be able to show systemic protection against virus infection. Although the *C. aculeatum* protein has been demonstrated to have very high antiviral activity and to be extremely useful as a plant immunizing agent, it has not been commercialized so far. The proteins from a few other non-host plants have also been recognized as good defence stimulants, but they have not been developed into products for disease control, because the industry finds it easier to patent newly synthesized compounds than natural plant products.

Singh (2006) reported the anti-proliferative property of phytoproteins from *B. diffusa* and *C. aculeatum* for the retardation of proliferation of human breast cancer cell lines as well as the inhibition of activity of the Semliki Forest virus in mice.

7.6 Conclusion

With the precipitous withdrawal of some of the toxic protectants, it may be profitable to explore natural plant products as alternatives, particularly against virus diseases where all other methods fail. The phytoproteins or their smaller peptides may prove valuable as 'lead structures' for the development of synthetic compounds. It would pay us to explore this rich source of antivirals more thoroughly. The value of these antiviral proteins is unlimited because they are quite safe, non-toxic even after repeated and prolonged use, and substantially enhance plant growth and yield. Antiviral phytoproteins may be useful if they are integrated with other strategies of virus disease management.

Botanical resistance inducers can be classified as ideal virus-suppressing agents, as they encompass all the characteristics of an ideal antiviral compound. Induced resistance operates through the activation of natural defence mechanisms of the host plant and induces systemic resistance to viral multiplication in plants. The active products present in these extracts have no direct effect on viruses; their antiviral activity is mediated by host cells in which they induce the antiviral state. Systemic resistance inducers obtained from plants have been shown to be effective against a wide range of viruses. Plant extracts or the semi-purified proteins from these plants stimulate the hosts to produce a virus inhibitory agent, which spreads to surrounding tissues and other plant parts.

References

Alexander, D., Goodman, R.M., Gut-Rella, M., Glascock, C., Weymann, K., Friedrich, L., Maddox, D., Ahl-Goy, P., Lunz, T., Ward, E. and Ryals, J. (1993) Increased tolerance of two oomycete pathogens in transgenic tobacco expressing pathogenesis-related protein 1a. *Proceedings of National Academy of Sciences USA* 90, 7327–7331.

Aron, G.M. and Irvin, J.D. (1980) Inhibition of herpes simplex virus multiplication by the pokeweed antiviral protein. *Antimicrobial Agents Chemotherapy* 17, 1032–1033.

Awasthi, L.P., Chowdhury, B. and Verma, H.N. (1984) Prevention of plant virus disease by *Boerhaavia diffusa* inhibitor. *Indian Journal of Tropical Plant Diseases* 2, 41–44.

Awasthi, L.P., Kluge, S. and Verma, H.N. (1989) Characterstics of antiviral agents induced by *Boerhaavia diffusa* glycoprotein in host plants. *Indian Journal of Virology* 3, 156–169.

Awasthi, L.P., Pathak, S.P., Gautam, N.C. and Verma, H.N. (1985) Control of virus diseases of vegetable crops by a glycoprotein isolated from *Boerhaavia diffusa*. *Indian Journal of Plant Pathology* 3, 311–327.

Awasthi, L.P. (1981) The purification and nature of an antiviral protein from *Cuscuta reflexa* plants. *Archives of Virology* 70, 215–223.

Baranwal, V.K. and Ahmad, N. (1997) Effect of *Clerodendrum aculeatum* leaf extract on tomato leaf curl virus. *Indian Phytopathology* 50, 297–299.

Barbieri, L. and Stirpe, F. (1982) Ribosome inactivating proteins from plants properties and possible uses. *Cancer Survey* 1, 489–520.

Barbieri, L., Batteli, M.G., and Stirpe, F. (1993) Ribosome-inactivating proteins from plants. *Biochimica et Biophysica Acta* 1154, 237–282.

Barbieri, L., Brigotti, M.G., Perocco, P., Carnicelli, D., Ciani, M., Mercatali, L. and Stirpe, F. (2003) Ribosome-inactivating proteins depurinate poly (ADP-ribosyl) ated poly (ADP-robose) polymerase and have transforming activity for 3T3 fibroblast. *FEBS Letters* 538, 178–182.

Baulcombe, D. (1994) Novel strategies for engineering virus resistance in plants. *Current Opinion in Biotechnology* 5,117-124.

Beachy, R.N., Loesch-Fries, S. and Tumer, N.E. (1990) Coat protein mediated resistance against virus infection. *Annual Review of Phytopathology* 28, 451–474.

Bolognesi, A., Polito, L., Lubelli, C., Barbieri, L., Parente, A. and Stirpe F. (2002) Ribosome-inactivating and adenine polynucleotide glycosylase activities in *Mirabilis jalapa* L. tissues. *Journal of Biological Chemistry* 277, 13709–13716.

Bozarth, R.F. and Ross, A.F. (1964) Systemic resistance induced by localized virus infection: extent of changes in uninfected plant parts. *Virology* 24, 446–455.

Chadha, K.C. and McNeil, D.H. (1969). An antiviral principle from tomatoes systemically infected with tobacco mosaic virus. *Canadian Journal of Botany* 47, 513–518.

Chen, Z.C., Antoniw, J.F. and White, R.F. (1993) A possible mechanism for the antiviral activity of pokeweed antiviral protein. *Physiology and Molecular Plant Pathology* 42, 249–258.

Chen, Z.C., White, R.F., Antoniw, J.F. and Lin, Q. (1991) Effect of pokeweed antiviral protein (PAP) on the infection of plant virus. *Plant Pathology* 40, 612–620.

Cho, H.I., Lee, S.J., Kim, S. and Kim, B.D. (2000) Isolation and characterization of cDNA encoding ribosome inactivating protein from *Dianthus sinensis* L. *Molecules and Cells* 10, 135–139.

Choudhary, N., Kapoor, H.C. and Lodha, M.L. (2008) Cloning and expression of antiviral / ribosome-inactivating protein from *Bougainvillea xbuttiana*. *Journal of Biosciences* 33, 91–101.

Cohen, Y., Gisi, U. and Niderman, T. (1993) Local and systemic protection against *Phytophthora infestans* induced in potato and tomato plants by jasmonic acid and jasmonic acids methyl ester. *Phytopathology* 83, 1054–1062.

Conejero, V., Picazo, I. and Segado, P. (1979) Citrus exocortis viroid (CEV): Protein alterations in different hosts following viroid infection. *Virology* 97, 454–456.

Cutt, J.R., Harpester, M.H., Dixon, D.C., Carr, J.P., Dunsmuir, P. and Klessig, F. (1989) Disease response to tobacco mosaic virus in transgenic tobacco plants that constitutively express the pathogenesis-related PR 1b gene. *Virology* 173, 89–97.

Durrant, W.E. and Dong, X. (2004) Systemic acquired resistance. *Annual Review of Phytopathology* 42, 185–209.

Edelbaum, O., Altman, A. and Sela, I. (1983) Polyinosinic-polycytidylic acid in association with cyclic nucleotides activities the antiviral factor (AVF) in plant tissue. *Journal of General Virology* 64, 211–214.

Endo, Y. and Tsurugi, K. (1987) RNA N-glycosidase activity of ricin A chain. Mechanism of action of the toxic lectin on eukaryotic ribosomes. *Journal of Biological Chemistry* 262, 8128–8130.

Endo, Y., Mitsui, K., Motizuki, M.Q. and Tusrugi, K. (1987) The mechanism of action of ricin and related toxic lectins on eukaryotic ribosomes. The site and the characteristics of the modification in 28S ribosomal RNA caused by the toxins. *Journal of Biological Chemistry* 262, 5908–5912.

Faccioli, G. and Capponi, R. (1983) An antiviral factor present in the plants of *Chenopodium amaranticolor* locally infected by tobacco necrosis virus. I. extraction, partial characterization, biological and chemical properties. *Phytopathology* 106, 289–301.

Farrag, E.S., Ziedan, E.S.H. and Mahmoud, S.Y. (2007) Systemic acquired resistance induced in cucumber plants against powdery mildew disease by preinoculation with tobacco necrosis virus. *Plant Pathology Journal* 6, 44–50.

Fraser, R.S.S. (1990) Evidence for the occurrence of the 'Pathogenesis-related' proteins in leaves of healthy tobacco plants during flowering. *Physiology and Plant Pathology* 19, 69–76.

Gera, A. and Loebenstein, G. (1983) Further studies of an inhibitor of virus replication fromTMV infected protoplast of a local lesion responding tobacco cultivar. *Phytopathology* 73, 111–115.

Gessner, S.L. and Irvin, J.D. (1980) Inhibition of pokeweed antiviral protein and ricin. *Journal of Biological Chemistry* 255, 3251–3253.

Gianinnazi, S., Martin, C. and Vallee, J.C. (1970) Hypersensibilite aux virus, temperature et proteins solubles chez le *Nicotiana xanthi* nc Appartition de nouvelles macromolecules lores de la repression de la synthese virale. *Comptes.-Rendus Academie des Sciences, Paris* 270, 2283–2386.

Gianinnazi, S., Ahl, P., Cornu, A., Scalla, R. and Cassini, R. (1980) First report of host P-Protein appearance in response to a fungal infection in tobacco. *Physiology and Plant Pathology* 16, 337–342.

Gilpatrick, J.D., and Weintraub, M. (1952) An unusual type of protection with carnation mosaic virus. *Science* 115, 701–702.

Gupta, R.K., (1999) Micropropagation of *B. diffusa* for producing the systemic resistance inducing protein (SRIP) for viral disease management. Ph.D. Thesis, Lucknow University, Lucknow, India.

Gupta, R.K., Srivastava, A., and Verma, H.N. (2004) Callus culture and organogenesis in *Boerhaavia diffusa*: A potent antiviral protein containing plant. *Physiology and Molecular Biology of Plant* 10, 263.

Habuka, N., Murakami, Y., Noma, M., Kudo, T, and Harikhosi, K. (1989) Amino acid sequence of *Mirabilis* antiviral protein, total synthesis of its gene and expression in *Escherichia coli*. *Journal of Biological Chemistry* 264, 6629–6637.

Hansen, J. (1989) Antiviral chemicals for plant disease control. *Critical Reviews in* Plant *Sciences* 8, 45–88.

Hartley, M.R., Legname, G., Osborn, R., Chen, Z. and Lord, J.M. (1991) Single chain ribosome-inactivating proteins from plants, depurinnate *Escherichia coli* 23S ribosomal RNA. *FEBS Letters* 290, 65–68.

Ikeda, T., Niino, K., Kataoka, J. and Matsumoto, T. (1987) Effects of nutritional and hormonal factors on the formation of anti-plant viral protein by highly producing cell strains of *Mirabilis jalapa* L. *Agricultural and Biological Chemistry* 51, 3119-3124.

Irvin, J.D. (1975) Purification and partial characterization of the anti viral protein from *Phytolacca Americana* which inhibits eukaryotic protein synthesis. *Archives of Biochemistry and Biophysics* 169, 522–528.

Irvin, J.D. (1983) Pokeweed antiviral protein. *Pharmacology and Therapeutics* 21, 371–387.

Jenns, A. and Kuc, J. (1977) Localized infection with TNV protects cucumber against *Colletotrichum lagenarium*. *Physiology and Plant Pathology* 11, 207–212.

Kato, S. and Misawa, T. (1976) Isolation and identification of a substance interfering with local lesions formation produced in cowpea leaves locally infected with cucumber mosaic virus. *Annual Phytopathological Society* 24, 450–452.

Kessmann, H., Staub, T., Hofmann, C., Maetzke, T., Herzog J. and Ward, E., Uknes, S. and Ryals, J. (1994) Induction of systemic acquired disease resistance in plants by chemicals. *Annual Review of Phytopathology* 32, 439–459.

Khan, M.M.A.A. and Verma, H.N. (1990) Partial characterization of an induced virus inhibitory protein associated with systemic resistance in *Cyamopsis tetragonoloba* (L.) Taub. Plants. *Annals of Applied Biology* 117, 617–623.

Kimmins, W.C. (1969). Isolation of a virus inhibitor from plants with localized infection. *Canadian Journal of Botany* 47, 1879–1886.

Kopp, M., Rouster, J., Fritig, B., Darvill, A. and Albersheim, P. (1989) Host-Pathogen interactions. XXXII. A fungal glucan preparation protects *nicotinase* against infection by viruses. *Plant Physiology* 90, 208–216.

Kubo, S., Ikeda, T., Imaizumi, S., Takanami, Y. and Mikami Y. (1990) A potent plant virus inhibitor found in *Mirabilis jalapa* L. *Annual Phytopathological Society,* 56, 481–487.

Kumar, D., Verma, H.N., Tuteja, N. and Tewari, K.K. (1997) Cloning and characterization of gene encoding an anti-viral protein from *Clerodendrum aculeatum* L. *Plant Molecular Biology* 33, 745–751.

Lamb, C.J. and Dixon, R.A. (1990) Molecular communication in interactions between plants and microbial pathogens. *Annual Review of Plant Physiology* 41, 439–463.

Lamb, F.I., Roberts, L.M. and Lord, J.M. (1985) Nucleotide sequence of cloned cDNA coding for preroricin. *Europian Journal of Biochemistry* 148, 265–270.

Legname, G, Bellosta, P, Gromo, G, Modena, D, Kun, J.N., Roberts, L.M. and Lord, J.M. (1991) Nucleotide sequence of cDNA coding for Dianthin 30, a ribosome inactivating protein from *Dianthus caryophyllus. Biochimica et Biophysica Acta* 1090, 119–122.

Loebenstein, G. (1963) Further evidence on systemic resistance induced by localized necrotic virus infections in plants. *Phytopathology* 53, 306–308.

Loebenstein, G. and Gera, A. (1981) Inhibitor of virus replication released from tobacco mosaic virus infected protoplasts of a local lesion responding tobacco cultivar. *Virology* 114, 132–139.

Lodge, K.J., Kaniewski, W.K. and Tumer, N.E. (1993) Board spectrum virus resistance in transgenic plants expressing pokeweed antiviral protein. *Proceedings of National Academy of Sciences* 90, 7089–7093.

Lubelli, C., Chalgilialoglu, A., Bolognesi, A., Strocchi, P., Colombatti, M. and Stripe, F. (2006). Detection of ricin and other ribosome inactivating proteins by an immune-polymerase chain reaction assay. *Annals of Biochemistry* 355, 102–109.

Mansouri, S., Nourollahzadeh, E. and Hudak K.A. (2006) Pokeweed antiviral protein depurinates the sarcin / ricin loop of the rRNA prior to binding of aminoacyl-tRNA to the ribosomal A site. *RNA* 12, 1683–1692.

Malamy, J., Carr, J.P., Klessig, D.F. and Raskin, I. (1990) Salicylic acid: A likely endogenous signal in the resistance response of tobacco to viral infection. *Science* 250, 1002–1004.

McIntyre, J.L., Dodds, J.A. and Hare, J.D. (1981) Effects of localized infections of *Nicotiana tabacum* by tobacco mosaic virus on systemic resistance against diverse pathogens and an insect. *Phytopathology* 71, 297–301.

Metraux, J.P. and Boller, T. (1986) Local and systemic induction of chitinase in cucumber plants in response to viral, bacterial and fungal infections. *Physiology and Molecular Plant Pathology* 56, 161–169.

Mozes, R., Antignus, Y., Sela, I. and Harpaz, I. (1978) The chemical nature of an antiviral factor (AVF) from virus infected plants. *Journal of General Virology* 38, 241–249.

Nagaich, B.B. and Singh, S. (1970) An antiviral principle induced by potato virus x inoculation in *Capsicam pendulum* wild. *Virology* 40, 203–205.

Nienhaus, F. and Babovic, M. (1978) Observation on an induced antiviral factor in leaves of grape vine inoculated with tomato ringspot virus. *Zeitschrift fur Pflanzenkrantheiten und Flanzenschutz* 85, 238–246.

Olsnes, S. and Pihl, A. (1982) Toxic lectins and related proteins, molecular action of toxins and viruses In: Cohen P. and Van Heyningen S. (eds) *Molecular Action of Toxins and Viruses.* Elsevier Biomedical Press, Amsterdam, 51–105.

Pennazzio, S. and Roggero, P. (1991) Systemic acquired resistance to virus infections and ethylene biosynthesis in asparagus beans. *Journal of Phytopathology* 131, 177–183.

Plobner, L. and Leiser, R.M. (1990) Induction of virus resistance by carnation protein. *VIIIth International congress of Virology,* Berlin (Abstract).

Prasad, V., Srivastava, S., Versha, and Verma, H.N. (1995) Two basic proteins isolated from *Clerodendrum inerme* Gaertn. Are inducers of systemic resistance in susceptible plants. *Plant Science* (UK) 110, 73–82.

Prasad, V. and Srivastava, S. (2001) Inducible mechanisms of plant resistance to virus infections. *Journal of Plant Biology 28*, 13–23.

Parveen, S., Tripathi, A., and Varma, A., (2001). Isolation and characterization of an inducer protein (Crip-31) from *Clerodendrum inerme* leaves responsible for induction of systemic resistance against viruses. *Plant Science* 161, 453–459.

Picard, Daniel C. Cheng Kao; Katalin A. Hudak, (2005). Pokeweed antiviral protein inhibits Brome Mosaic Virus replication in plant cells. *Journal of Biological Chemistry* 280, 20069–20075.

Ragetli, H.W.J. (1957). Behaviour and nature of virus inhibitor occurring in *Dianthus caryophyllus* L. *Tijdschr. Planteziekten, 61*, 245–344.

Ragetli, H.W.J. and Weintraub, M. (1962) Purification and characteristics of a virus inhibitor from *Dianthus caryophyllus* I. Purification and activity. *Virology,* 18, 232–240.

Ragetli, H.W.J. and Weintraub, M. (1962) Purification and characteristics of a virus inhibitor from *Dianthus caryophyllus* II. Characterization and mode of action. *Virology,* 18, 241–248.

Rajamohan, F., Venkatachalam, T.K., Irvin, J.D. and Uckun, F.M. (1999) Pokeweed antiviral protein isoforms PAP-I, PAP-II and PAP-III depurinate RNA of human immunodeficiency virus (HIV)-I. *Biochemical and biophysical research communications* 260, 453–458.

Reddy, M.P., Brown, D.T. and Robertus, J.D. (1986) Extra cellular localization of pokeweed antiviral protein. *Proceedings of National Academy of Sciences USA* 83, 5053–5056.

Reisbig, R.R. and Bruland, O. (1983) Diathin 30 and 32 from *Dianthus caryophyllus*: two inhibitors of plant protein synthesis and their tissue distribution. *Archives of Biochemistry and Biophysics* 224, 700–706.

Rodes, T.L. and Irvin, J.D. (1981) Reversal of the inhibitory effects of the pokeweed antiviral protein upon protein upon protein synthesis. *Biochimica et Biophysica Acta* 652, 160–167.

Ross, A.F. (1961a) Localized acquired resistance to plant virus infection in hypersensitive host. *Virology* 14, 329–339.

Ross, A.F. (1961b) Systemic acquired resistance induced by localized virus infections in plants. *Virology* 14, 340–358.

Ryals, J., Uknes, S. and Ward, E. (1994) Systemic Acquired Resistance. *Plant Physiology* 104, 1109–1112.

Sela, I., and Applebaum, S.W. (1962) Occurrence of antiviral factor in virus infected plants. *Virology* 17, 543–548.

Sela, I., Harpez, I. and Birk, Y. (1964) Seperation of a highly active antiviral factor from virus infected plants. *Virology* 22, 446–451.

Sela, I., Harpez, I. and Birk, Y. (1966) Identification of the active component of an antiviral factor isolated from virus infected plants. *Virology* 28, 71-78.

Singh, A.K. (2006) Comparative studies on control of certain ailments of plants, mice and cancer cell lines and *in vitro* stimulation of growth of plants and virus resistance using phytoproteins from *Boerhaavia diffusa* and *Clerodendrum aculeatum*. Ph.D. Thesis, Lucknow University, Lucknow, India.

Srivastava, A. (1999) Micropropagation of *Clerodendrum aculeatum* and isolation of virus inhibitory substance from culture cells inducing systemic resistance in susceptible host. Ph.D. Thesis, Lucknow University, Lucknow, India.

Srivastava, A., Gupta, R.K. and Verma, H.N. (2004) Micro Propagation of *Clerodendrum aculeatum* through adventitious shoot induction and production of consistent amount of virus resistance inducing protein. *Indian Journal of Experimental Biology* 42, 1200–1207.

Srivastava, A., Trivedi, S., Krishna, S., Verma, H.N. and Prasad, V. (2008) Suppression of papaya ringspot virus infection in *Carica papaya* with CAP. *European Journal of Plant Pathology,* DOI: 10.1007 / s10658-008-9358-2.

Srivastava, A., Trivedi, S., Krishna, S.K., Verma, H.N. and Prasad, V. (2009). Supression of *Papaya ring spot virus* infection in *Carica papaya* with CAP-34, a systemic antiviral resistance inducing protein from *Clerodendrum aculeatum. European Journal of Plant Pathology* 123, 241–246.

Srivastava N. and Padhya M.N. (1995) Punarnavine profile in the regenerated roots of *Boerhaavia diffusa* L. from leaf segments. *Current Science* 68, 653–656.

Stevens, W.A., Spurdon, C., Onyon, L.J., and stripe, F. 1981. Effect of inhibitors of protein synthesis from plants on tobacco mosaic virus infection. *Experientia* 37, 257–259.

Sticher, L., Mauch-Mani, B. and Metraux, J.P. (1997) Systemic acquired resistance. *Annual Review of Phytopathology* 35, 235–270.

Stintzi, A., Heitz, T., Prasad, V., Wuedemann-Merdinoglu, S., Kaufmann, S., Geoffroy, P., Legrand, M. and Fritig, B. (1993) Plant pathogenesis related proteins and their role in defense against pathogens. *Bicochimie* 75, 687–706.

Stirpe, F., Barbieri, L., Battelli, M.G., Soria, M. and Douglas, A. (1992) Ribosome-inactivating proteins from plants: Present status and future prospectus. *Biotechnology* 10, 405–412.

Stirpe, F., Williams, D.G., Onyon, L.J., Leggg, R.F. and Stevens, W. A. (1981) Dianthus, ribosome damaging proteins with antiviral properties from *Dianthus caryophyllus* L. (carnation) *Biochemical Journal* 195, 399–405.

Strocchi, P., Barbieri, L. and Stirpe, F. (1992) Immunological properties of ribosome inactivating proteins and a saporin immunotoxin. *Journal of Immunological Methods* 155, 57–63.

Smirnov, S., Shulaev, V. and Tumer, N.E. (1997) Expression of pokeweed antiviral protein in transgenic plants induces virus resistance in grafted wild type plants independently of salicylic acid accumulation and pathogenesis-related protein synthesis. *Plant Physiology* 114, 1113–1121.

Tomlinson, J.A., Walker, V.M., Flewett, T.H. and Barclay, G.R. (1974) The inhibition of infection of cucumber mosaic virus and influenza virus by extracts from *Phytolacca Americana. Journal of General Virology* 22, 225–231.

Ukness, S., Winter, A., Delaney, T., Vernooji, B., Friedrich, L., Nye, G., Potter, S., Ward, E. and Ryals, J. (1993) Biological induction of systemic acquired resistance in *Arabidopsis. Molecular Plant Microbe Interactions* 6, 692–698.

Ussery, M.A., Irvin, J.D. and Hardesty, B. (1977) Inhibition of polio virus replication by a plant antiviral peptide. *Annals of New York Academy of Sciences* 284, 431–440.

Van Damme, E.J.M., Hao, Q., Chen, Y., Barre, A., Vandenbussche, F., Desmyter, S., Rouge, P. and Peumans, W.J. (2001) Ribosome-inactivating proteins: a family of plant proteins that do more than inactivate ribosomes. *Critical Reviews in Plant Sciences* 20, 395–465.

Van Kammen, A., Noordam, D. and Thing, T.H. (1961) The mechanism of inhibition of infection with tobacco mosaic virus by an inhibitor from carnation sap. *Virology* 14, 100–108.

Van Loon, L.C. and Van Kammen, A. (1970) Polyacrylamide disc electrophoresis of the soluble leaf proteins from *Nicotiana tabacum* var. 'Samsun' and 'Samsun' NN II. Changes in protein constitution after infection with tobacco mosaic virus. *Virology* 40, 119–211.

Varma, A., dhar, A.K. and Mandal, B. (1992) MYNV transmission and control in India. In: Green, S.K. and Kim, D. (eds) *Mungbean yellow mosaic disease*. AVRDC, Taipei, pp. 8–20.

Verma, H.N. (1982) Inhibitor of plant viruses from higher plants. In: Singh, B.P. and Raychoudhary, S.P. (eds) *Current Trends in Plant Virology*. Today and Tomorrow's Printers and Publishers, New Delhi, India, 151–159.

Verma H.N. and Awasthi L.P. (1979) Antiviral activity of *Boerhaavia diffusa* root extract and the physical properties of the virus inhibitor. *Canadian Journal of Botany* 57, 926–932.

Verma, H.N. and Awasthi, L.P. (1979) Prevention of virus infection and multiplication by leaf extract of *Euphorbia hirta* and the properties of the virus inhibitor. *New Botanist* 6, 49–59.

Verma, H.N. and Awasthi, L.P. (1980) Occurrence of a highly antiviral agent in plants treated with *Boerhaavia diffusa* inhibitor. *Candian Journal of Botany* 58, 2141–2144.

Verma, H.N. and Baranwal, V.K. (1999) Antiviral phytoproteins as biocontrol agents for efficient management of plant virus diseases. In: Rajak, R.L. and Upadhyay, R.K. (eds.) *Biocontrol Potential and their Exploitation in Crop Pest and Disease Management.* Aditya Book Pvt. Ltd., New Delhi, pp. 7–79.

Verma, H.N. and Baranwal, V.K. (1983) Antiviral activity and the physical properties of the leaf extracts of *Chenopodium ambrossoides*. L. *Proceedings of Indian Academy of Sciences (Plant Science)* 92, 461–465.

Verma, H.N. and Dwivedi, S.D. (1983) Preventions of plant virus disease in some economically important plant by *Bougainvillea* leaf extract. *Indian Journal Plant Pathology* 1, 97–100.

Verma, H.N. and Dwivedi, S.D. (1984) Properties of a virus inhibiting agent, isolated from plants following treatment with *Bougainvillea spectabilis* leaf extract. *Physiology and Plant Pathology* 25, 93–101.

Verma, H.N. and Khan, M.M.A.A. (1984) Management of plant virus diseases by *Pseuderanthemum bicolor* leaf extract. *Journal of plant Diseases and Protection* 91, 266–272.

Verma, H.N. and Khan, M.M.A.A. (1985) Occurrence of strong virus interfering agent in susceptible plants sprayed with *Pseuderanthemum atropurpurem tricolor* Leaf extract. *Indian Journal of Virology* 1, 26–34.

Verma, H.N. and Kumar, V. (1980) Prevention of plant virus diseases by *Mirabilis jalapa* leaf extract. *New Botanist* 7, 87–91.

Verma, H.N. and Mukherjee, K. (1975) Brinjal leaf extract induced resistance to virus infection in plants. *Indian Journal of Experimental Biology* 13, 416–417.

Verma, H.N. and Mukherjee, K. (1977) Properties of the interfering agent from brinjal leaves inducing resistance against tobacco mosaic virus. *New Botanist,* 5: 49–59.

Verma, H.N. and Prasad, V. (1992) Virus inhibitors and inducers of resistance: potential avenues for biological control of viral diseases. In: Mukerji, K.G., Tewari, J.P., Arora, D.K. and Saxena, G. (eds) *Recent Development in Biocontrol of Plant Diseases.* Aditya Books Pvt. Ltd, New Delhi, India, pp. 81.

Verma, H.N. and Srivastava, A. (1985) Natural occurrence of a severe mosaic disease of *Argemone mexicana* L. *Indian Journal of Plant Pathology* 3, 42–46.

Verma, H.N. and Srivastava, A. (1985) A potent systemic inhibitor of plant virus infection from *Aerva sanguinolenta* Blume. *Current Science* 54, 526–528.

Verma, H.N. and Varsha (1994) Induction of durable resistance by primed *Clerodendrum aculeatum* leaf extract. *Indian Phypathology* 47, 19–22.

Verma, H.N. and Varsha (1994) Prevention of natural occurrence of tobacco leaf curl disease by primed *Clerodendrum aculeatum* leaf extract. In: Verma, J.P. (ed.) *Detection of plant Pathogens and their Management.* Angor Publications (P) Ltd. New Delhi, pp. 202–206.

Verma, H.N. Choudhary, B. and Rastogi, P. (1984) Antiviral activity in leaf extracts of different *Clerodendrum* species. *Zeitschrift fur Pflanzenkrantheiten und Flanzenschutz* 91, 34–41.

Verma, H.N., Awasthi, L.P. and Mukherjee, K. (1979) Induction of systemic resistance by anti-viral plant extracts in non-hypersenstive hosts. *Journal of Plant Diseases and Protection* 87, 735–740.

Verma, H.N., Awasthi, L.P. and Saxena, K.C. (1979) Isolator of the virus inhibitor from root extract of *Boerhaavia diffusa* inducing systemic resistance in plants. *Candian Journal of Botany* 57, 11214–1217.

Verma, H.N., Baranwal, V.K. and Srivastava, S. (1998) Antiviral substances of plant origin. In: Hadidi, A., Khetarpal, R.K. and Kozenzava, A. (eds) *Plant virus disease control,* Americana Phytopathological Society (APS) Press, St. Paul, USA, pp. 154–162.

Verma, H.N., Choudhary, B. and Rastogi, P. (1984) Antiviral activity of different *Clerodendrum* L. species. *Zeitschrift fur Pflanzenkrantheiten und Flanzenschutz* 91, 34–41.

Verma, H.N. and Khan, M.M.A.A. (1984) Management of plant virus diseases by *Pseuderanthemum bicolor* leaf extracts. *Zeitschrift fur Pflanzenkrantheiten und Flanzenschutz* 91, 266–272.

Verma, H.N., Khan, M.M.A.A. and Dwivedi, S.D. (1985) Biological properties of highly antiviral agents present in *Pseudoranthemum atropurpureum* and *Bougainvillea apectabilis* extract. *Indian Journal of Plant Pathology* 3, 13–20.

Verma, H.N., Rastogi, P., Prasad, V. and Srivastava, A. (1985) Possible control of natural virus infection on *Vigna radiate* and *Vigna Mungo* by plant extracts. *Indian Journal of Plant Pathology* 3, 21–24.

Verma, H.N., Srivastava, A., Gupta, R.K. (1998a) Seasonal variation in systemic resistance inducing basic protein isolated from leaves of *Clerodendrum aculeatum*. *Indian Journal of Plant Pathology* 16, 9–13.

Verma, H.N., Srivastava, S., Varsha, and Kumar, D. (1996) Induction of systemic resistance in plants against viruses by a basic protein from *Clerodendrum aculeatum* leaves. *Phytopathology* 86, 485–492.

Verma, H.N., Varsha and Baranwal, V.K. (1995a) Endogenous virus inhibitors from plants: Their physical and Biological properties. In: Chessin, M., De Borde, D. and Zipf, A. (eds) *Antiviral Proteins in Higher Plants*. CRC Press, Boca Raton, pp. 1–21.

Verma, H.N. and Khan, M.M.A.A. (1991) A soluble protein from a non-host plant stimulating resistance mechanisms in systemic hosts. In: Sen, C. and Datta, S. (eds) *Biotechnology in Crop Protection*. Bidhan Chandra Krishi Vishwavidyalay, Kalyani, W.B., India, pp. 149–161.

Ward, E.R., Uknes, S.J., Williams, S.C., Dincher, S.S., Weiderhold, D.L., Alexander, D.C., Ahl-Goy, P., Metraux, J.P. and Ryals, J.A. (1991) Coordinate gene activity in response to agents that induce systemic acquired resistance. *Plant Cell*, 3, 1085–1094.

Weiringa-Brants, D.H. and Dekker W.C. (1987) Induced resistance in hypersensitive tobacco against tobacco mosaic virus infection of intercellular fluid from tobacco plants with systemic acquired resistance. *Journal of Phytopathology* 118, 165–170.

Xin-Yun, L., Huang-fang, L. and Wei-fan, C. (1988) Studies on the application of virus tolerance inducer-NS-83, In: *Proceedings of International Symposium of Plant Pathology*, Beijing, China.

Yalpani, N., Silverman, P., Wilson, T.M.A., Kleier, D.A. and Raskin, I. (1991) Salicylic acid is a systemic signal and an inducer of pathogenesis related proteins in virus infected tobacco. *Plant Cell* 3, 809–818.

Ye, X.S., Pan, S.Q. and Kuc, J. (1989) Pathogenesis-related proteins and systemic resistance to blue mold and tobacco mosaic virus induced by tobacco mosaic virus, *Peronospora tabacina* and asprin. *Physiology and Molecular Plant Pathology* 35, 161–175.

Zarling, J.M., Moran, P.A., Haffer, O., Sias, J., Rchman, D.D., Spina, C.A., Myers, D.A., Kuebelbeck, V., Ledbetter, J.A. and Uckun, F.M. (1990) Inhibition of HIV replication by pokeweed antiviral protein targeted to CD4+ cells by monoclonal antibodies. *Nature* 347, 92–95.

Zoubenko, O., Hudak, K. and Tumer, N.E. (2000) A non-toxic pokeweed antiviral protein mutant inhibits pathogen infection via a novel salicylic acid-independent pathway. *Plant Molecular Biology* 44, 219–229.

Zhang, D. and Halaweisch, F.T. (2007) Isolation and characterization of ribosome inactivating proteins from cucurbitaceae. *Chemistry and Biodiversity* 4, 431–442.

8 Phytochemicals as Natural Fumigants and Contact Insecticides Against Stored-product Insects

MOSHE KOSTYUKOVSKY AND ELI SHAAYA

Agricultural Research Organization, the Volcani Center, Israel

Abstract

For centuries, traditional agriculture in developing countries has used effective methods of insect pest control using botanicals. In order to make them a cheap and simple means of insect control for users, their efficacy and optimal use still need to be assessed. Currently, the measures to control pest infestation in grain, dry stored food and cut flowers rely heavily on toxic fumigants and contact insecticides. In recent years, the number of pesticides has declined as health, safety and environmental concerns have prompted authorities to consider restricting the use of toxic chemicals in food. Lately a new field is developing on the use of phytochemicals in insect pest management, such as edible and essential oils and their constituents. The aim here has been to evaluate the potential use of edible oils obtained from oil seeds and essential oils, and their constituents obtained from aromatic plants, as fumigants and contact insecticides for the control of the legume pest *Callosobruchus maculatus* F. (Coleoptera: Bruchidae). The most active edible oils as contact insecticides were crude oils from rice, maize, cottonseed and palm, and the fatty acids capric acid and undecanoic acid. The essentials oils and their constituents were found to have higher activity as fumigants than contact insecticides. From our studies, to elucidate the mode of action of essential oils, it was possible to postulate that essential oils may affect octopaminergic target sites.

8.1 Introduction

Insect damage in stored grains and other durable commodities may amount to 10–40% in developing countries, where modern storage technologies have not been introduced (Raja *et al.*, 2001). Currently, food industries rely mainly on fumigation as an effective method for insect pest control in grain and other dry food commodities. At present only two fumigants are still in use: methyl bromide and phosphine. The first one is mostly being phased out in developed countries due to its ozone depletion effects (WMO, 1995;

Shaaya and Kostyukovsky, 2006). In addition, there have been repeated indications that certain insects have developed resistance to phosphine, which is widely used today (Nakakita and Winks, 1981; Mills, 1983; Tyler *et al.*, 1983). It should be mentioned that although effective fumigants and contact synthesized insecticides are available, there is global concern about their negative effects on non-target organisms, pest resistance and pesticide residues (Kostyukovsky *et al.*, 2002a; Ogendo *et al.*, 2003). In recent years, attention has been focused on the use of botanicals as possible alternatives to toxic insecticides. Lately, there has been a growing interest in the use of plant oils and their bioactive chemical constituents for the protection of agriculture products due to their low mammalian toxicity and low persistence in the environment (Raja *et al.*, 2001; Papachristos and Stamopoulos, 2002). Plant oils have repellent and insecticidal (Shaaya *et al.*, 1997; Kostyukovsky *et al.*, 2002a; Papachristos and Stamopoulos, 2002), nematicidal (Oka *et al.*, 2000), antifungal (Paster *et al.*, 1995; Srivastava *et al.*, 2009), antibacterial (Matasyoh *et al.*, 2007), virucidal (Schuhmacher *et al.*, 2003), antifeedant and reproduction inhibitory (Raja *et al.*, 2001; Papachristos and Stamopoulos, 2002) effects.

Numerous plant species have been reported to have insecticidal properties capable of controlling insects (Grainge and Ahmed, 1988; Arnason *et al.*, 1989). The toxicity of a large number of essential oils and their constituents has been evaluated against a number of stored-product insects. Essential oils extracted from *Pogostemon heyneanus, Ocimum basilicum* and *Eucaluptus* showed insecticidal activities against *Sitophilus oryzae, Stegobium paniceum, Tribolium castaneum* and *Callosobruchus chinensis* (Deshpande *et al.*, 1974; Deshpande and Tipnis, 1977). Toxic effects of the terpnenoids d-limonene, linalool and terpineol were observed on several coleopterans damaging postharvest products (Karr and Coats, l988; Coats *et al.*, 1991; Weaver *et al.*, 1991). Fumigant toxic activity and reproductive inhibition induced by a number of essential oils and their monoterpenenoids were also evaluated against the bean weevil *Acanthoscelides obtectus* and the moth *Sitotroga cerealella* (Klingauf *et al.*, 1983; Regnault-Roger and Hamraoui, 1995). Our earlier investigations on the effectiveness of the essentials oils extracted from aromatic plants, showed great promise for the control of the major stored-product insects. Several of them were found to be active fumigants at low concentrations against these insects (Shaaya *et al.*, 1991, 1993, 1994).

The use of edible oils as contact insecticides to protect grains, especially legumes, against storage insects is traditional practice in many countries in Asia and Africa. The method is convenient and inexpensive for the protection of stored seeds in households and in small farms. Many different edible oils have been studied as stored grain protectants against insects (Oca *et al.*, 1978; Varma and Pandey, 1978; Pandey *et al.*, 1981; Santos *et al.*, 1981; Messina and Renwick, 1983; Ivbijaro, 1984; Ivbijaro *et al.*, 1984; Pierrard, 1986; Ahmed *et al.*, 1988; Don Pedro, 1989; Pacheco *et al.*, 1995).

This chapter evaluates the efficacy of essential and edible oils as fumigants and contact insecticides to suppress populations of the stored-product insects, mainly the legume pest insect *Callosobruchus maculatus*. The pulse beetle, *Callosobruchus maculatus* F. (Coleoptera: Bruchidae), is one of the major pests of stored cowpea, lentils and green and black gram in the tropics (Sharma, 1984; Raja *et al.*, 2007). The infestation often begins in the field on the dry ripe seeds before harvest. The damage to the seeds can reach up to 50% after 6 months of storage (Caswell, 1980) and up to 90% annually, according to the International Institute of Tropical Agriculture (IITA, 1989).

8.2 Studies with Edible Oils and Fatty Acids: Biological Activity and Repellency

The biological activity of a number of crude and distilled edible oils and a number of straight-chain fatty acids, which contain from C5 to C18, was evaluated in laboratory tests against the common legume pest *C. maculatus*. All the edible oils tested were found to have different degrees of activity at a concentration of 1 g/kg (= 1 kg/ton) chickpea seeds (Table 8.1).

The most active oils were crude oils from rice, maize, cottonseeds and palm. Of the eggs laid, 90–96% did not develop to larvae and only 0–1% developed to F_1 adults in the seeds treated with these oils (Table 8.1). In addition, some of the oils tested were found to prevent oviposition. The most active oil in this regard was the rice crude oil: only 50 eggs were laid on seeds treated with 1 g/kg oil compared to 287 eggs on the control seeds (Table 8.1).

Table 8.1. Biological activity of various edible oils against *Callosobruchus maculatus*.

Oil	Number of eggs laid	Egg mortality (%)	Adult emergence Number	%
Crude rice	50	100	0	0
Crude maize	137	96	1	1
Refined maize	150	97	3	2
Crude cotton seed	212	99	2	1
Refined cotton seed	275	89	22	8
Crude palm	375	99	4	1
Refined palm	237	84	38	16
Crude soya bean	212	85	21	10
Refined soya bean	262	96	10	4
Crude coconut	125	95	6	5
Distilled peanut	150	95	8	5
Distilled safflower	287	97	9	3
Crude olive	175	74	46	26
Refined olive	300	86	18	6
Refined sunflower	187	82	30	16
Distilled kapok	387	88	39	10
Control	287	5	270	94

Each oil was used at a concentration of 1 g/kg chickpea. The required amount of the oil was first mixed with acetone (50 ml/kg seeds) and the acetone was evaporated under a hood. Five males and five females were introduced to 5 g seeds. The data are the average of three experiments, each one was in triplicate.

To obtain an insight into the nature of the activity of the oils, the activity of straight-chain fatty acids ranging from C5 to C18 at a concentration of 4 g/kg was studied against *C. maculatus* (Table 8.2). The results showed that C9–C11 acids were the most active in preventing oviposition at this concentration, with the C11 acid the most active: only 13, 13 and 0 eggs were found on the treated seeds, respectively (Table 8.2).

C12–C16 fatty acids were less effective and activity was remarkably decreased for the lower C5–C7 and higher C17–C18 acids. At lower concentrations of 1.6, 0.8 and 0.4 g/kg, the C11 acid was found to be the most active; fewer eggs were laid than with C9 or C10 and no eggs developed to adults. The data presented in Table 8.3 show clearly that C9–C11 acids are strongly repellent to *C. maculatus*, but they have no lethal effect on the adults.

Field tests using crude palm kernel and rice bran oils showed that both oils were effective in controlling *C. maculatus* infestation providing full protection for the first 4–5 months of storage at a rate of 1.5–3.0 g/kg seeds. They persisted in controlling insect infestation for up to 15 months. The numbers of adult insects found in the treated seeds after 15 months of storage were only about 10% of that in the control samples (results not shown).

8.3 Studies with Essential Oils as Contact Insecticides

The efficacy of a large number of essential oils has been evaluated for the control of *C. maculatus* at a concentration of 400 ppm. The various oils tested were found to have a toxic effect on the eggs laid and on the development of the eggs to adults (Table 8.4).

Table 8.2. Biological activity of straight-chain fatty acids C5–C18 against *Callosobruchus maculatus*.

Fatty acid	Number of eggs laid	Adult emergence	
		Number	%
Pentanoic(C5)	185	110	59
Hexanoic (C6)	130	92	71
Heptanoic (C7)	112	66	59
Octanoic (C8)	105	22	21
Nanonoic (C9)	13	0	0
Decanoic (C10)	13	0	0
Undecanoic (C11)	0	0	0
Dodecanoic (C12)	32	22	69
Tridecanoic (C13)	32	20	62
Tetradecanoic (C14)	62	45	73
Pentadecanoic (C15)	72	56	77
Hexadecanoic (C16)	65	37	57
Heptadecanoic (C17)	120	88	73
Octadecanoic (C18)	140	115	82
Control	287	272	95

Each fatty acid was used at a concentration of 4 g/kg chickpea. The data are the average of three experiments, each of which were in triplicate.

Table 8.3. Toxicology and repellency of the fatty acids C9–C11 tested against *Callosobruchus maculatus*.

Fatty acid	Concentration (g/kg)	Number of eggs laid	Undeveloped eggs (%)	Adult emergence Number	Adult emergence %
C9	1.6	60	100	0	0
	0.8	67	95	3	4
	0.4	100	72	18	18
Control	0	360	3	345	96
C10	1.6	65	100	0	0
	0.8	27	92	1	4
	0.4	62	78	7	11
Control	0	300	5	280	93
C11	1.6	20	100	0	0
	0.8	20	100	0	0
	0.4	22	100	0	0
Control	0	300	3	287	96

The data are the average of three experiments, each of which were in triplicate.

Table 8.4. Contact activity of a number of essential oils on egg laying, egg development and F_1 of *Callosobruchus maculatus*.

Essential oil	Eggs laid Number	Eggs laid % of control value	Eggs developed to larvae (%)	No. of adults F_1	Eggs developed to adults (%)
Syrian marjoram	171	83	87	0	0
Lemon grass	144	70	88	5	3.5
Geranium	119	34	14	8	6.7
Vistria	157	57	77	15	9.6
Basil	51	78	93	23	45
Clary sage	40	22	75	22	55
Orange	67	34	84	45	67
Grapefruit	120	64	77	49	41
Lemon	172	88	91	50	29
Caraway	97	51	79	62	64
Cumin	148	59	74	55	37
Thyme	120	69	68	46	38
Celery	239	179	79	169	71
Thyme leaved savory	98	68	89	75	77
Ruta	109	75	76	51	47
Rosemary	123	68	79	64	32
Peppermint	159	92	83	54	34
SEM76 oil	67	28	93	27	17
Control	150–350	–	85–95	120–280	80–85

Each essential oil was tested at a concentration of 400 ppm and an exposure time of 24 h.
Twenty unsexed adults were introduced to the treated and untreated seeds 24 h after treatment.

In the case of seeds treated with Syrian marjoram, the number of eggs laid on the treated seeds was 83% of the control, but 0% of the eggs developed to adults. Other oils were found to cause a reduction in the number of eggs laid on the treated seeds but were less effective on egg to adult development. In the case of clary sage oil, only 40 eggs were laid on the treated seeds (= 22% of control), but 55% of the eggs developed to adults. In contrast, the essential oil SEM76, obtained from *Labiata* plant species, was found effective on both the reduction of number of eggs laid (28% of control) and on the number of eggs that developed to adults (17% of control) (Table 8.4).

8.4 Studies with Essential Oils as Fumigants Against Stored-product Insects

In order to isolate active essential oils we firstly screened a large number of essential oils extracted from aromatic plants and isolated their main constituents, using space fumigation (Shaaya *et al.*, 1991, 1993, 1994). The most active compounds are summarized in Table 8.5.

Among the tested compounds, the essential oil SEM76 and its main constituent were found to be the most potent fumigants. To obtain the LC_{90} of all insect species tested at adult stage, SEM76 oil was required at the concentration of 0.6–1.2 μl/l air (Table 8.5). The main constituent of the oil, which accounts for approximately 80% of the oil, was also found to have high activity, a little higher than the oil, as expected (Table 8.5). Space fumigation studies with SEM76 against various developmental stages of *C. maculatus* showed that eggs and young larvae before they penetrated into the seeds were the most susceptible to the compound. A concentration of 0.5 μl/l air was enough to cause 100% mortality of the eggs and the first instar larvae. After the larvae penetrated inside the seeds, they became more tolerant (Table 8.6). Only pupae 1–2 days before adult emergence again became sensitive to the compound (Table 8.6).

Table 8.5. Fumigant toxicity of the most active monoterpenes tested on stored-product insect pests in space fumigation tests.

Compound	Oryzaephilus surinamensis LC_{50}	Oryzaephilus surinamensis LC_{90}	Rhizopertha dominica LC_{50}	Rhizopertha dominica LC_{90}	Sitophilus oryzae LC_{50}	Sitophilus oryzae LC_{90}	Tribolium castaneum LC_{50}	Tribolium castaneum LC_{90}
1,8-Cineol	3.1	7.3	2.5	4.0	7.2	14.2	7.5	8.5
Carvacrol	–	–	>15.0	–	>15	–	>15.0	–
Limonene	>15.0	–	6.7	10.3	>15	–	7.6	8.6
Linalool	3.0	6.0	6.0	8.5	10.1	19.8	>15	–
Pulegone	1.7	2.8	2.8	4.5	0.7	1.4	2.5	3.2
SEM-76	–	<1.0	0.4	0.6	–	<0.5	0.7	1.2
SEM-76 oil	–	<1.0	0.5	0.8	–	<1.0	0.9	1.4
Terpinen-4-ol	3.9	11.4	1.3	2.0	1.2	5.2	2.5	3.3
Terpineol	4.3	12.7	>15.0	–	>15	–	>15	–

LC_{50} and LC_{90} values are μl/l air. Exposure time was 24 h.

Table 8.6. Fumigant activity of the essential oil SEM76 extracted from Labiatae species on various developmental stages of *Callosobruchus maculatus*.

Stage treated	Age	Concentration (µl/l)	No. of eggs	F_1	
				No. of adults	Eggs developed to adults (%)
Egg	20–24 h	0.5	20	0	0
Larvae outside the seed	0–1 day	0.5	20	0	0
Larvae inside the seed	2 days	1.0	20	3.5	17
		1.5	20	4.5	22
Larvae	3 days	1.5	20	10.5	53
		3.0	20	8	40
Larvae	7 days	1.5	20	14	70
		3.0	20	13	63
Larvae	11 days	1.5	20	17.5	88
		3.0	20	16.5	83
Pupae	4–5 days before emergence	1.5	20	12	60
		3.0	20	11	55
Pupae	1–2 days before emergence	0.5	20	4	20
		1.5	20	1	5
		3.0	20	0	0
Control		0	20	17	85

Twenty unsexed adults were used for each test. The data are an average of three duplicate experiments. Exposure time was 24 h.

8.5 Distribution of α-Terpineol in the Fumigation Chamber

Most of the research to study the potential of essential oils and their constituents (terpenes) for the control of insect pests was done using space fumigation, without paying attention to the amount of the fumigant available to the treated insects. As a model, we studied the distribution of the terpenoid α-terpineol in the fumigation chamber among the air in the chamber space, filter paper and the flask walls. Using concentrations of 3, 5, 10 and 15 mg/l air and stirring at 20°C, only minor differences in the amount of the terpenoid in the chamber space was measured: 0.710, 0.676, 0.782 and 0.897 mg/l air, respectively. The rest was found on the filter paper and the flask walls (Table 8.7). At a higher temperature of 25°C, a higher amount of the terpenoid, 50–60%, was recovered in the chamber space compared to the amount recovered at 20°C (Table 8.7). No pronounced change of the terpenoid in the chamber space was measured at a different intensity of stirring or when no stirring was applied (Table 8.8). It should be mentioned that stirring caused a higher concentration of the terpenoid on the glass walls compared to the filter paper, and the opposite if no stirring was applied (Table 8.8). Using stirring at 20°C, we could show that after 2 h of fumigation the air in the fumigation chamber was already saturated (Table 8.9).

Table 8.7. The distribution of α-terpineol in the chamber space, filter paper and the fumigation chamber walls at different temperatures, using stirring.

Concentration used (mg/l air)	Temperature (°C)	Amount of α- terpineol in the chamber space (mg/l air)	Total of α- terpineol recovered (%)	Amount recovered (%)		
				Chamber space	Filter paper	Flask walls
3	20	0.710	85.0	27.8	10.7	61.6
			(2.682)	(0.745)	(0.287)	(1.65)
5	20	0.676	81	16.6	18.9	64.5
			(4.265)	(0.710)	(0.805)	(2.75)
5	25	1.106	84	26.4	6.1	67.4
			(4.366)	(1.16)	(0.266)	(2.96)
10	20	0.782	105	7.4	41.1	51.5
			(10.932)	(0.820)	(4.54)	(5.68)
10	25	1.165	91	12.8	32.8	54.4
			(9.53)	(1.22)	(3.13)	(5.18)
15	20	0.897	112	5.2	62	33
			(17.591)	(0.921)	(10.89)	(5.78)

The numbers in brackets are the value of α-terpineol in mg, recovered after 24 h of space fumigation. The volume of the fumigation chamber is 1050 ml. The sensitivity of the GC measurements was ± 15%.

Table 8.8. Effect of stirring on the distribution of α-terpineol in the chamber space, filter paper and fumigation chamber walls.

Concentration used (mg/l) air	Stirring time and intensity	Amount of α- terpineol in the chamber space (mg/l air)	Total of α- terpineol recovered (%)	Amount recovered (%)		
				Chamber space	Filter paper	Flask walls
5	24 h moderate	0.965	85.0	22.7	6.7	71
			(4.549)	(1.010)	(0.278)	(3.160)
	24 h slow	0.856	93	16.6	18.9	64.5
			(4.265)	(0.710)	(0.805)	(2.750)
	15 min slow	0.800	99	16.2	63.5	20.4
			(5.200)	(0.840)	(3.30)	(1.060)
	0	0.858	103	16.9	76.7	6.4
			(5.320)	(0.901)	(4.080)	(0.339)
15	24 h moderate	0.684	107	4.3	51.5	44.3
			(16.898)	(0.718)	(8.70)	(7.480)
	0	0.639	111	3.8	90.9	5.3
			(17.490)	(0.671)	(15.90)	(0.919)

The numbers in brackets are the value of α-terpineol in mg, recovered after 24 h of space fumigation. The volume of the fumigation chamber was 1050 ml. The fumigation was conducted at a temperature of 20°C. The sensitivity of the GC measurements was ± 15%.

Table 8.9. Amount of α-terpineol measured in the fumigation chamber space at various time intervals of fumigation, using stirring.

Concentration used (mg/l) air	The amount of α-terpineol (mg) measured in the fumigation chamber space, hours following fumigation			
	2	4	6	24
3	0.685	0.678	0.865	0.731
5	0.877	0.804	–	0.701
10	0.900	0.922	–	0.821
15	0.838	0.769	0.767	0.857

The volume of the fumigation chamber was 1050 ml.
The fumigation was conducted at a temperature of 20°C.

8.6 Possible Mode of Action of Essential Oils

The essential oil SEM76 and pulegone were found to be highly toxic against all stored-product insects tested (Shaaya *et al.*, 2001, 2002). The effect of these two compounds on acetylcholinesterase and the octopamine systems in insects was studied in order to elucidate their mode of action.

In our studies with acetylcholinesterase (Greenberg-Levy *et al.*, 1993; Kostyukovsky *et al.*, 2002b), using acetylcholinesterase extracted from *Rhizopertha dominica*, a stored-product insect, we showed that the inhibitory activity of these two terpenoids on acetylcholinesterase was only evident when high concentrations of these compounds (10^{-3} M) were applied. These doses were at such high levels that they cannot account for the toxic effects observed *in vivo* by these compounds on the same insect species (Shaaya *et al.*, 2001, 2002), which were obtained at much lower concentration (10^{-6} M = 1.5 μl/1 of air). The failure of these biologically active compounds to produce a stronger enzyme inhibition indicated that acetylcholinesterase was probably not the main site of action of essential oils.

Another possible target for neurotoxicity of essential oils is the octopaminergic system in insects. Octopamine is a multifunctional, naturally occurring biogenic amine that plays a key role as a neurotransmitter, neurohormone and neuromodulator in invertebrate systems (Evans, 1981; Fig. 8.1), with a physiological role analogous to that of norepinephrine in vertebrates.

The octopaminergic system in insects represents a biorational target for insecticidal action and has been targeted by various insecticides in the past, e.g., formamidines (Haynes, 1988; Perry *et al.*, 1998). Many of the physiological functions of octopamine appear to be mediated by a class of octopamine receptors, specifically linked to a protein coupled to the enzyme adenylate cyclase. These physiological actions have been shown to be associated with elevated levels of cyclic-AMP (Evans, 1984). Using an insect *in-vitro* cuticular tissue preparation, octopamine was found to induce a significant increase in the intracellular messenger cyclic AMP (Rafaeli and Gileadi, 1995). SEM76 and pulegone directly affect the intracellular response of these tissues by mimicking the octopamine response, thereby increasing the production of

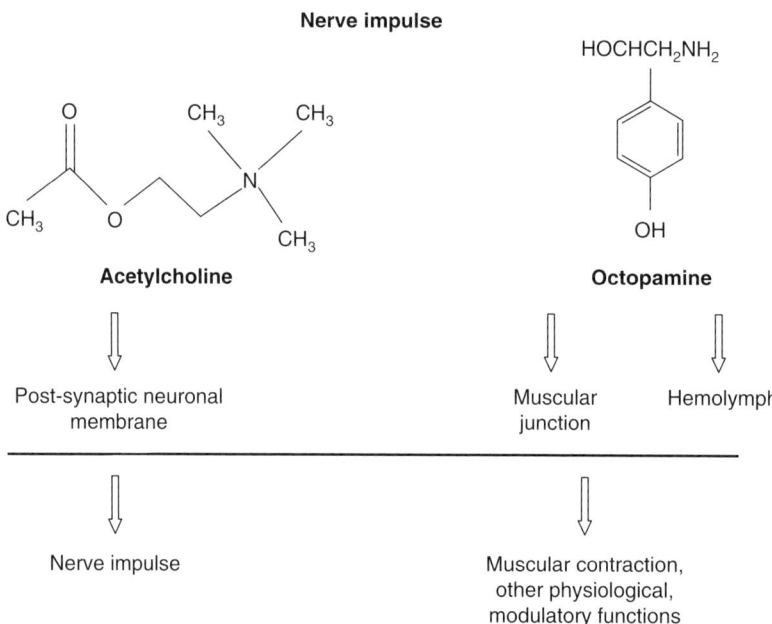

Fig. 8.1. Neurotransmitters in insect systems as possible sites of action of essential oil toxicity.

Table 8.10. Effect of essential oils in the absence and presence of the octopamine antagonist, phentolamine (10^{-5} M), on intracellular cyclic-AMP levels.

Treatment	Concentration (M)	Intracellular cyclic – AMP levels pmol/abdominal segment			
		Octopamine	SEM76	Pulegone	d-limonene
Without	0			0.2	
phentolamine	10^{-8}	0.39	0.027	0.023	–
	10^{-7}	0.58	0.42	0.88	0.02
With	0			0.14	
phentolamine	10^{-8}	0.18	0.17	–	–
	10^{-7}	0.05	0.17	0.2	–
	10^{-6}	0.11	0.01	–	–

cyclic-AMP at low physiological concentrations of 10^{-8} M (Table 8.10). Moreover, addition of the octopamine antagonist phentolamine strongly antagonized the response to octopamine, as well as to the essential oil SEM76 and pulegone, were strongly antagonized and no elevation in cyclic-AMP was observed (Shaaya, *et al.*, 2001; Kostyukovsky *et al.*, 2002b). These effects were observed at low dilutions of the essential oils (estimated at 10^{-7} and 10^{-8} M), levels that induced the overt behavioural toxicity responses *in vitro* (Table 8.10; Kostyukovsky *et al.*, 2002b). It should be mentioned that d-limonene, which was found to have very low toxicity to stored-product insects, did not show any effect on the production of cyclic-AMP (Table 8.10).

Because the essential oil response is strongly insect specific, as is octopamine neurotransmission, the essential oils can mimic the action of octopamine at low doses, and phentolamine, an octopamine inhibitor, has an inhibitory action on the essential oils, we can postulate that the essential oils may affect octopamonergic target sites.

8.7 Discussion

Studies with edible oils and fatty acids showed that crude oils of rice, maize, cottonseed and palm were found the most potent against *C. maculatus* compared to the other oils tested. The purified forms of these oils were found to be less effective. In field studies, rice and palm crude oils at concentrations of 1.5 and 3 g/kg, protected chickpeas completely from insect infestation for a period of 4–5 months and partially for up to 15 months. Studies by Khaire *et al.* (1992) showed that adult emergence of *C. chinensis* was completely prevented for up to 100 days in pigeonpea treated with 1% neem oil or karanj oil. In addition, theses oils were found to have no adverse effect on seed germination. Boeke *et al.* (2004) showed that a number of botanical products might provide effective control of *C. maculatus* in cowpea.

The mode of action of edible oils was first attributed to interference with normal respiration, resulting in suffocation of the insects. The action of the oil, however, is more complex, since insects deprived of oxygen survived longer than those treated with oils (Gunther and Jeppson, 1960). From this study and others (Wigglesworth, 1942; Ebeling and Wagner, 1959; Ebeling, 1976; Shaaya and Ikan, 1979), it may be postulated that the biological activity of the edible oils is attributed to both their physical and chemical properties such as viscosity, volatility, specific gravity and hydrophobicity.

Among the straight-chain fatty acids ranging from C5 to C18 carbon atoms, it was found that C9–C11 acids were the most effective in preventing oviposition of *C. maculatus* on the treated seeds, but have no lethal effect on the adults. In earlier studies (Shaaya *el al.*, 1976), we showed that wheat seeds treated with C10 fatty acid, at a concentration of 4 g/kg, repelled *Sitophilus oryzae* L, but that forced contact of the beetles with the treated seeds was found to have no effect on mortality rates of the beetles (Shaaya *et al.*, 1976). It should be mentioned that *C. maculatus* adults do not consume food and live approximately 1 week only, whereas *S. oryzae* adults feed as long as they live – several months. We postulate, therefore, that fatty acids act as repellents and the beetles die because of starvation.

Essential oils as contact insecticide were found to have low toxicity. A high concentration of 400 ppm of the most active essential oil from Syrian marjoram was needed to prevent the development from egg to adult. On the contrary, essential oils and their constituents as fumigants showed high activity in controlling young larvae and pupae prior to emergence of *C. maculatus*. The contact and fumigant toxicity of five essential oils, cardamom, cinnamon, clove, eucalyptus and neem were also investigated against the cowpea weevil (Mahfuz and Khalequzzaman, 2007). In space fumigation, at a concentration of 1 µl/l air, *Ocimum gratissimum* oil and its constituent eugenol caused 100% mortality of *C. chinensis* 24 h after treatment (Ogendo *et al.*, 2008). The essential oils of citrodora and lemongrass were found to have insecticidal and ovicidal activities against adults and eggs of *C. maculatus* (Raja and William, 2008). Van Huis (1991), Adabie-Gomez *et al.* (2006), Henning (2007), Boateng and Kusi (2008) have reported the efficacy of *Jatropha* seed oil as contact insecticide against *C. maculatus*. Similarly, Kéita *et al.* (2001) reported that kaolin powder aromatized with pure oil of *Ocimum* spp. at 6.7 µl/g of pea seeds provided complete protection over 3 months of storage.

In space fumigation studies to evaluate the potency of the essential oils and their constituents against insect pests, special attention has to be given to not using a higher amount of the fumigant than needed to saturate the air in the fumigation chamber at constant temperature. It should be mentioned that the amount of the various test materials needed to saturate the air in the fumigation chamber might be different, depending on the physical and chemical properties of the fumigant. Our findings, as well as results of other researchers, suggest that certain essential oils, and particularly their terpenoids, are highly selective to insects, since they are probably targeted to an insect-selective octopaminergic receptor, which is a non-mammalian target. The worldwide availability of plant essential oils and their terpenoids, and their use as flavouring agents in food and beverages is a good indication of their relative safety to humans. In our earlier studies, we showed that essential oils have a high potential as alternative fumigants to methyl bromide for the control of stored-product insects and cut flowers (Shaaya, 1998; Wilson and Shaaya, 1999; Kostyukovsky and Shaaya, 2001; Kostyukovsky *et al.*, 2002). The reduction in the use of conventional synthetic toxic pesticides gives to these bioactive chemicals the potential to be used as a complementary or alternative method in crop production and integrated pest management.

References

Adabie-Gomez, D.A., Monford, K.G., Agyir-Yawson, A., Owusu-Biney, A. and Osae, M. (2006) Evaluation of four local plant species for insecticidal activity against *Sithophilus zeamais* Motsch. (Coleoptera: Curculionidae) and *Callosobruchus maculatus* F. (Coleoptera: Bruchidae). *Ghana Journal of Agricultural Science* 39, 147–154.

Ahmed, K., Khalique, F., Afzal, M., Malik, B.A. and Malik, M.R. (1988) Efficacy of vegetable oils for protection of greengram from attack of bruchid beetle. *Pakistan Journal of Agricultural Research* 9, 413–416.

Arnason, E.T., Philogene, B.J.R. and Morand, P. (1989) *Insecticides of Plant Origin*. American Chemical Society Symposium series 387, 213.

Boateng, B.A. and Kusi, F. (2008) Toxicity of *Jatropha* seed oil to *Callosobruchus maculatus* (Coleoptera: Bruchidae) and its parasitoid, *Dinarmus basalis* (Hymenoptera: Pteromalidae). *Journal of Applied Sciences Research* 4, 945–951.

Boeke, S.J., Baumgart, I.R., Van Loon, J.J.A., van Huis, A., Dicke, M. and Kossou, D.K. (2004) Toxicity and repellence of African plants traditionally used for the protection of stored cowpea against *Callosobruchus maculatus*. *Journal of Stored Products Research* 40, 423–438.

Caswell, G.H. (1980) A review of the work done in the Entomology section of the Institute for Agricultural Research on the Pests of Stored Grain. Samaru Misc. Paper 99, Zaria, Nigeria.

Coats, J.R., Karr, L.L. and Drewes, C.D. (1991) Toxicity and neurotoxic effects of monoterpenoids in insects and earthworms. In: Heiden, P.A. (ed.) *Naturally Occurring Pest Bioregulators*. American Chemical Society Symposium series, 449, pp. 305–316.

Deshpande, R.S. and Tipnis, H.P. (1977) Insecticidal activity of *Ocimum basilicum* Linn. *Pesticides* 11, 11–12.

Deshpande, R.S., Adhikary, P.R. and Tipnis, H.P. (1974) Stored grain pest control agents from *Nigella sativa* and *Pogostemon heyneanus*. *Bulletin of Grain Technology* 12, 232–234.

Don Pedro, K.N. (1989) Mechanisms of action of some vegetable oils against *Sitophilus zeamais* Motsch. (Coleoptera: Curculionidae) on wheat. *Journal of Stored Products Research* 25, 217–223.

Ebeling, W. (1976) Insect integument: a vulnerable organs system. In: Hepburn, H.R. (ed.) *The Insect Integument*. Elsevier, Amsterdam, pp. 383–399.

Ebeling, W. and Wagner, R.E. (1959). Rapid desiccation of drywood termites with inert sorptive dusts and other substances. *Journal of Economic Entomology* 52, 190–207.

Evans, P.D. (1981) Multiple receptor types for octopamine in the locust. *Journal of Physiology* (London) 318, 99–122.

Evans, P.D. (1984) A modulatory octopaminergic neurone increases cyclic nucleotide levels in locust skeletal muscle. *Journal of Physiology* (London) 348, 307–324.

Grainge, M. and Ahmed, S. (1988) *Handbook of Plants with Pest-control Properties*. Wiley and Sons, New York, USA.

Greenberg-Levy, S., Kostjukovysky, M., Ravid, U. and Shaaya, E. (1993) Studies to elucidate the effect of monoterpenes on acetycholinesterase in two stored-product insects. *Acta Horticulturae* 344, 138–146.

Gunther, F.A. and Jeppson, L.R. (1960) *Modern Insecticides and World Food Production*, Wiley, New York, pp. 284.

Haynes, K.F. (1988) Sublethal effects of neurotoxic insecticides on insect behaviour. *Annual Review of Entomology* 33, 148–168.

Henning, R.K. (2007) *Jatropha curcas* L. [Internet] Record from Protabase. van der Vossen, H.A.M. and Mkamilo, G.S. (eds). PROTA (Plant Resources of Tropical Africa), Wageningen, Netherlands, available at http://database.prota.org/search.htm.

IITA, 1989 International Institute of Tropical Agriculture. *Annual Report 1988/89*. Ibadan, Nigeria.

Ivbijaro, M.F. (1984) Toxic effects of groundnut oil on the rice weevil *Sithophilus oryzae* (L.). *Insect Science and its Application* 5, 251–252.

Ivbijaro, M.F., Ligan, C. and Youdeowei, A. (1984) Comparative effects of vegetable oils as protectants of maize from damage of the rice weevil *Sithophilus oryzae* (L.). *Proceedings of 17th International Congress of Entomology*, Hamburg, Germany, p. 643.

Karr, L. and Coats, R.J. (1988) Insecticidal properties of d-limonene. *Journal of Pesticide Science* 13, 287–290.

Keita, S.M., Vincent, C., Schmit, J.P. Arnason, J.T. and Bélanger, A. (2001) Efficacy of essential oil of *Ocimum basilicum* L. and *O. gratissimum* L. applied as an insecticidal fumigant and powder to control *Callosobruchus maculatus* (Fab.) (Coleoptera: Bruchidae). *Journal of Stored Products Research* 37, 339–349.

Khaire, V.M., Kachare, B.V. and Mote, U.N. (1992) Efficacy of different vegetable oils as grains protectants against pulse beetle *Callosobruchus chinensis* L. in increasing the storability of pigeonpea. *Journal of Stored Products Research* 28, 153–156.

Klingauf, F., Bestmann, H.J., Vostrowsky, O. and Michaelis, K. (1983) Wirkung von altherishcen Olen auf Schadinsekten. *Mitteilung Deutsche Gesselschaft fuer*

Allgemeine und Angewandte Entomologie 4, 123–126.

Kostyukovsky, M. and Shaaya, E. (2001) Quarantine treatment on cut flowers by natural fumigants. *Proceedings of the 7th International Conference on Controlled Atmosphere and Fumigation in Stored Products 29 October to 3 November 2000.* Fresno, USA, pp. 821–827.

Kostyukovsky, M., Ravid, U. and Shaaya, E. (2002a) The potential use of plant volatiles for the control of stored product insects and quarantine pests in cut flowers. *Acta Horticulturae* 576, 347–358.

Kostyukovsky, M., Rafaeli, A., Gileadi, C., Demchenko, N. and Shaaya E. (2002b) Activation of octopaminergic receptors by essential oil constituents isolated from aromatic plants: possible mode of activity against insect pests. *Pest Management Science* 58, 1–6.

Mahfuz, I. and Khalequzzaman, M. (2007) Contact and Fumigant Toxicity of Essential Oils against *Callosobruchus maculatus*. *University Journal of Zoology. Rajshahi University* 26, 63–66.

Matasyoh, L.G., Matasyoh, J.C., Wachira, F.N., Kinyua, M.G., Muigai, A.T.M. and Mukiama, T.K. (2007) Chemical composition and antimicrobial activity of the essential oil of *Ocimum gratissimum* L. growing in eastern Kenya. *African Journal of Biotechnology* 6, 760–765.

Messina, F.J. and Renwick, J.A.A. (1983) Effectiveness of oils in protecting stored cowpea from the cowpea weevil (Coleoptera: Bruchidae). *Journal of Economic Entomology* 76, 634–635.

Mills, K.A. (1983) Resistance to the fumigant hydrogen phosphine in some stored-product species associated with repeated inadequate treatments. *Mitteilung Deutsche Gesselschaft fuer Allgemine Angewandle Entomologie* 4, 98–101.

Nakakita, H. and Winks, R.G. (1981) Phosphine resistance in immature stages of a laboratory selected strains of *Tribolium castaneum* (Herbst.). *Journal of Stored Products Research* 17, 43–52.

Oca, G.M., Garcia, F. and Schoonhoven, A.V. (1978) Efecto de cuatro aceites vegetales sobre *Sithophilus oryzae* y *Sitotroga cerealella* in maiz, sorgo y trigo almacenados. *Revista Colombiana de Entomología* 4, 45–49.

Ogendo, J.O., Belmain, S.R., Deng, A.L. and Walker, D.J. (2003) Comparison of toxic and repellent effects of *Lantana camara* L. with *Tephrosia vogelii* Hook and a synthetic pesticide against *Sitophilus zeamais* Motschulsky in maize grain storage. *Insect Science and Its Application* 23, 127–135.

Ogendo, J.O., Kostyukovsky, M., Ravid, U., Matasyoh, J.C., Deng, A.L., Omolo, E.O., Kariuki, S.T. and Shaaya, E. (2008) Bioactivity of *Ocimum gratissimum* L. oil and two of its constituents against five insect pests attacking stored food products. *Journal of Stored Products Research 44*, 328–334.

Oka, Y., Nacar, S., Putievsky, E., Ravid, U., Yaniv, Z. and Spiegel, Y. (2000) Nematicidal activity of essential oils and their constituents against the root-knot nematode. *Phytopathology* 90, 710–715.

Pacheco, A.I., De Castro, F., De Paula, D., Lourencao, A., Bolonhezi, S.and Barbieri, K.M. (1995) Effecacy of soybean and caster oils in the control of *Callosobruchus maculatus* (F.) and *Callosobruchus phaseoli* (Gyllenhal) in stored chickpeas (*Cicer arietinum* L.). *Journal of Stored Products Research* 19, 57–62.

Pandey, G.P., Doharey, R.B. and Varma, B.K. (1981) Efficacy of some vegetable oils for protecting greengram against the attack of *Callosobruchus maculatus* (Fabr.). *Indian Journal of Agricaltural Science* 51, 910–912.

Papachristos, D.P. and Stamopoulos, D.C. (2002) Repellent, toxic and reproduction inhibitory effects of essential oil vapours on *Acanthoscelides obtectus* (Say) (Coleoptera: Bruchidae). *Journal of Stored Products Research* 38, 117–128.

Paster, N., Menasherov, M., Ravid, U. and Juven, B. (1995) Antifungal activity of *Oregano* and *Thyme* essential oils applied as fumigants against fungi attacking stored grain. *Journal of Food Protection* 58, 81–85.

Perry, A.S., Yamamoto, I., Ishaaya, I. and Perry, R. (1998) *Insecticides in Agriculture and Environment - Retrospects and Prospects.*

Springer, Berlin, Heidelberg, New York, pp. 261.

Pierrard, G. (1986) Control of the cowpea weevil, *Callosobruchus maculatus* (F), at the farmer level in Senegal. *Tropical Pest Management* 32, 197–200.

Rafaeli, A. and Gileadi, C. (1995) Modulation of the PBAN-stimulated pheromonotropic activity in *Helicoverpa armigera*. *Insect Biochemistry and Molecular Biology* 25, 827–834.

Raja, M. and JohnWilliam, S. (2008) Impact of volatile oils of plants against the Cowpea Beetle *Callosobruchus maculatus* (Fab.) (Coleoptera: Bruchidae). *International Journal of Integrative Biology* 2, 62.

Raja, N., Albert, S., Ignacimuthu, S. and Dorn, S. (2001) Effect of plant volatile oils in protecting stored cowpea *Vigna unguiculata* (L.) Walpers against *Callosobruchus maculatus* (F.) (Coleoptera: Bruchdae) infestation. *Journal of Stored Products Research* 37, 127–132.

Raja, M., John William, S. and Jayakumar, M. (2007) Repellent activity of plant extracts against pulse beetle *Callosobruchus maculatus* (Fab.) (Coleoptera: Bruchidae). *Hexapoda* 14, 142–145.

Regnault-Roger, C. and Hamraoui, A. (1995) Fumigant toxic activity and reproductive inhibition induced by monoterpenes upon *Acanthoscelides obtectus* Say (Coleoptera), bruchid of kidney bean (*Phaseolus vulgaris* L). *Journal of Stored Products Research* 31, 291–299.

Santos, J.H.R., Beleza, M.G.S, and Silva, N.L. (1981) A mortalidade do *Callosobruchus maculatus* (F) em graos de *Vigna sinensis*, tratados com oleo de algodao. *Revista Ciência Agronômica* 12, 45–48.

Schuhmacher, A., Reichling, J. and Schnitzler, P. (2003) Virucidal effect of peppermint oil on the enveloped viruses herpes simplex virus type 1 and type 2 *in vitro*. *Phytomedicine* 10, 504–510.

Shaaya E. (1998) Phyto-oils control insects in stored products and cut flowers. Methyl Bromide Alternatives Newsletter, USDA 4, 6–7.

Shaaya, E. and Ikan, R. (1979) Insect control using natural products. In: Gessbuhler, H. (ed.) *Advances in Pesticide Science, part 2*. Pergamon Press. Oxford, pp. 303–306.

Shaaya, E., Grossman, G, and Ikan, R. (1976) The effect of straight chain fatty acids on growth of *Sithophilus oryzae*. *Israel Journal of Entomology* 11, 81–91.

Shaaya, E., Paster, N., Juven, B., Zisman, U. and Pisarev V. (1991) Fumigant toxicity of essential oils against four major stored-product insects. *Journal of Chemical Ecology* 17, 499–504.

Shaaya, E., Ravid, U., Paster, N., Kostjukovsky, M. and Plotkin, S. (1993) Essential oils and their components as active fumigant against several species of stored-product insects and fungi. *Acta Horticulturae* 344, 131–137.

Shaaya, E., Kostjukovsky, M. and Ravid, U. (1994) Essential oils and their constituents as effective fumigants against stored-product insects. *Israel Agrisearch* 7, 133–139.

Shaaya, E., Kostjukovsky, M., Eilberg, J. and Sukprakarn, C. (1997) Plant oils as fumigants and contact insecticides for the control of stored-product insects. *Journal of Stored Products Research* 33, 7–15.

Shaaya, E., Kostyukovsky, M., Ravid, U. and Rafaeli, A. (2001) The activity and mode of action of plant volatiles for the control of stored product insects and quarantine pests in cut flowers. In: Özgüven, M. (ed.) *Proceedings of a Workshop on Agricultural and Quality Aspects of Medicinal and Aromatic Plants*, Adana, Turkey, pp. 65–78.

Shaaya, E., Kostjukovsky, M. and Rafaeli, A. (2002) Phyto-chemicals for controlling insect pests. Abstract of paper presented at the Second-Israel-Japan Workshop: *Ecologically sound new plant protection technologies*. The Japan Israel Binational Committee for Plant Protection, Tokyo, Japan. *Phytoparasitica* 30, pp. 2.

Shaaya, E. and Kostyukovsky, M. (2006) Essential oils: potency against stored product insects and mode of action. *Stewart Postharvest Review* 2, 4.

Sharma, S.S. (1984) Review of literature of the loss caused by *Callosobruchus* spp. (Coleoptera: Bruchidae) during storage of pulses. *Bulletin of Grain Technology* 22, 62–71.

Srivastava, B., Singh, P., Srivastava, A.K., Shukla, R. and Dubey, N.K. (2009) Efficacy of *Artabotrys odoratissimus* oil as a plant based antimicrobial against storage fungi and aflatoxin B1 secretion. *International Journal of Food Science and Technology* 44, 1909–1915.

Tyler, P.S., Taylor, R.W. and Rees, D.P. (1983) Insect resistance to phosphine fumigation in food warehouses in Bangladesh. *International Pest Control* 25, 10–13.

Van Huis, A. (1991) Biological methods of bruchid control in the tropics: a review. *Insect Science and its Application* 12, 87–102.

Varma, B.K. and Pandey, G.P. (1978) Treatment of stored green gram seed with edible oils for protection from *Callosobruchus maculatus* (Fabr.). *Indian Journal of Agricultural Science* 48, 72–75.

Weaver, D.K., Dunkel, F.V., Ntezurubaza, L., Jackson, L.L. and Stock, D.T. (1991) The efficacy of linalool, a major component of freshly milled *Ocimum canum* Sims *(Lamiaceae)* for protection against postharvest damage by certain stored product Coleoptera. *Journal of Stored Products Research* 27, 213–220.

Wigglesworth, V.B. (1942) Some notes on the integument of insects in relation to the entry of contact insecticides. *Bulletin of Entomological Research* 33, 205–218.

Wilson, L. and Shaaya, E. (1999) Natural plant extracts might sub for methyl bromide. *Agricultural Research* 47, 14–15.

WMO (1995) Scientific assessment of ozone depletion: 1994. *World Meteorological Organization Global Ozone Research and Monitoring Projects. Report no. 37,* WMO, Geneva, Switzerland.

9 Prospects of Large-scale Use of Natural Products as Alternatives to Synthetic Pesticides in Developing Countries

D.B. Olufolaji

Department of Crop, Soil and Pest Management, The Federal University of Technology, Akure, Nigeria

Abstract

The incidence of pests and diseases on crops in developing countries has caused a lot of food insecurity and negatively affected the agro-allied industries. The use of synthetic pesticides has been the only means of combating the pests and diseases to achieve large-scale food production in developing countries due to lack of appropriate technology to formulate their own pest control measures. The use of these chemicals is associated with a lot of problems. The problems of mammalian toxicity, pollution of the environment and high cost of the few available chemical pesticides are enormous. The Environmental Protection Agency (EPA) worldwide reacts against these problems and encourages research into a better alternative to these synthetic pesticides. The natural products as alternatives are available at all times of the year due to their location in the tropical regions, they are easy to formulate, are not toxic to mammals, are less costly and are environmentally friendly. These attributes assist in the production of cheap and safe food, so the botanicals are assets to developing countries and encourage their large-scale usage. Extracts from plants such as *Azadirachta indica,* and *Ocimum* spp. are used for the control of various crop pests such as cowpea beetles (*Callosobruchus maculatus*), and diseases such as wet rot of *Amaranthus* spp. (*Choanephora curcubitarum*) and anthracnose of soybean (*Colletotrichum truncatum*).

9.1 Introduction

About one third of the world's agricultural produce in terms of plants and animals are destroyed by a combination of both pests and diseases. In Africa, where most of the developing countries are located, most of the food crops are attacked by one pest or another, thereby causing high crop loss (Table 9.1; Ogbalu, 2009).

Table 9.1. Estimate of crop losses due to insect attacks in some African countries.

Crop	Country	Pests	Estimated crop loss (%)
Cereals			
Sorghum	Chad	*Busseola fusca*	54
	Nigeria	*Busseola fusca*	47
	Ethiopia	Stemborers	25
Rice	Nigeria	Stemborers / weevils	35–60
	Ghana	Stemborers	25–30
	Chad	Stemborers	35–55
Maize	Ethiopia	Stemborers	32–55
		Weevils	48–55
	Nigeria	Weevils	68
		Maruca testulalis	43–68
		Busseola fusca	24–55
Maize tassels	Nigeria	Thrips	55
	Ethiopia	Stemborers	37
	Cameroun	Stemborers	24
Legumes			
Cowpea	Nigeria	Weevils	30–58
		Pod borers	60–72
	Swaziland	Pod borers	20–48
	Ethiopia	Hemipterans	38–57
	Kenya	*Ophiomyia phaseoli*	28–43
	Chad	Pod borers	48–54
Roots and tubers			
Cocoyam	Nigeria	Scale insects	34
Cassava	Nigeria	*Zonocerus variegatus*	28–31
	Ghana	*Zonocerus variegatus*	15–27
Yam	Chad	Yam beetles	24–48
	Nigeria	Yam beetles	54
Sweet potatoes	Nigeria	*Cylas puncticollis*	67–76
Vegetables			
Tomatoes	Nigeria	Dipterous pests	45–90
Peppers	Nigeria	Dipterous / bugs	35–53
Okra	Nigeria	Flea beetles	45–60
Bitterleaf	Nigeria	*Lixus cameranus*	54–68
		Diacrisia maculosa	55–87
Fluted pumpkin	Nigeria	*Z. variegatus*	55–82
		D. maculosa	

Source: Ogbalu, 2009.

Knowledge of natural products or botanicals as pesticides is as old as agriculture and local usage of the plants for both medicinal and pesticidal purposes also predates history. However, the advent of chemical pesticides in the mid-20th century has relegated the use of botanicals to the background, the striking negative side effect of these synthetic pesticides not withstanding (Lale, 2007). Until now the synthetic pesticides still record very formidable success in their control attributes on both pests and diseases in plants. The discovery of residues of synthetic pesticides in the food chain and their effect on the environment have led to agitation by environmentalists, ecologists and the Environmental Protection Agencies (EPA) worldwide to look inward for a better alternative in order to produce safe food for everybody, and maintain a friendly environment. The natural products which can also be referred to as botanicals and their products have and will continue to play important roles in alleviating human suffering and also control the losses of farm products experienced in most developing countries (Edeoga and Eriata, 2001; Aworinde *et al.,* 2008). Crop production and protection are being supported by the use of botanicals in developing countries in Africa and some third-world countries (Lale, 2007; Isman 2008). Researchers in Nigeria and some other African countries have investigated several plant species for their efficacy as pesticides for protecting their crops against pest degradation especially with the storage pests and diseases (Boeke *et al.,* 2001; Ofuya *et al.,* 2007; Olufolaji, 2008).

Until recently, conventional pesticides had been used to curb these pests and, in most cases, quite efficiently. However, the technicalities involved in the formulation, standardization and distribution of the botanicals as pesticides to reach the teeming population in the developing countries still remain problems for their usage (Lale, 2007). There are the attendant problems of resistance, environmental pollution, and adverse effect on health and climate change traceable in part to the high levels of chloro-fluoro-carbon being used as carriers in aerosol pesticides, which is affecting the ozone layer. This led to caution in the continued use of persistent synthetic pesticides, especially those in the class of chlorinated hydrocarbons and the highly toxic organophosphorus pesticides (Olaifa, 2009).

9.2 History of Natural Products in the Developing Countries and their Use by Local Farming Communities

Developing countries have been endowed by nature with large quantities and diverse collections of many such natural products, which are more of plant origin (botanicals). The population of inhabitants in developing countries is fed by farm produce from the local peasant farmers. Thus the major users of botanical pesticides are the local farmers who are less educated, very poor and most of them cannot afford the expensive synthetic pesticides.

The tropical forests, characteristic of most developing countries, are the abode of naturally occurring genera and species of most of these botanicals. Over centuries, evolution of these botanicals has occurred to produce various biotypes that are in good use today as medicines and pesticides (Table 9.2). A survey of the use of the botanicals in local farms revealed that around 50% of about 500 peasant farmers in India and Pakistan who stored food grains for more than 6 months used neem (*Azadirachta indica*) leaves to get rid of storage pests, whereas the relatively affluent and better educated farmers used neem cake for nematode control (Ahmed and Stoll, 1996). These farmers buy neem cake in the market because they find no synthetic pesticides to be as equally potent, cheap and readily available. One of the few reported cases of the use of the botanical pesticidal materials for pest management in Nigeria is the application of chilli pepper (*Capsicum* spp.) and tobacco (*Nicotiana tabacum*) leaf extracts by some farmers in the south western states of Nigeria (Nworgu, 2006), and the use of *Chromolaena odorata* to solve some pest and disease problems e.g. anthracnose of soybean caused by *Colletotrichum truncatum* has also been reported (Ajayi and Olufolaji, 2007). There are many other plants (*Eichhornia crassipes*, *Vernonia amygdalina*, *Piper guineense*, etc.) used by farmers as botanical pesticides for various pests and diseases, even in their crude form, and they have been found to be very effective (Olufolaji, 1999a, 2006, 2008; Ajayi and Olufolaji, 2007). Those farmers have adopted the use of botanicals in the control of pests and diseases on their crops due to desperation for survival, and since they are dependent on an agrarian economy, they evolved various method of formulating and using the botanicals without sourcing the orthodox expertise and technicalities that are either not within their reach or expensive to adopt.

Table 9.2. Some bioactive plants reported for the control of agricultural pests.

Pests	Bioactive plants	References
Maize weevil	Neem seed	Ivbijaro, 1983
Sitophilus orazae (L.)	*Eugenia aromatica*	Lale, 1992, 1994; Aranilewa *et al.*, 2002
Callosobruchus maculatus	*Piper guineense*	Olaifa and Erhum,1988
	Acrae epomia	Ofuya and Dawodu, 2002
Dysdercus superstittiosus	Several plants	Adedire and Lajide, 1999
Ootheca mutabilis	*Monodora tenuifolia*	Adedire *et al.*, 2003
Riptortus dentipes	*Zanthoxylum zanthoxyloide*	Ogunwolu and Idowu, 1994
Callosobruchus maculatus	*Azadirachta indica*	Lale and Abdulrahman, 1999
Callosobruchus maculatus	*Lippia adoensis*	Olaifa *et al.*, 1987
Acrae epomia		
Dysdercus superstittiosus		
Ootheca mutabilis		
Riptortus dentipe		
Grasshopper	*Clausena anisata*	Okunade and Olaifa, 1987
(*Zonocerus variegates*)	*Azadirachta indica*	Olaifa and Akingbohungbe,
	Piper guineense	1987a
	Azadirachta indica	Olaifa and Adenuga,
	Azadirachta indica	1988a,b
		Olaifa *et al.*, 1991a
Tetrapteura tetraptera	*Azadirachta indica*	Olaifa *et al.*, 1991b
Field pests of cowpea, maize, sorghum, cassava, banana	*Tephrosia vogelli*	Adebayo, *et al.*, 2007

Source: Olaifa, 2009.

9.3 Availability of Natural Products in Large and Required Quantities

Most countries in the developing world are favoured by very good climatic conditions and agro-ecosystems that support the growth and development of most of the botanicals used for the production of the botanical pesticides. The farmers have a locational advantage in that the botanicals are available in abundance in their ecosystem and exist throughout the year in both rainy and dry seasons. For example, most of these popular plants grow in the wild and are regarded as weeds, and some of them are not useful for food or raw materials for the industries (Table 9.2).

However, some are now being cultivated to prevent the extinction of useful species. In recent times, most countries in the developing world find it easy to incorporate the botanicals into their agro-forestry programmes and this has increased and maintained the existence of the botanicals in large quantities (Ogunnika, 2007). This has further strengthened the large-scale use of the botanicals as a means of pest and pathogen control in the sub-region. Furthermore, germination of the seeds of most of these botanicals is not tied down by an unnecessarily long dormant period, which is characteristic of some botanicals in developed world (Ogunnika, 2007). The activities of the farming population in the tropics have included intercropping and mix-cropping of main crops with the botanicals. These involve planting *Manihot esculenta* (cassava) with *Ocimum gratissimum* (camphor basil), *Theobromae cacao* (cocoa) with *Carica papaya* (pawpaw) and so on (Ogunnika, 2007). It is worth noting here that some of these botanicals when mixed or intercropped with the main crop could serve as a control for pests and diseases in the form of allelopathy. *N. tabaccum* and some other plant species, especially grasses, have been found to possess this attribute (Florentine *et al.*, 2003, Ogunnika, 2007). *Eucalyptus camaldulensis* and some other botanicals are also used as weed control in farmlands to maximize yields (Oyun and Agele, 2009). Many farm crops are protected by this means without the need to carry out formulation and application processes for pest and disease control. There is the availability of rain and moisture throughout the year and no occurrence of snow or completely dry period in most developing countries close to the tropics. Thus, plants can grow throughout the year and there is no threat of extinction of most of these botanicals, both wild and cultivated species.

The fact that almost half of the world's forest that houses these botanicals is in Africa is an indication of the availability of the much needed plants for botanical pesticides (Owolabi and Olanrewaju, 2007). Farnsworth and Soejarto (1991) estimated that 28% of the total inventories of the world's plant species are used in ethnomedicines. Groombridge (1992), and Lange and Schippmann (1997) reported that over 21,000 plant taxa (including synonyms) are being used worldwide for pesticidal purposes, which amounted to about US$800 million in 1995–1996. If these plants had mainly resided in the developed world, then developing countries would not benefit immensely from this since they are mostly poor. Despite the indiscriminate deforestation in developing countries in which these botanicals exist, rapid and large recovery of most of these botanicals still occurs. It has been observed that most of the botanicals obtained in the tropical forests of developing countries sprout easily and survive through the little water stress experienced during the dry season. These attributes support the possibility of large-scale use of these botanicals as pesticides.

9.4 Formulations of Botanicals in Developing Countries

Local formulation by peasant farmers

Methods adopted by farmers in the developing countries for the formulation and delivery of botanical pesticides vary depending on the parts of plants used and the target pests. The plant parts could be either pulverized fresh with a pestle and mortar or grinding stones before soaking in water (hot or cold) or sun-dried before pulverizing into powdered form. They are then soaked and extracted before use. The powdered form could also be used as dust to prevent incursion of pests and diseases on the stored farm produce or for on-farm dusting for pest and disease control. The formulated botanicals can be buried near the crop plants such that they can diffuse into the soil and protect the root or move in a systemic way to control pests and diseases on the farm. Termites, soil insects and some other soil pathogens such as *Pythium* spp. could be controlled through this method.

Some botanicals are effective when they are burnt and the fumes are directed into storage silos or into the soil as in the orthodox fumigation practice.

Scientific formulations

Researchers in most of the developing countries have investigated more than 100 different species of botanicals for their efficacy in crop protection against pest degradation, especially the storage pests and diseases (Boeke *et al.*, 2001; Ofuya *et al.*, 2007; Olufolaji, 2008). Some researchers have suggested that a few botanicals used as pesticides in Nigerian flora with confirmed pesticidal attributes against stored-product insects merit scientific formulations (Lale, 2001; Ofuya, 2003). However, pesticidal formulations of most botanicals which have been found effective and may be recommended for use in crop protection in Nigeria are rather simple, and easy to make in solid and liquid formulations (Ofuya, 2009). Solid formulations are mainly powders, whereas liquid formulations include oils and the crude extracts prepared in water and organic solvents. These methods of formulation of the botanicals as pesticides also make their large-scale use easy and expansive in the developing countries because not much technicality is involved.

Methods of formulation

Formulations of the botanicals in developing countries are also of immense importance in their large-scale use. Due to low-level scientific technicality and non-availability of sophisticated equipment associated with the production of synthetic pesticides, the type of formulation associated with the production of the botanical pesticides in developing countries is the type that the majority of the agrarian community and scientists will be able to handle and utilize for effective botanical pesticide formulation.

Powders

These are prepared by harvesting the plant materials, which are then sun-dried and pulverized into fine powder. The powders have been investigated undiluted for stored-product protection against insects and fungi. The required quantity of powder is admixed with an appropriate quantity of commodity prior to storage. Powders have also been extracted with water (water-extractable powders), filtered and applied as aqueous solutions for protecting field crops and grains (Table 9.3; Ofuya, 2009).

Crude water extracts

These are crude extracts obtained by using water as a solvent, and may be obtained simply by pressing out juices and then diluting in water or through maceration (steeping in water for prolong periods). They may also be obtained by infusion (the immersion of plant parts in boiling water for prolonged periods). Such aqueous extracts or solutions have mostly been investigated against field-crop insect pests and diseases (Table 9.3) (Olufolaji, 2006; Ofuya, 2009).

Oils

Oils are usually extracted from pulverized medicinal plant parts by means of organic solvents and are used very much like concentrate liquids. Organic

Table 9.3. Some medicinal plants in Nigeria and potential formulations in crop protection.

Botanical	Potential formulation
Azadirachta indica	Dust, water-extractable powder, emulsifiable concentrate, liquids
Piper guineense	Dust, water-extractable powder, emulsifiable concentrate, liquids
Eugenia aromatica	Dust, water-extractable powder, emulsifiable concentrate, liquids
Dementia tripetalia	Dust, water-extractable powder, emulsifiable concentrate, liquids
Tephrosia vogelli	Water-extractable powder, emulsifiable concentrate, liquids
Nicotaina tabacum	Dust, water-extractable powder
Allium sativum	Water-extractable powder, emulsifiable concentrate, liquids
Zingiber officinale	Water-extractable powder, emulsifiable concentrate, liquids

Source: Ofuya, 2009.

solvents, particularly methanol, ethanol, acetone, hexane, petroleum ether, diethyl ether, chloroform or methyl chloride, have been used (Table 9.3) (Olufolaji, 2006; Ofuya, 2009).

Mixed formulations

Using herbal mixtures for crop protection is traditional with some farmers (Kitch *et al.*, 1997). Efficacy of mixed formulations of medicinal plant powders for stored-grain protection has been subjected to empirical verification (Ogunwolu and Idowu, 1994; Dawodu and Ofuya, 2000; Emeasor *et al.*, 2007). Overall pesticidal activity of each material was not mitigated by mixing the two against the test insects. However, synergistic or additive effects would be desirable to enhance efficacy. Lale (2002) reported that mixing different essential oils from plants, in some cases, provided much better control than single use (Table 9.3) (Ofuya, 2009; Olufolaji, 2006, 2008).

Use of adjuvants

Ofuya *et al.* (2007) have demonstrated the possibility of using organic flours from yam, cassava and plantain as diluents in the formulation of insecticidal dusts from buds of *Eugenia aromatica* and dry fruits of *Piper guineense*. Filtrates that are obtained in the preparation of aqueous neem extracts can be improved after dilution by adding brown sugar, which improves the adherence of the filtrate on leaves and other plant parts (Jackai, 1993).

9.5 Attributes of the Botanicals Versus Conventional Synthetic Pesticides

The botanicals have an array of good attributes over the synthetic pesticides, which is encouraging because the low technicality required for use has enabled the developing world to achieve desirable advantages from their use. Ahmed and Stoll (1996) highlighted the various problems caused by synthetic pesticides including toxicity to non-target organisms such as man and beneficial macro/microbes, pollution in the agro-ecosystem and development of resistance by the target organisms. However, the botanical pesticides possess some special characteristics and may be used as better alternatives in managing different agricultural pests:

- The materials used for the production of the botanical pesticides are easily available since they are usually weeds in the agro-ecosystem.
- Botanical pesticides are usually inherently less harmful than conventional pesticides.
- Botanical pesticides generally affect only the target pest and closely related organisms, in contrast to broad-spectrum conventional pesticides that may affect organisms as different as birds, insects and mammals.

- Botanical pesticides are often effective in very small quantities and often decompose quickly, thereby resulting in lower exposure and largely avoiding the pollution problems caused by conventional pesticides.
- When used as a component of integrated pest management (IPM) programmes, botanical pesticides can greatly decrease the use of conventional pesticides, while crop yields remain high.

9.6 The Views of Environmental Protection Agency and the Policy of Most Developing Countries

The view of the EPA at national and international levels is not to support the use of synthetic pesticides. These views have made some advanced and developed countries to promulgate laws against the usage of some of these synthetic pesticides. In Nigeria, the body vested with the authority of pesticides registration and control is the National Agency for Food and Drug Adminstration and Control (NAFDAC) (Olaifa, 2009). Despite this, some pesticides such as DDT and some mercuric-based pesticides, which had been banned from use due to the residue left on crops for consumption and because they constitute a health hazard, are still found in pesticides markets. The presence of heavy metals in most pesticides have caused untold hardship and death to the consumer of the crops on which it has been used.

Most of the developing countries have formulated various policies on the usage of the synthetic pesticides. The policy emanated from the effect which the pesticides have unleashed on the crops due to their phytotoxicity, mammalian toxicity and carcinogenic effects on both the crops and man.

In 1994, the Botanical Pesticides and Pollution Prevention Division was established in the Office of Pesticide Programmes to facilitate the registration of botanical pesticides. This division promotes the use of safer pesticides, including botanical pesticides, as components of IPM programmes. The division also coordinates the Pesticide Environmental Stewardship Programme (PESP).

Since botanical pesticides tend to pose fewer risks than conventional pesticides, the EPA generally requires much less data to register a botanical pesticide than to register a conventional pesticide. In fact, new botanical pesticides are often registered in less than a year, compared with an average of more than 3 years for conventional pesticides.

While botanical pesticides require less data and are registered in less time than conventional pesticides, the EPA always conducts rigorous reviews to ensure that pesticides will not have adverse effects on human health or the environment. For the EPA to be sure that a pesticide is safe, the agency requires that registrants submit a variety of data about the composition, toxicity, degradation and other characteristics of the pesticide.

9.7 The Attitudes of the Poor Resource Farmers in the Developing Countries to the Use of Botanical Pesticides

The poor resource farmers in the developing countries have been relieved by the use of botanicals as it affords them the advantages of low pest control cost, easy pest control strategies, minimum pollution in their agro-ecosystem and, ultimately, higher yield and profit. They are now in search of any available information regarding new innovations on botanicals to control the pests on their farm lands. They are also ready to pay for the services because botanicals cost less than the conventional synthetic pesticides with all the safe attributes to food production in the developing world. The farmers are already trying to combine the research findings of scientists with their age-long practice to arrive at formidable control measures on their farm lands. Interestingly, the farmers are now adopting the principle of IPM in order to maximize yields on their farms. Moreover, the botanists who engage in floriculture and horticulture are included in the use of botanicals to preserve most of their important species of plants from the scourge of pests and diseases. The farmers have already formed cooperatives and through them approach government extension agencies for more enlightenment on the use of botanicals for their benefit.

9.8 The Future of the Use of Botanicals

The future of botanicals as alternatives to synthetic pesticides is bright, in that more people are using them due to the availability of the raw materials, accessibility of the raw materials and easy use of the botanicals, with minimum demand for specialized application and equipment that is characteristic of synthetic pesticides. Most of the research on pest and disease control in both developed and developing countries are now centred on the use of botanicals as alternatives to synthetic pesticides. The World Health Organization has also encouraged the use of botanicals because there is a prevalence of diseases traceable to chemicals used for the control of plant diseases because the consumed plant materials possess high degree of residues that cause carcinogenic diseases and other health issues. It is also an issue with environmentalists concerned about the threat of global warming because most of the pollution accredited to cause global warming is traced to the pollutants produced and discharged in the developing countries (Olufolaji, 2006). The present economic recession has even made the purchase of synthetic pesticides more expensive and thus difficult for the poor and even the enlightened farmers in the developing countries, hence the adoption of an alternative, which is the use of botanical pesticides.

9.9 Conclusion

The botanicals have solved enormous problems of farmers in the developing world who are faced with an array of pests and diseases that reduce their yields and make farming a harder process.

Acknowledgement

The author is grateful for the opportunity given by the coordinator of this book collection, Prof. N.K. Dubey, to be a contributor to this book. I also appreciate the management of The Federal University of Technology, Akure, Nigeria, with whom I am residing to carry out this research work.

References

Adebayo, T.A., Olaniran, O.A. and Akanbi, W.B. (2007) Control of insect pests of cowpea in the field with allelochems from *Tephrosia vogelii* and *Petiveria alliacea* in the southern guinea Savannah of Nigeria. *Agricultural Journal* 2, 365–369.

Adedire, C.O. and Lajide, L. (1999) Toxicity and oviposition deterrency of some plant extracts on cowpea storage bruchid *Callosobruchus maculatus* (F.) (Coleoptera:Bruchidae). *Journal of Plant Disease and Protection* 106, 647–653.

Adedire, C.O., Adebowale, K.O. and Dansu, O.M. (2003) Chemical composition and insecticidal properties of *Monodora tenuifolia* seed oil (Annonaceae). *Journal of Tropical Forest Products* 9 (1&2), 15–25.

Ahmed, S. and Stoll, G. (1996) Biopesticides, In: Binders, J. Haverkort, B., Hiemstra, W. (eds) *Biotechnology: Building on Farmers' Knowledge*. MacMillan Education Ltd. London and Basingstoke, pp. 52–59.

Ajayi, A.M. and Olufolaji, D.B. (2007) The Biofungicidal Attributes of some plant extracts on *Colletotrichum capsici;* the fungal pathogen of brown blotch disease of cowpea. *Nigerian Journal of Mycology* 1, 59–65.

Aranilewa, S.T., Odeyemi, O.O. and Adedire, C.O. (2002) Effects of medicinal plant extract and powder on the maize weevil, *Sitophilus zeamais* Mots. (Coleoptera: Curculionidae). *Annals of Agricultural Science* 3, 1–10.

Aworinde, D.O., Abeegunrin, T.A. and Ogundele, A.A. (2008) Ethnobotanical survey of plants of medicinal importance in south western Nigeria. *Applied Tropical Agriculture* 13, 24–33.

Boeke, S.J., Van Loon, J.J.A., Van Huis, A., Kossou, D.K. and Dickle, M. (2001) The use of plant materials to protect stored leguminous seeds against seed beetles; a review. *Wageningen University Papers 2001–2003*, Backhuys Publishers, The Netherlands, p. 108.

Dawodu, E.O. and Ofuya, T.I. (2000) Effects of mixing fruit powders of *Piper guineense* Schum and Thonn. and *Dennetia tripetala* Baker on oviposition and adult emergence of *Callosobruchus maculatus* (F.) (Coleoptera: Bruchidae). *Applied Tropical Agriculture* 5, 158–162.

Edeoga, H.O. and Eriata, D.O. (2001) Alkaloid, tannin and saponin content of some Nigerian medicinal plants. *Journal of Medicinal and Aromatic Plants* 23, 344–349.

Emeasor, K.C., Emosairue, S.O. and Ogbuji, R.O. (2007) Preliminary laboratory evaluation of the efficacy of mixed seed powders of *Piper guineense* (Schum and Thonn) and *Thevetia peruviana* (Persoon) Schum against *Callosobruchus maculatus* (F.) (Coleoptera: Bruchidae). *Nigerian Journal of Entomology* 24, 114–118.

Farnsworth, N.R. and Soejarto, D.D. (1991) Global importance of medicinal plants. In: Akerele, O., Heywood, V. and Synge, H. (eds) The conservation of medicinal plants. Cambridge University Press, Cambridge, pp. 25–51.

Florentine, S.K., Duhan, J.S. and Fox, J.E. (2003) Allelopathic species and grasses. *Allelopathic Journal* 11, 77–83.

Groombridge, B. (1992) Global *Biodiversity Status of the Earth's Living Resources*. Chapman and Hall, London, Glasgow, New York, pp. 585.

Isman, M.B. (2008) Perspective of botanical insecticides: for richer, for poorer. *Pest Management Science* 64, 8–11.

Ivbijaro, M.F. (1983) Toxicity of neem seed, *Azadirachta indica* A. Juss. to *Sitophilus oryzae* (L.) in stored maize. *Protection Ecology* 5, 353–357.

Jackai, L.E.N. (1993) The use of neem in controlling cowpea pests. *International Institute of Tropical Agriculture (IITA) Research* 7, 5–11.

Kitch, L.W., Bottenberg, H. and Wolfson, J.L. (1997) Indeginous knowledge and cowpea pest management in sub-Saharan Africa. In: Singh, B.B., Mohan-raj, D.R., Dashiell, K.E. and Jackai, L.E.N. (eds) *Advances in Cowpea Research Copublication of IITA/JIRCAS, IITA,* Ibadan, Nigeria, pp. 292–301.

Lale, N.E.S. (1992) A Laboratory study of the comparative toxicity of products from three species of the maize weevil. *Postharvest Biology and Technology* 2, 61–64.

Lale, N.E.S. (1994) Laboratory assessment of the effectiveness and persistence of powders of four spices on cowpea bruchid and maize weevil in an air-tight storage facilities. *Samaru Journal of Agricultural Research* 11, 79–84.

Lale, N.E.S. and Abdulrahman, H.T. (1999) Evaluation of neem (*Azadirachta indica* A. juss.) seed oil obtained by different methods and neem powder for the management of *Callosobruchus maculatus* (F.) (Coleoptera: Bruchidae) in stored cowpea. *Journal of Stored Products Research* 35, 135–143.

Lale, N.E.S. (2007) Overview of use of medicinal plants in crop production and protection. In: *Proceedings of the Humboldt Kellog / 3rd Annual Conference of School of Agriculture and Agricultural Technology,* School of Agriculture and Agricultural Technology, The Federal University of Technology, Akure. Ondo State, Nigeria, pp 270–274.

Lale, N.E.S. (2001) The impact of storage insect pests on post-harvest losses and their management in the Nigerian Agricultural system. *Nigerian journal of Experimental and Applied Biology* 2, 231–239.

Lale, N.E.S. (2002) *Stored Products Entomology and Acarology in Tropical Africa.* Mole Publications, Maiduguri, Nigeria, pp. 204.

Lange, D. and Schippmann, U. (1997) Trade survey of Medicinal Plants in Germany, A contribution to International Plants Species Conservation. Bonn-Bad Godesberg (Bundesamt fur Naturschutz), pp. 128.

Nworgu, F.C. (2006) Prospects and Pitfalls of Agricultural Production in Nigeria. Blessed Publications, Ibadan, Oyo State. Nigeria, pp. 181.

Ofuya, T.I. (2003) Beans, Insect and Man. *Inaugural Lecture Series 35,* The Federal University of Technology, Akure. Nigeria, pp. 45.

Ofuya, T.I. (2009) Formulation of medicinal plants for crop protection in Nigeria. In: *Proceeding of the Humboldt Kellog / 5th SAAT Annual Conference of formulations of Medicinal Plants in Plant and Animal Production in Nigeria.* School of Agriculture and Agricultural Technology, The Federal University of Technology, Akure, Ondo State, Nigeria, pp. 1–6

Ofuya, T.I. and Dawodu, E.O. (2002) Aspects of insecticidal action of *Piper guineense* Schum and Thom. fruit powders against *Callosobruchus maculatus* (F.) (Coleoptera: Bruchidae). *Nigerian Journal of Entomology* 20, 40–50.

Ofuya, T.I., Olotuah, O.F. and Aladesanwa, R.D. (2007) Potentials of dusts of *Eugenia aromatica* Baill, dry flower buds and *Piper guineense* Shum and Thonn dry fruits formulated with three organic flours for controlling *Callosobruchus maculatus* Fabricus (Coleoptera : Bruchidae). *Nigerian Journal of Entomology* 24, 98–106.

Ogunnika, C.B. (2007) Medicinal plants: A Potential agroforestry components in Nigeria. In: *Proceedings of the Humboldt Kellog / 3rd Annual Conference of School of Agriculture and Agricultural Technology,* School of Agriculture and Agricultural Technology, The Federal University of

Technology, Akure, Ondo State, Nigeria, pp. 2–7.

Ogbalu, O.K. (2009). A review of estimate of crop losses in field and stored conditions from different locations. *Niger Delta Biologia* 9, 1–18.

Ogunwolu, O. and Idowu, O. (1994) Potential of powdered *Zanthoxylum zanthoxyloides* root, bark and *Azadirachta indica* seeds for control of the cowpea seed bruchid *Callosobruchus maculatus* in Nigeria. Journal of Africa Zoology 108, 521–528.

Okunade, A.I. and Olaifa, J.I. (1987) Extragole an acute toxic principle from the volatile oil of the leaves of *Clausena anisata. Journal of Natural Products* 50, 1990–1991.

Olaifa, A.I. and Akingbohungbe, A.E. (1987a) Antifeedant and insecticidal effects of *Azadirachta indica, Petiveria alliacea* and *Piper guineense* on the variegated grasshopper *Zonocerus* variegates. *Proceedings of the 3rd International Neem Conference,* Nairobi, Kenya, 1986, GTZ, Eschborn, Germany, pp. 405–416.

Olaifa, J.I. (2009) Formulation of bioactive plants for plant and animal production in Nigeria; Realities and challenges. *Proceeding of the Humboldt Kellog / 5th SAAT Annual Conference of formulations of Medicinal Plants in Plant and Animal Production in Nigeria.* School of Agriculture and Agricultural Technology, The Federal University of Technology, Akure. Ondo State, Nigeria. pp. 10 18.

Olaifa, J.I. and Erhum, W.O. (1988) Laboratory evaluation of *Piper guineense* for the protection of cowpea against *Callosobruchus maculates. Insect Science and its Application* 9, 55–59.

Olaifa, J.I., Erhum, W.O. and Akingbohungbe, A.E. (1987) Insecticidal activity of some Nigerian Plants. *Insects Science and its Application* 8, 221–224.

Olaifa, J.I. and Adenuga, A.O. (1988a) Neem products for protecting field cassava from grasshopper. *Insect Science and its Application* 9, 267–270.

Olaifa, J.I. and Adenuga, A.O. (1988b) Protecting field crops with locust lotion;

a natural- locally sourced insecticide. *Farmer Advisory Bulletin, July 1988. Federal Department of Livestock and Pest control Services, Federal Ministry of Agriculture and Natural Resources Kaduna, Nigeria,* pp. 15.

Olaifa, J.I., Adedokun, T.A. and Adenuga, A.O. (1991a) Antifeedant and growth-regulating effects of neem products on the variegated grasshopper *Zonocerus variegates* L. (Orthoptera:Pyrgomorphidae). *Journal of African Zoology (Belgium)* 105, 157–162.

Olaifa, J.I., Adenuga, A.O. and Kanu, K.M. (1991b) Relative efficacy of a neem formulation and four conventional insecticides in the protection of some arable crop against Acridoid grasshopper. *Discovery and Innovation* 3, 115–121.

Olufolaji, D.B. (1999a) Control of wet rot disease of *Amaranthus* cause by *Chromolaena curcubitarum* using neem (*Azadirachta indica*) extract. *Journal of Sustainable Agriculture and Environment* 4, 68–77.

Olufolaji, D.B. (2006) Effects of crude extracts of *Eichhornia crassipes* and *Chromolaena odorata* on the control of red-rot disease of sugarcane in Nigeria. In: Yang-Rui, L. and Solomon, S. (eds) *Proceedings of the International Conference of Professionals in Sugar and Integrated Technologies,* International Association of Professionals in Sugar and Integrated Technologies, Nanning, Guangxi – 530007, P.R. China, pp. 337–342.

Olufolaji, D.B. (2008) Combination of plant extracts of *Chromolaena odorata* and *ocimum gratissimum* in the control of pineapple disease of sugarcane in Nigeria. In: Yang-Rui Li, Nasr, M.I., Solomon, S. and Rao, G.P. (eds) *Proceedings of the International Conference of Professionals in Sugar and Integrated Technologies.* International Association of Professionals in Sugar and Integrated Technologies, Nanning, PR China, pp. 426–428.

Owolabi, O.O. and Olanrewaju, J. (2007) Imperative conservation approach: Our last option to save the plants that save lives in Nigeria. In: *Proceedings of the*

Humboldt Kellog / 3rd Annual Conference of School of Agriculture and Agricultural Technology, School of Agriculture and Agricultural Technology, The Federal University of Technology, Akure, Nigeria, pp. 8–13.

Oyun, M.B. and Agele, S.O. (2009) Testing of extracts from *Eucalyptus camaldulensis, Acacia auriculiformis Gliricidia sepium* for weed suppression in a maize farm. In: *Proceeding of the Humboldt Kellog/ 5th SAAT Annual Conference of formulations of Medicinal Plants in Plant and Animal Production in Nigeria,* School of Agriculture and Agricultural Technology, The Federal University of Technology, Akure. Ondo State, Nigeria, pp. 89–93.

10 Current Status of Natural Products in Pest Management with Special Reference to *Brassica carinata* as a Biofumigant

María Porras

Department of Crop Protection, IFAPA Centro Las Torres – Tomejil, Sevilla, Spain

Abstract

This chapter covers the use of natural products in plant disease management in the form of fungicides and bactericides, and personal experience of the development and optimization of biofumigation with *Brassica carinata* and soil solarization as alternatives to the traditional use of chemicals in strawberry production. The potential of biofumigation as a component of the integrated management of soil pathogens has been demonstrated in various agricultural systems. The presence of high amounts of glucosinolates, and of the enzyme myrosinase that catalyses their hydrolysis, linked to the high biocidal activity of some glucosinolate enzymatic hydrolysis derivative products have been suggested for a practical use of amending soil with these natural biocidal compounds through cultivation and green manure of selected species of the family Brassicaceae.

10.1 Introduction

Chemical pesticides such as methyl bromide are being phased out globally because of their impact on the ozone layer (European Parliament, 2000). Since 2005, the use of methyl bromide has been banned in European Union countries. Chemical, physical and biological alternative methods for pathogen control have been evaluated in crop production to replace the compounds lost due to the new registration requirements (Duniway, 2002; Martin and Bull, 2002; Porras *et al.*, 2007b). Natural-product-based pesticides can sometimes be specific to the target species and have unique modes of action with little mammalian toxicity (Duke *et al.*, 2003). Modern instrumentation and improved methods should increase interest in natural-product-based pesticide discovery research (Duke *et al.*, 2002).

Naturally occurring substances found in fungi, bacteria and plants are important sources of molecules with different biological properties. They may be developed either as products per se or used as starting points for synthesis to optimize specific properties. Low mammalian toxicity, low environmental impact, low levels of residues in food and compatibility with integrated pest management (IPM) programmes are important considerations in the election of fungicides for development (Knight *et al.*, 1997).

This chapter covers the use of natural products in plant disease management (fungicides and bactericides), and personal experience of *Brassica carinata* as a biofumigant in strawberry fields.

10.2 Natural Products Used as Fungicides and Bactericides in Agriculture

The natural products used as agrochemicals have been well documented (Knight *et al.*, 1997; Warrior, 2000; Copping and Duke, 2007; Kim and Hwang, 2007; Dayan *et al.*, 2009). Microorganisms can synthesize secondary metabolites of versatile chemical structures with diverse biological activities that exceed the scope of synthetic organic chemistry (Porter, 1985). Pesticidal potency is not the only factor for formulating a plant-based product as a commercial pesticide. In-field chemical stability adequate to reduce the application times to an economical level, and lowered volatility for sufficient retention on the surface of host plants are also important factors for microbial fungicides to be efficiently applied in agriculture (Kim and Hwang, 2007). Therefore, only a few microbial metabolites have been successfully developed into commercial fungicides. The excellent fungicidal activity of these microbial metabolites and their potential as lead candidates for further fungicide development continue to stimulate research and screening for antifungal microbial metabolites (Kim and Hwang, 2007; Dubey *et al.*, 2009). Table 10.1 summarizes the microbial metabolites, and plant- or animal-derived products used as fungicides or bactericides in agriculture.

Biological control agents (including both bacteriological or fungus origin), plants, natural compounds and preparations have been described with activity against bacterial or fungal plant pathogens. Furthermore, plants protect themselves from microbial attacks with both constitutive antimicrobials and compounds induced by the attacking pathogens (phytoalexins). Natural products have been used to protect plants indirectly from pathogens by inducing systemic acquired resistance (SAR). The SAR-inducing compounds are termed elicitors. Pathogens cannot evolve resistance directly to the elicitor because the activity is indirect, making such products excellent candidates for integrated disease management. Nevertheless, elicitors are generally not as effective as chemical fungicides, partly because the timing of elicitor application and threat to the crop by a pathogen is crucial, but difficult to maximize (Dayan *et al.*, 2009).

Disease suppression by biocontrol agents is the sustained manifestation of interactions between the plant, the pathogen, the biocontrol agent, and the

Table 10.1. A summary of natural products used as agrochemicals.

Species	Products	Target pathogens	References
Bacteria	Bacteria-derived products		
Streptomyces griseochromogenes	Blasticidin-S	*Pyricularia oryzae* Cavara; perfect stage *Magnaporthe grisea* (Hebert) Barr	Kim and Hwang, 2007
Erwinia amylovora (Burrill) Winslow	Harpin protein	*Xanthomonas campestris* (Pammell) Dowson *Pseudomonas syringae* Van Hall *Pseudomonas solanacearum* (Smith) Smith *Fusarium* sp. *Phytophthora* sp. *Magnaporthe salvinii* (Cattaneo) Krause & Webster *Rhizoctonia solani* Kühn (Pellicularia sasakii Ito) *Venturia inaequalis* (Cooke) Winter *Erwinia amylovora* *Botrytis* sp. *Guignardia bidwellii* (Ellis) Viala & Rivas *Diplocarpon rosae* Wolf	Copping and Duke, 2007
Streptomyces kasugaensis Hamad *et al.*	Kasugamycin	*P. oryzae* *Cercospora* spp. *Venturia* spp. *Cladosporium fulvum* *Erwinia atroscptica* *Xanthomonas campestris*	Anon, 2005; Copping and Duke, 2007
Streptoverticillium rimofaciens	Mildiomycin	*Erysiphe* spp. *Uncinula necator* (Schwein) Burrill *Podosphaera* spp. *Sphaerotheca* spp.	Copping and Duke, 2007
Streptomyces natalensis Struyk, Hoette, Drost, Waisvisz, Van Eek & Hoogerheide *S. chattanoogensis* Burns & Holtman	Natamycin	*Fusarium oxysporum* Schlecht	Copping and Duke, 2007

Continued

Table 10.1. Continued.

Species	Products	Target pathogens	References
Streptomyces rimosus Sobin *et al.*	Oxytetracycline	*Erwinia amylovora* *Pseudomonas* spp. *Xanthomonas* spp.	Copping and Duke, 2007
Streptomyces cacoai var. asoensis Isono *et al.*	Polyoxins:		Copping and Duke, 2007
	polyoxin B	*Sphaerotheca* spp. and other powdery mildews *Botrytis cinerea* Pers. *Sclerotinia sclerotiorum* De Bary *Corynespora melonis* Lindau *Cochliobolus miyabeanus* (Ito & Kuribay) Drechsler ex Dastur *Alternaria alternata* (Fr.) Keissler and other *Alternaria* spp.	
	polyoxorim (polyoxin D)	*R. solani* *Nectria galligena* Bresadola (*Diplodia pseudodiplodia* Fuckel) *Drechslera* spp. *Bipolaris* spp. *Curvularia* spp. *Helminthosporium* spp.	
Streptomyces griseus (Krainsky) Waksman & Henrici	Streptomycin	Bacteria, particularly effective against: *Xanthomonas oryzae* Dowson *X. citri* Dowson *Pseudomonas tabaci* Stevens *P. lachrymans* Carsner	Copping and Duke, 2007
Streptomyces hygroscopicus (Jensen) Waksman & Henrici var. limoneus	Validamycin	*Rhizoctonia* spp.	Copping and Duke, 2007
Pseudomonas fluorescens	Pyrrolinitrin	*B. cinerea* *M. grisea* (*In vitro* and in the greenhouse, not in field because of photosensitivity)	Haas and Keel, 2003; Kim and Hwang, 2007

Continued

Table 10.1. Continued.

Species	Products	Target pathogens	References
	Fenpiclonil and Fludioxonil (Phenylpyrroles derivates to enhanced photostability)	*Fusarium graminearum* Schwabe *Gerlachia nivalis* Gams & Müll *Botrytis* spp. *Monilinia* spp. *Sclerotinia* spp.	
Fungi	Fungus-derived products		
Saccharomyces cerevisiae Meyer ex Hansen	Yeast extract hydrolysate	ni	Copping and Duke, 2007
Strobilurus tenacellus	Strobilurin A		Ishii *et al.*, 2001;
Oudemansiella mucida	Oudemansin A		Balba, 2007
	From which synthetic strobilurins such as azoxystrobin and kresoxim-methyl have been developed	Resistance to strobilurins has already evolved	
Plants	Plant-derived products		
Cassia tora L. (also known as *Cassia obtusifolia* L.)	Cinnamaldehyde	*Verticillium* spp. *Rhizoctonia* spp. *Pythium* spp. *Sclerotinia homeocarpa* Bennett *Fusarium moniliforme* var. subglutinans Wollenw. & Reinking	Copping and Duke, 2007
The two active ingredients are found in virtually all living organisms and in their pure form	L-glutamic acid and γ-aminobutyric acid (Auxien has registered the combination product containing both active indredients)	Powdery mildew	Copping and Duke, 2007
Simmondsia californica Nutt. and *S. chinensis* Link.	Jojoba oil	Powdery mildew	Copping and Duke, 2007
Laminaria digitata (Hudson)	Laminarine	Fungal pathogens of cereals, particularly septoria and powdery mildews	Copping and Duke, 2007

Continued

Table 10.1. Continued.

Species	Products	Target pathogens	References
Reynoutria sachalinensis (Fr. Schm.) Nakai	Extract of giant knotweed	*Botrytis* spp. Powdery mildews *Xanthomonas* spp.	Copping and Duke, 2007
Macleaya cordata R. Br.	Pink plume poppy extract	The target pathogens are those causing foliar fungal diseases, such as powdery mildew, alternaria leaf spot and septoria leaf spot	Copping and Duke, 2007
Rosmarinus officinalis	Rosemary oil	ni	Dayan *et al.*, 2009
Thymus vulgaris	Thyme oil	ni	Dayan *et al.*, 2009
Azadirachta indica	Clarified hydrophobic extract of neem oil	ni	Dayan *et al.*, 2009
Gossypium hirsutum and *Allium sativum*	Cottonseed oil with garlic extract	ni	Dayan *et al.*, 2009
Animals	Animal-derived products		
Cow	Milk	Powdery mildews	Copping and Duke, 2007
Dried, crushed crustacean exoskeletons	Poly-D-glucosamine or chitosan	Powdery mildews *Botrytis* spp.	Copping and Duke, 2007

ni, information not provided in reference.

microbial community on and around the plant and the physical environment (Handelsman and Stabb, 1996). Relevant biocontrol microorganisms of the rhizosphere include *Pseudomonas* spp., *Bacillus* spp., *Streptomyces* spp., *Trichoderma* spp., and non-pathogenic *Fusarium* spp. (Cook, 1993; Paulitz and Belanger, 2001; Whipps, 2001). Different species of *Trichoderma* are studied primarily for their ability to control plant disease through antagonism, rhizosphere competence, enzyme production, induction of defence responses in plants, metabolism of germination stimulants, and beneficial growth of the host following root colonization (Weindling, 1934; Zimand *et al.*, 1996; Bailey and Lumsden, 1998; Howell, 2003; Porras *et al.*, 2007a,b). Determination of these effects depends on many interactions that take place in the soil among *Trichoderma* spp., other microorganisms, the plant root, and the soil environment (Bailey and Lumsden, 1998).

10.3 Biofumigation with *Brassica*

The term biofumigation has been defined as the 'suppression of soil-borne pathogens and pests by *Brassica* rotation or green manure crops, through liberation of isothiocyanates from hydrolysis of the glucosinolates that characterize the *Brassicaceae*' (Kirkegaard and Matthiessen, 2004).

Biofumigation is based on the action of volatile compounds, especially isothiocyanates, produced by the hydrolysis of glucosinolates. Glucosinolates are secondary metabolites produced by plants belonging to the *Capparales* order (Gimsing and Kirkegaard, 2009). Of the many hundreds of cruciferous species investigated, all are able to synthesize glucosinolates, and at least 500 species of non-cruciferous dicotyledonous angiosperms have been reported to contain them. Most glucosinolate-containing genera are clustered within the *Brassicaceae*, *Capparaceae* and *Caricaceae*. Among the *Brassicaceae*, the genus *Brassica* contains a large number of the commonly consumable species (Fahey *et al.*, 2001). The enzyme myrosinase, produced by all plants that produce glucosinolates, is physically separated from the glucosinolates in the intact plant tissue. Glucosinolates and myrosinase are reported to be located in vacuoles, and are not colocalized in the same cells (Andréasson *et al.*, 2001) (Fig. 10.1). Upon tissue disruption the glucosinolates and the myrosinase come into contact (Fig. 10.2) and the myrosinase hydrolyses the glucosinolates to form hydrolysis products such as isothiocyanate, nitrile and thiocyanate (Gimsing and Kirkegaard, 2009) (Fig. 10.3).

The different chemical side-chain structures of the isothiocyanates determine important chemical and physical properties such as hydrophobicity and volatility, and their toxicity (Brown and Morra, 1997). Biofumigation can be achieved by incorporating fresh plant material (green manure), seed meals (a by-product of seed crushing for oil), or dried plant material treated to preserve isothiocyanate activity (Kirkegaard and Matthiessen, 2004; Lazzeri *et al.* 2004; Matthiessen and Kirkegaard, 2006).

A field experiment conducted in Huelva (South-west Spain; the most important area of strawberry production in Europe) contributes to the development and optimization of biofumigation with *Brassica* and soil solarization as alternatives to the traditional use of chemicals in strawberry production (Porras *et al.*, 2008).

Solarization is a process that employs solar radiation to heat soil under a transparent plastic film to temperatures that are detrimental to soilborne pathogens. Increased soil temperatures can decrease populations of plant pathogens (Katan, 1981). Solarization can enhance the effectiveness of other pest management approaches (Katan, 1981; Ben-Yephet *et al.*, 1987; Ramirez-Villapudua and Munnecke, 1988; Porras *et al.*, 2007a,b) and has the additional advantages of being a non-chemical alternative method for pathogen control (Batchelor, 2002).

The soil in the field experiment had never been previously treated with methyl bromide, and was naturally infested by *Phytophthora* spp. Treatments were soil solarization (S), biofumigation + solarization (B + S), and the untreated control (C). Treatments were applied in the same plots each year

○ Myrosinase

△ Glucosinolate

Fig. 10.1. Intact vegetal tissue with the myrosinase enzymes and the glucosinolates physically separated (within vacuoles).

(2005–2006 and 2006–2007). Biofumigation was done with *Brassica carinata* in the siliqua stage that showed the major suppressive effect in previous *in vitro* experiments (Romero *et al.*, 2008). *B. carinata* was cut, because upon tissue disruption the substrate and the enzyme come into contact and the myrosinase hydrolyses the glucosinolates to form hydrolysis products. The chopped material was incorporated into the soil to a depth of 10 cm using a rotary tiller. The soil was covered using clear 50-μm low-density polyethylene mulch, and drip-irrigated from July to September (Fig. 10.4).

Plots were covered to avoid the escape of isothiocyanates from the soil by volatilization and to increase soil temperatures. Drip irrigation was applied to enhance the formation of isothiocyanates in soil (Matthiessen *et al.* 2004).

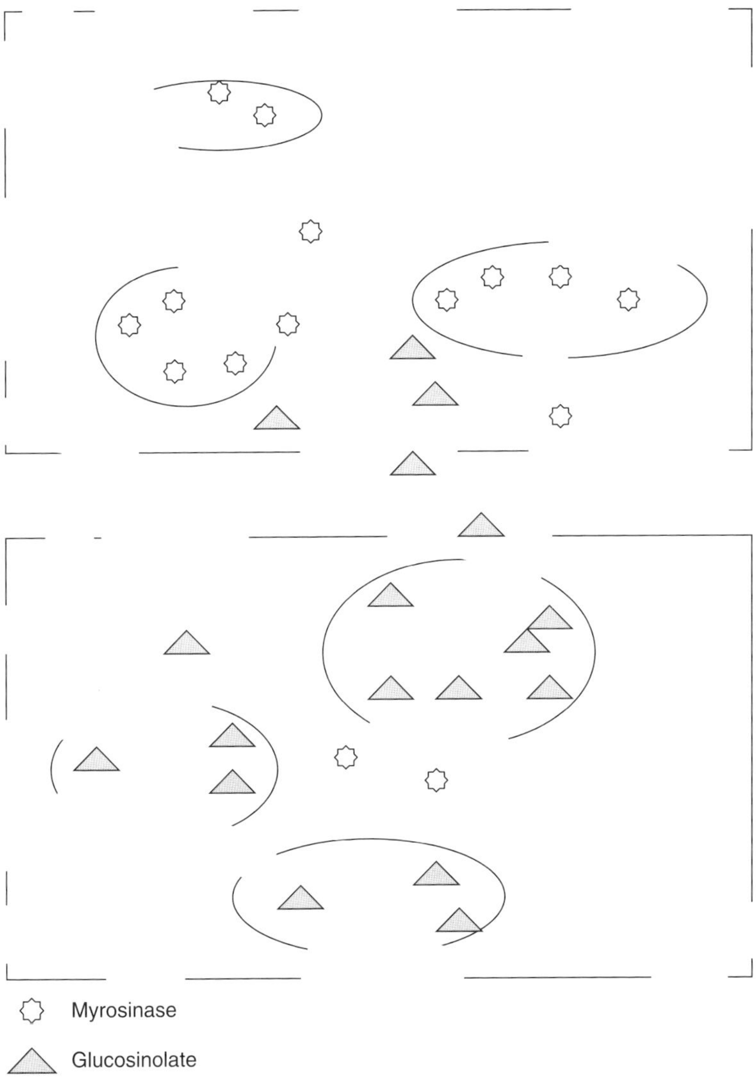

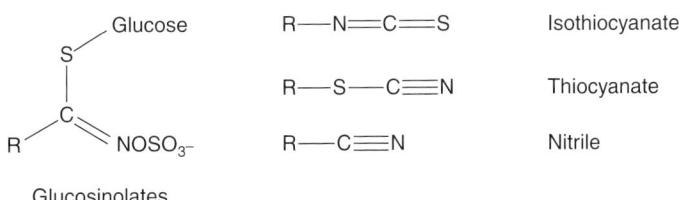

○ Myrosinase

△ Glucosinolate

Fig. 10.2. Upon tissue disruption the glucosinolates and the myrosinases come into contact and the myrosinase hydrolyses the glucosinolates to form hydrolysis products.

Glucose

$$S$$

$$C$$

$$R \qquad NOSO_3^-$$

Glucosinolates

R—N$=$C$=$S   Isothiocyanate

R—S—C$\equiv$N   Thiocyanate

R—C$\equiv$N   Nitrile

Fig. 10.3. Glucosinolates and hydrolysis products originated by the activity of the enzyme myrosinase.

Fig. 10.4. *Brassica carinata* (a) used as biofumigant was chopped (b), buried (c), and covered and drip-irrigated (d).

Biofumigation and solarization increased foliar surface and straw-berry yield the most each year, and differences were observed relative to solarization alone and the untreated control (Tables 10.2 and 10.3).

Biofumigation and solarization reduced *Phytophthora* spp. soil popula-tion relative to the untreated control, but it did not totally eliminate the pathogen. According to previous experiments (Porras, 2005; Porras *et al.*, 2007a), the results indicate that solarization may serve as a component in an integrated, sustainable approach to the management of *Phytophthora* spp. in strawberry.

The presence of high amounts of glucosinolates, and of the enzyme myrosinase that catalyses their hydrolysis, linked to the high biocidal activity of some glucosinolate enzymatic hydrolysis derivative products (mainly isothiocyanates and nitriles) have suggested the practical possibility of amending soil with these natural biocidal compounds by the cultivation and green manure of selected species of the family *Brassicaceae* (Lazzeri *et al.*, 2004). The potential of biofumigation as a

Table 10.2. Foliar surface (cm^2) in December and January 2005–2006 and 2006–2007.

	2005–2006		2006–2007	
	16th December	13th January	19th December	9th January
B + S	295.90 a	502.49 a	367.04 a	435.14 a
S	229.91 b	413.58 b	301.07 b	346.31 b
C	204.74 b	350.57 c	208.32 c	228.03 c
$F_{2,374}$	29.00*	20.83*	36.10*	53.48*

Letters indicate a significant difference, Tukey HSD test; *, $P<0.001$.

Table 10.3. Total accumulated strawberry yield (g/plant) until 31st March 2006 and 2007.

	2005–2006	2006–2007
B + S	349.31 a	318.66 a
S	300.17 b	261.45 b
C	252.84 c	148.13 c
$F_{2,38}$	11.12*	63.70*

Letters indicate a significant difference, Tukey's multiple range test; ***, $P<0.001$.

component of the integrated management of soil pathogens has been demonstrated in various agricultural systems (Gimsing and Kirkegaard, 2009).

10.4 Conclusion

Naturally occurring substances found in fungi, bacteria and plants are important sources of molecules with pesticidal properties. They may be developed either as products per se or used as starting points for synthesis to optimize specific properties. Low mammalian toxicity, low environmental impact, low levels of residues in food, and compatibility with IPM programmes are important considerations in formulation of plant-based products as pesticides. In addition, chemical pesticides such as methyl bromide are being phased out globally because of the impact on the ozone layer. Chemical, physical and biological alternative methods for pathogen control have been evaluated in crop production to replace the compounds lost due to the new registration requirements. Natural-product-based pesticides can sometimes be specific to the target species and have unique modes of action with little mammalian toxicity. Modern instrumentation and improved methods would increase interest in natural-product-based pesticide discovery research.

References

Andréasson, E., Jorgensen, L.B., Höglund, A.S., Rask, L. and Meijer, J. (2001) Different myrosinase and idioblast distribution in *Arabidopsis* and *Brassica napus*. *Plant Physiology* 127, 1750–1763.

Anonymous (2005) Available at http://www.epa.gov/opprd001/factsheets/kasuga-mycin.pdf

Bailey, B.A. and Lumsden, R.D. (1998) Direct effects of *Trichoderma* and *Gliocladium* on plant growth and resistance to pathogens. In: Harman, G.E. and Kubicek, C.P. (eds) *Trichoderma and Gliocladium: Enzymes, biological control and commercial applications*. Taylor and Francis, London, United Kingdom, pp. 185–204.

Balba, H. (2007) Review of strobilurin fungicide chemicals. *Journal of Environmental Science and Health part B - Pesticides Food Contaminants and Agricultural Wastes* 42, 441–451.

Batchelor, T.A. (2002) International and European Community controls on methyl bromide and the status of methyl bromide use and alternatives in the European Community. In: Batchelor, T.A. and Bolivar, J.M. (eds) *Proceedings of International Conference on Alternatives to Methyl Bromide*, Seville, Spain, pp. 28–33.

Ben-Yephet, Y., Stapleton, J.J., Wakeman, R.J. and DeVay, J.E. (1987) Comparative effects of soil solarization with single and double layers of polyethylene film on survival of *Fusarium oxysporum* f. sp. *vasinfectum*. *Phytoparasitica* 15, 181–185.

Brown, P.D. and Morra, M.J. (1997) Control of soil-borne plant pests using glucosinolate-containing plants. *Advances in Agronomy* 61, 167–231.

Cook, R.J. (1993) Making greater use of introduced microorganisms for biological control of plant pathogens. *Annual Review of Phytopathology* 31, 53–80.

Copping, L.G. and Duke S.O. (2007) Natural products that have been used commercially as crop protection agents. *Pest Management Science* 63, 524–554.

Dayan, F.E., Cantrell, C.L. and Duke, S.O. (2009) Natural products in crop protection. *Bioorganic and Medicinal Chemistry* 17, 4022–4034.

Dubey, N.K., Kumar, A., Singh, P., and Shukla, R. (2009) Exploitation of natural compounds in ecofriendly management of plant pests. In: Gisi, U., Chet, I. and Gullino, M.L. (eds) *Recent Developments in Management of Plant Diseases*, Springer Netherlands, pp. 181–198.

Duke, S.O., Baerson, S.R., Dayan, F.E., Rimando, A.M., Scheffler, B.E., Tellez, M.R., Wedge, D.E., Schrader, K.K., Akey, D.H., Arthur, F.H., De Lucca, A.J., Gibson, D.M., Harrison, H.F., Peterson, J.K., Gealy, D.R., Tworkoski, T., Wilson, C.L. and Morris, J.B. (2003) United States Department of Agriculture - Agricultural Research Service research on natural products for pest management. *Pest Management Science* 59, 708–717.

Duke, S.O., Dayan, F.E., Rimando, A.M., Schrader, K.K., Aliotta, G., Oliva, A. and Romagni J.G. (2002) Chemicals from nature for weed management. *Weed Science* 50, 138–151.

Duniway, J.M. (2002) Status of chemical alternatives to methyl bromide for pre-plant fumigation of soil. *Phytopathology* 92, 1337–1343.

European Parliament (2000) Regulation EC2037/00 of the European Parliament and of the Council of 29 June 2000 on Substances that Deplete the Ozone Layer. *Official Journal of European Communities*, 29 September 2000, L244, 1–24.

Fahey, J.W., Zalcmann, A.T. and Talalay, P. (2001) The chemical diversity and distribution of glucosinolates and isothiocyanates among plants. *Phytochemistry* 56, 5–51.

Gimsing, A.L. and Kirkegaard, J.A. (2009) Glucosinolates and biofumigation: fate of glucosinolates and their hydrolysis products in soil. *Phytochemistry Reviews* 8, 299–310.

Haas, D. and Keel, C. (2003) Root-colonizing *Pseudomonas* spp. and relevance for biological control of plant disease. *Annual Review of Phytopathology* 41, 117–153.

Handelsman, J. and Stabb, E.V. (1996) Biocontrol of soilborne plant pathogens. *The Plant Cell* 8, 1855–1869.

Howell, C.R. (2003) Mechanisms employed by *Trichoderma* species in the biological control of plant diseases: the history and evolution of current concepts. *Plant Diseases* 87, 4–10.

Ishii, H., Fraaije, B.A., Sugiyama, T., Noguchi, K., Nishimura, K., Takeda, T., Amano, T., and Hollomon, D.W. (2001) Occurrence and molecular characterization of strobilurin resistance in cucumber powdery mildew and downy mildew. *Phytopathology* 91, 1166–1171.

Katan, J. (1981) Solar heating (solarization) of the soil for control of soilborne pests. *Annual Review of Phytopathology* 19, 211–236.

Kim, B.S. and Hwang, B.K. (2007) Microbial fungicides in the control of plant diseases. *Journal of Phytopathology* 155, 641–653.

Kirkegaard, J.A. and Matthiessen, J.N. (2004) Developing and refining the biofumigation concept. *Agroindustria* 3, 233–239.

Knight, S.C., Anthony, V.M., Brady, A.M., Greenland, A.J., Heaney, S.P., Murray, D.C., Powell, K.A., Schulz, M.A., Spinks,C.A., Worthington, P.A. and Youle, D. (1997) Rationale and perspectives on the development of fungicides. *Annual Review of Phytopathology* 35, 349–372.

Lazzeri, L., Leoni, O. and Manici, L.M. (2004) Biocidal plant dried pellets for biofumigation. *Industrial Crops and Products* 20, 59–65.

Martin, F.N. and Bull, C.T. (2002) Biological approaches for control of root pathogens of strawberry. *Phytopathology* 92, 1356–1362.

Matthiessen, J.N. and Kirkegaard, J.A. (2006) Biofumigation and enhanced biodegradation: opportunity and challenge in soilborne pest and disease management. *Critical Reviews in Plant Science* 25, 235–265.

Matthiessen, J.N., Warton, B. and Shackleton, M.A. (2004) The importance of plant maceration and water addition in achieving high *Brassica*-derived isothiocyanate levels in soil. *Agroindustria* 3, 277–280.

Paulitz, T.C. and Belanger, R.R. (2001) Biological control in greenhouse systems. *Annual Review of Phytopathology* 39, 103–133.

Porras, M. (2005) Respuesta de patógenos de fresa (*Fragaria* x *ananassa* Duch.) a la utilización de *Trichoderma* spp. y solarización. Ph.D. Thesis. University of Seville, Spain.

Porras, M., Barrau, C., Arroyo, F.T., de los Santos, B., Blanco, C. and Romero, F. (2007a) Reduction of *Phytophthora cactorum* in strawberry fields by *Trichoderma* and soil solarization. *Plant Diseases* 91, 142–146.

Porras, M., Barrau, C. and Romero, F. (2007b) Effects of soil solarization and *Trichoderma* on strawberry production. *Crop Protection* 26, 782–787.

Porras, M., Barrau, C., Romero, E., Zurera, C. and Romero, F. (2008) Biofumigant effect of *Brassica carinata* on *Phytophthora* spp. in strawberry fields. *Journal of Plant Pathology* 90, S2.337.

Porter, N. (1985) Physicochemical and biophysical panel symposium biologically active secondary metabolites. *Pesticide Science* 16, 422–427.

Ramirez-Villapudua, R.J. and Munnecke, D.E. (1988) Effect of solar heating and soil amendments of cruciferous residues on *Fusarium oxysporium* f. sp. *conglutinans* and other organisms. *Phytopathology* 78, 289–295.

Romero, E., Zurera, C., Porras, M., Barrau, C. and Romero, F. (2008) Comparative study of different *Brassica* species to control *Phytophthora cactorum* and *Verticillium dahlie. Journal of Phytopathology* 90, S2. 337.

Warrior, P. (2000) Living systems as natural crop-protection agents. *Pest Management Science* 56, 681–687.

Weindling, R. (1934) Studies on a lethal principle effective in the parasitic action of *Trichoderma lignorum* on *Rhizoctonia solani* and other soil fungi. *Phytopathology* 24, 1153–1179.

Whipps, J.M. (2001) Microbial interactions and biocontrol in the rhizosphere. *Journal of Experimental Botany* 52, 487–511.

Zimand, G., Elad, Y. and Chet, I. (1996) Effect of *Trichoderma harzianum* on *Botrytis cinerea* pathogenicity. *Phytopathology* 86, 1255–1260.

11 Fungal Endophytes: an Alternative Source of Bioactive Compounds for Plant Protection

R.N. KHARWAR[1] AND GARY STROBEL[2]

[1]Department of Botany, Banaras Hindu University, India; [2]Department of Plant Sciences, Montana State University, Bozeman, USA

Abstract

Endophytes are a group of microorganisms that represent an abundant and dependable source of bioactive and chemically novel compounds with potential for exploitation in a wide variety of applications. The mechanisms through which endophytes exist and respond to their surroundings must be better understood in order to be more predictive about which higher plants to seek, study and employ in isolating their microfloral components. This may facilitate the natural product discovery process. Endophytic fungi are now attracting great interest from researchers for an alternative way of controlling plant pathogens.

11.1 Introduction

The bioactive compounds of natural origin have been under ever-increasing demand to solve various problems related to human health and agriculture. Plants have provided humans with resources for healing purposes for millennia. Some representative and well known medicines derived from plants are quinine, digitalin, taxol, aspirin, reserpine and many more (Wani *et al.*, 1971; Hoffman *et al.*, 1998). Some plants have been under great threat as a result of the enormous pressure brought upon them by virtue of their healing and biocontrol properties.

In some cases, plant-associated fungi are able to make the same bioactive compounds as the host plant itself and one of the best examples of this is the discovery of the gibberellins in *Fusarium fujikuroi* in the early 1930s (Borrow *et al.*, 1955). Eventually it was learnt that the gibberellins are one of only five classes of phytohormones that are to be found in virtually all plants. Endophytic (endon = inner; phyton = plants) fungi are a group of microorganisms that reside in healthy and functional inner tissues of the plants without causing any detectable symptoms (Petrini, 1991).

© CAB International 2011. *Natural Products in Plant Pest Management* (ed. N.K. Dubey)

The observation that fungi make the same compounds as their host led Stierle *et al.* (1993) to examine the prospect that endophytic fungi associated with *Taxus brevifolia* may also produce taxol, the most promising natural bioactive molecule discovered against cancer. This significant discovery established that the endophytic fungi which reside in the living healthy tissues of the plant, may produce one or more bioactive principles or natural products with either medicinal/therapeutic or biocontrol potential which were previously known to be produced by the host plant (Wani *et al.*, 1971; Stierle *et al.*, 1993; Dimetry *et al.*, 1995; Hoffman *et al.*, 1998; Gahukar, 2005). Taxol itself is the world's first billion dollar anticancer drug and its main source is *Taxus* spp. Potentially, a fungal source of taxol would reduce its price and save the plant, in some areas, from extinction. The success of finding fungal taxol has produced a paradigm for other bioactive compounds to be found in endophytic fungi. This might have happened during the course of evolution of symbiosis between fungus and host plant, where reciprocal gene transfer has occurred. As observed in the case of *Agrobacterium tumefaciens*, the t-DNA of the bacterial plasmid was transferred and incorporated into the genome of the host plant. This is a fine example of the transaction of genetic materials between prokaryote and eukaryote. However, in case of the fungal endophyte–host relationship, it is a eukaryote to eukaryote transaction.

As the fungi are a less studied group of microbes, only a few studies can be exemplified. Alkaloids synthesized by *Neotyphodium* sp. in its grass hosts have been implicated in fescue toxicosis in rangeland animals. The chemistry and biology of this and other grass endophytes are reviewed elsewhere. Unfortunately, because this work was so comprehensive, one may be led to the conclusion that endophytes only produce toxic compounds in their respective hosts and that they hold no promise for any medicinal applications. It turns out that this is simply not the case. Endophytes examined from a plethora of sources show that an overwhelming number of them produce natural products with promising potential for many applications.

Bacon and White (2000) gave an inclusive and widely accepted definition of endophytes: 'Microbes that colonize living, internal tissues of plants without causing any immediate, overt negative effects' (Fig. 11.1). While the usual lack of adverse effect nature of endophyte occupation in plant tissue has prompted a focus on symbiotic or mutualistic relationships between endophytes and their hosts, the observed biodiversity of endophytes suggests they can also be aggressive saprophytes or opportunistic pathogens (Promputtha *et al.*, 2007).

Both fungi and bacteria are the most common microbes existing as endophytes. It would seem that other microbial forms most certainly exist in plants as endophytes such as mycoplasmas, rickettsia, streptomycetes and archebacteria. In fact, it may be the case that the majority of microbes existing in plants are not culturable with common laboratory techniques, making their presence and role in plants even more intriguing. The most frequently isolated endophytes are the fungi (Redlin and Carris, 1996). However, at the

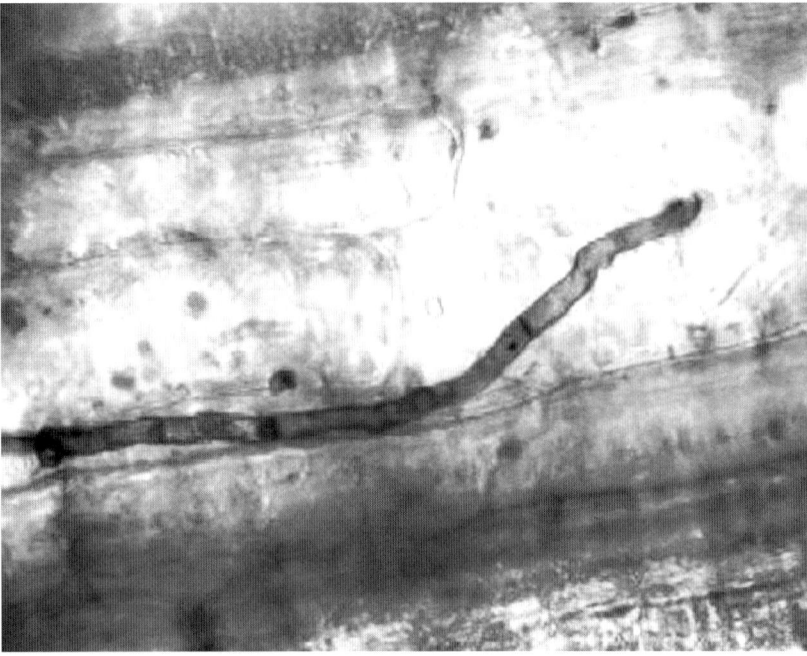

Fig. 11.1.   A prominent endophytic fungal mycelium lying between the cellular spaces.

outset, it is important to note that the vast majority of plants have not been studied for any endophytic association. Thus, enormous opportunities exist for the recovery of novel fungal forms, including genera, biotypes, as well as species in the myriad of plants yet to be studied. Hawksworth and Rossman (1987) estimated there may be as many as 1.5 million different fungal species, yet only about 100,000 have been described. As more evidence accumulates, estimates keep rising as to the actual number of fungal species. For instance, Dreyfuss and Chapela (1994) have estimated at least 1 million species of endophytic fungi alone. It seems obvious that endophytes are a rich and reliable source of genetic diversity and may represent many previously undescribed species.

Out of 300,000 higher plants that exist on the earth, only a handful, primarily grass species, have been studied relative to their endophytic biology. Consequently, there is enormous opportunity to find new and interesting endophytic fungi amongst the myriad of plants in different settings and ecosystems. The intent of this chapter is to provide insights into the occurrence of endophytes in nature, the products that they make, and how some of these organisms show their potential for human and plant use. The chapter discusses the rationale for study, methods used, and examples of some of the endophytes isolated and studied over the course of time. This chapter, however, also includes some specific examples that illustrate the work done in this emerging field of bioprospecting and sourcing the endophytic fungi.

11.2 Diversity in Endophytes

Of the myriad of ecosystems on Planet Earth, those having the greatest general biodiversity of life seem to be the ones that also have the greatest number and most diverse endophytes. Tropical and temperate rainforests are the most biologically diverse terrestrial ecosystems. The most threatened of these spots cover only 1.44% of the land's surface, yet they harbour over 60% of the world's terrestrial biodiversity. In addition, each of the 20–25 areas identified as supporting the world's greatest biodiversity also support unusually high levels of plant endemism (Mittermeier, 1999). As such, one would expect, that with high plant endemism, there should also exist specific endophytes that may have evolved with the endemic plant species. Biological diversity implies chemical diversity because of the constant chemical innovation that is required to survive in ecosystems where the evolutionary race to survive is most active. Tropical rainforests are a remarkable example of this type of environment. Competition is great, resources are limited and selection pressure is at its peak. This gives rise to a high probability that rainforests are a source of novel molecular structures and biologically active compounds (Redell and Gordon, 2000).

Bills *et al.* (2002) describe a metabolic distinction between tropical and temperate endophytes through statistical data which compares the number of bioactive natural products isolated from endophytes of tropical regions to the number of those isolated from endophytes of temperate origin. Not only did they find that tropical endophytes provide more active natural products than temperate endophytes, but they also noted that a significantly higher number of tropical endophytes produced a larger number of active secondary metabolites than did fungi from other substrata. This observation suggests the importance of the host plant as well as the ecosystem in influencing the general metabolism of endophytic microbes.

11.3 Techniques for Isolation, Preservation and Storage of Endophytes

Cultures for product isolation

Detailed techniques for the isolation of microbial endophytes are outlined in a number of reviews and technical articles (Strobel *et al.*, 1993; Strobel, 2002; Strobel and Daisy, 2003; Strobel *et al.*, 2004; Castillo *et al.*, 2005). If endophytes are being obtained from plants growing in polar regions, the dry tropics, or some temperate areas of the world, one can expect to acquire from none to only one or two endophytic cultures per plant sample (0.5–10.0 cm limb piece). However, from the wet tropics this number may rise to 20–30 or even more microbes per plant piece. Given limited fermentation capabilities, it is critical to label and store cultures of freshly isolated microbes for work in the future and for patent and publication purposes.

Generally, preservation in an aqueous 15% glycerol solution at –70°C is an exceedingly good procedure for saving cultures until work on them can proceed at a later date (Strobel *et al.*, 1993; Strobel, 2002; Strobel and Daisy, 2003; Strobel *et al.*, 2004).

Growing fungal cultures on sterilized barley or other grains with subsequent storage at –70°C is an effective alternative to the glycerol storage solution.

Plant selection norms and ethics

It is important to understand that the methods and rationale used to guide plant selection seem to provide the best opportunities to isolate novel endophytic microorganisms at the genus, species or biotype level. Since the number of plant species in the world is so great, creative and imaginative strategies must be used to quickly narrow the search for endophytes displaying bioactivity (Mittermeier *et al.*, 1999).

A specific rationale for the collection of each plant for endophyte isolation and natural product discovery is used. Several hypotheses govern this plant selection strategy:

1. Plants from unique environmental settings, especially those with an unusual biology, and possessing novel strategies for survival are seriously considered for study.
2. Plants that have an ethnobotanical history (use by indigenous peoples) that are related to the specific uses or applications of interest are selected for study. These plants are chosen either by direct contact with local people or via local literature. This approach may also provide benefits to the local/ tribal people by documenting their help under a participatory research programme. The ethics of science restrict the researcher drawing the monetary and intellectual benefits alone, especially when knowledge of local people is used. However, it often happens that at the time of documentation, scientists don't give proper importance to the knowledge source or the local people, and this must be checked to maintain the sanctity of the research community. Ultimately, it may be learned that the healing powers of the botanical source, in fact, may have nothing to do with the natural products of the plant, but of an endophyte inhabiting the plant.
3. Plants that are endemic, having an unusual longevity, or that have occupied a certain ancient land mass, such as Gondwanaland, are also more likely to lodge endophytes with active natural products than other plants.
4. Plants growing in areas of great biodiversity, it follows, also have the prospect of housing endophytes with great biodiversity.

Distinct environmental settings are therefore considered to be a promising source of novel endophytes and their compounds, and so too are plants with an unconventional biology. For example, an aquatic plant, *Rhyncholacis penicillata*, was collected from a river system in south-west Venezuela where that plant faced constant beating by virtue of rushing waters, debris, and tumbling rocks and pebbles. These environmental insults created many portals through which common phytopathogenic oomycetes could enter the plant. Still, the plant population appeared to be healthy, possibly due to the protection provided by an endophytic product. This was the environmental biological clue used to choose this plant for a comprehensive study of its endophytes.

An example of plants with an ethnobotanical history is the snakevine *Kennedia nigriscans*, from the Northern Territory of Australia, which was selected for study since its sap has traditionally been used as a bush medicine. In fact, this area was selected for plant sampling as it has been home to the world's longest standing human civilization, the Australian Aborigines. The snakevine is harvested, crushed and heated in an aqueous brew by local Aborigines in South-west Arnhemland to treat cuts, wounds and infections. As it turned out, the plant contains an entire suite of streptomycetes (Castillo *et al.*, 2005). One in particular, designated *Streptomyces* NRRL 30562, has unique partial 16S rDNA sequences when compared to those in GenBank (Castillo *et al.*, 2002). It produces a series of actinomycins including actinomycins D, Xo_β, and X_2. It produces novel, broad-spectrum peptide antibiotics called munumbicins. This seems to be an excellent example illustrating the potential benefits of knowledge of indigenous peoples and researchers.

Furthermore, some plants generate bioactive natural products and have associated endophytes that produce the same natural products. Such is the case with taxol, a highly functionalized diterpenoid and famed anticancer agent that is found in *Taxus brevifolia* and other yew species (*Taxus* spp.) (Wani *et al.*, 1971). In 1993, a novel taxol-producing fungus *Taxomyces andreanae* was isolated and characterized from *Taxus brevifolia* (Strobel *et al.*, 1993).

While combinatorial synthesis produces compounds at random, secondary metabolites, defined as low molecular weight compounds not required for growth in pure culture, are produced as an adaptation for specific functions in nature (Demain, 1981). Schulz (2001) points out that certain microbial metabolites seem to be characteristic of certain biotopes, both at an environmental as well as an organism level. Accordingly, it appears that the search for novel secondary metabolites should centre on organisms that inhabit unique biotopes. Thus, it behoves the investigator to carefully study and select the biological source before proceeding, rather than to take a totally random approach in selecting the source material. Careful study also indicates that organisms and their biotopes that are subjected to constant metabolic and environmental interactions should produce even more secondary metabolites (Schulz *at al.*, 2002). Endophytes are microbes that inhabit such biotopes, namely higher plants, which is why they are currently considered as a wellspring of novel secondary metabolites offering the potential for exploitation of their medical benefits (Tan and Zou, 2001). Moreover, novel compound-screening experiments have already proved that endophytes are considerably ahead of soil isolates, with 51% new structure compounds compared with 38%, respectively (Schulz *et al.*, 2002).

Molecular techniques provided unexpected and thrilling results with *Pestalotiopsis microspora*

It is of some compelling interest how the genes for taxol production may have been acquired by endophytes such as *Pestalotiopsis microspora* (Long *et al.*, 1998). Although the complete answer to this question is not at hand, some other relevant genetic studies have been performed on this organism. *P. microspora* Ne 32 is one of the most easily genetically transformable fungi that has been studied to date. *In vivo* addition of telomeric repeats to foreign DNA generates chromosomal DNAs in this fungus (Long *et al.*, 1998). Repeats of the telomeric sequence 5'-TTAGGG-3' were appended to non-telomeric-transforming DNA termini. The new DNAs, carrying foreign genes and the telomeric repeats, replicated independently of the chromosome and expressed the information carried by the foreign genes. The addition of telomeric repeats to foreign DNA is unusual among fungi. This finding may have important implications in the biology of *P. microspora* Ne 32 since it explains at least one mechanism as to how new DNA can be captured by this organism and eventually expressed and replicated. Such a mechanism may begin to explain how the enormous biochemical variation may have arisen in this fungus. Also, this initial work represents a framework to aid in the understanding of how this fungus may adapt itself to the environment of its plant hosts and suggests that the uptake of plant DNA into its own genome may occur. In addition, the telomeric repeats have the same sequence as human telomeres which points to the possibility that *P. microspora* may serve as a means to make artificial human chromosomes, a totally unexpected result.

11.4 Natural Products and Phytochemicals from Endophytes

It has been suggested that the reason some endophytes produce certain phytochemicals, originally characteristic of the host, might be related to a genetic combination of the endophyte with the host that occurred in evolutionary time (Tan and Zou, 2001). This is a concept that was originally proposed as a mechanism to explain why *Taxomyces andreanae* may be producing taxol (Stierle *et al.*, 1993). All aspects of the biology and interrelationship of endophytes with their respective hosts is a vastly under-investigated and exciting field (Strobel, 2002; Strobel and Daisy, 2003; Strobel *et al.*, 2004). Thus, more background information on a given plant species and its microorganismal biology would be exceedingly helpful in directing the search for bioactive products (Lane *et al.*, 2000). Presently, no one is quite certain of the role of endophytes in nature and what appears to be their relationship to various host plant species. While some endophytic fungi appear to be ubiquitous (e.g. *Fusarium* spp., *Pestalotiopsis* spp., *Aspergillus* spp. and *Xylaria* spp.), one cannot definitively state that endophytes are truly host specific or even systemic within plants any more than assume that their associations are chance encounters (Promputtha *et al.*, 2007). Frequently, many endophytes of the same species are isolated from the same plant, and only one or a few biotypes of a given fungus will produce a highly biologically active compound in culture. A great deal of uncertainty also exists about what an endophyte produces in culture and what it may produce in nature. It does seem possible that the production of certain bioactive compounds by the endophyte *in situ* may facilitate the domination of its biological niche within the plant or even provide protection to the plant from harmful invading pathogens, insects and environmental stresses (Omacini, *et al.*, 2001; Rodriguez, *et al.*, 2008). Furthermore, little information exists on the biochemistry and physiology of the interactions of the endophyte with its host plant. It would seem that many factors changing in the host depending on the season, and other factors including age, environment, and location, may influence the biology of the endophyte. Indeed, further research at the molecular level must be conducted in the field to study endophyte interactions and ecology. All of these interactions are probably chemically mediated for some purpose in nature. An ecological awareness of the role these organisms play in nature will provide the best clues for targeting particular types of endophytic bioactivity with the greatest potential for bioprospecting.

There is a general call for new antibacterial, antifungal, antiviral and chemotherapeutic agents that are highly effective and possess low toxicity. This new approach of search is driven by the development of resistance in infectious microorganisms (e.g. *Staphylococcus, Mycobacterium, Streptococcus, Bacillus, Aspergillus* spp.) to existing drugs and by the menacing increase of naturally resistant organisms. The ingress to the human population of disease-causing agents such as AIDS, Ebola and SARS requires the discovery and development of new drugs to combat them. Not only diseases such as AIDS require drugs that target them specifically, but new therapies are needed for treating ancillary infections that are a consequence of a weakened immune system. Furthermore, others who are immunocompromised (e.g. cancer and organ-transplant patients) are at high risk of infection by opportunistic pathogens, such as *Aspergillus, Cryptococcus* and *Candida* sp., that are not normally major problems in the human population. In addition, more drugs are needed to treat efficiently parasitic protozoan and nematodal infections such as malaria, leishmaniasis, trypanomiasis, incephalitis and filariasis. Malaria alone is more dangerous in causing human death each year than any other single infectious agent with the exception of AIDS and tuberculosis (NIAID, 2001). However, the enteric diseases as a group claim the most lives each year of any other disease complex and, unfortunately, the victims are mostly children (NIAID, 2001).

Endophytic fungi are now attracting great interest from researchers as an alternative source in controlling plant pathogens. The control of late blight, caused by *Phytophthora infestans*, is very difficult by chemical means, because of the high virulence of the pathogen and increasing resistance to available fungicides (Griffith *et al.*, 1992). The fermentation broths of 52.3% of endophytic fungi displayed growth inhibition of at least one pathogenic fungus, such as *Neurospora* sp., *Trichoderma* sp. and *Fusarium* sp. (Huang *et al.*, 2001). In a similar study, fermentation broths of nine (4.8%) out of the 187 endophytic fungi isolated from mainly woody plants were highly active against

P. infestans in tomato plants (Park *et al.*, 2005). Induced resistance was generated in tomato plant against fusarium wilt by endophytic *Fusarium oxysporum* (Duijff *et al.*, 1998). *Sclerotinia sclerotiorum* is a common root-, crown- and stem-rot-causing pathogen in cabbage, common bean, citrus, celery, coriander, melon, squash, soybean, tomato, lettuce, cucumber and so on. Cyclosporine is characterized as a major antifungal substance against *S. sclerotiorum* from the fermentation broth of *F. oxysporum* S6 (Rodriguez *et al.*, 2006). *F. oxysporum* strain EF119, isolated from roots of red pepper, showed the most potent *in vivo* anti-oomycete activity against tomato late blight and *in vitro* anti-oomycete activity against several oomycete pathogens (Kim *et al.*, 2007). Out of 510 endophytic fungi screened, 64 strains exhibited antifungal activities against *Candida albicans*, *Candida glabrata*, *Candida krusei*, *Cryptococcus neoformans*, *Aspergillus fumigatus*, *Aspergillus flavus*, *Rhizopus oryzae*, *Trichophyton rubrum* and *Microsporum canis* (Anke *et al.*, 2003).

Narisawa *et al.* (2000) found that the root endophytic hyphomycete *Heteroconium chaetospira* suppressed Verticillium yellows in Chinese cabbage in the field. Verticillium wilt is one of the most destructive diseases of aubergine. Eleven out of 123 isolates of endophytic fungi, especially *Heteroconium chaetospira*, *Phialocephala fortinii*, *Fusarium*, *Penicillium*, *Trichoderma* and mycelium radicis atrovirens (MRA), after being inoculated onto axenically reared aubergine seedlings, almost completely suppressed the pathogenic effects of a post-inoculated, virulent strain of *Verticillium dahliae* (Narisawa *et al.*, 2000). Out of 39 endophytes of *Artemisia annua* investigated, 21 showed *in vitro* antifungal activity against crop-threatening fungi *Gaeumannomyces graminis* var. *tritici*, *Rhizoctonia cerealis*, *Helminthosporium sativum*, *Fusarium graminearum*, *Gerlachia nivalis* and *Phytophthora capsici* (Liu *et al.*, 2001). The extracts of endophytic *Alternaria* sp., isolated from medicinal plants of Western Ghats of India, inhibited the growth of *C. albicans* (Raviraja *et al.*, 2006). About 41.2% of all the isolated endophytic fungi from rice plants showed antagonism against *Magnaporthe grisea*, *Rhizoctonia solani* and *Fusarium moniliforme*. *Colletotrichum gloeosporioides* (Penz) Penz & Sacc. was isolated as an endophyte from healthy leaves of *Cryptocarya mandioccana* and had antifungal activity against phytopathogenic fungi *Cladosporium cladosporioides* and *Cladosporium sphaerospermum* (Inacio *et al.*, 2006). Fungal endophytes *Chaetomium* spp., *Phoma* sp., isolated from asymptomatic leaves of wheat, reduced the number of pustules and the area of pustules of *Puccinia recondita* f. sp. *Tritici* (Dingle and Mcgee, 2003). An *in vitro* study showed 40%, 65% and 27% antagonistic interaction by endophytic morphospecies against cacao pathogens *Moniliophthora roreri*, *Phytophthora palmivora* and *Crinipellis perniciosa*, respectively, while in-field endophytic *C. gloeosporioides* produced a significant decrease in pod loss (Mejia *et al.*, 2008).

For novel natural products, exciting possibilities exist for those who are willing to venture into the wild and unexplored territories of the world to experience the excitement and thrill of engaging in the discovery of endophytes, their biology, and potential usefulness. It is evident that natural-product-based compounds have an immense impact on modern medicine. For instance, about 40% of prescription drugs are based on them, and well over 50% of the new chemical products registered by the US Food and Drug Administration (FDA) as anticancer agents, antimigraine agents and antihypertensive agents from 1981 to 2002 are natural products or their derivatives (Newman *et al.*, 2003). Excluding biologics, between 1989 and 1995, 60% of approved drugs and pre-new-drug-application (NDA) candidates were of natural origin. From 1983 to 1994, over 60% of all approved and pre-NDA stage cancer drugs were of natural origin, as were 78% of all newly approved antibacterial agents (Concepcion *et al.*, 2001). The discovery and development of taxol is a modern example of a natural product that has made an enormous impact on medicine (Wani *et al.*, 1971; Suffness, 1995; Schulz and Christine, 2005).

Antibiotics from endophytic fungi

Generally, the most commonly isolated endophytic fungi are in the group of Fungi Imperfecti or Deuteromycotina. Also, it is quite common to isolate endophytes that produce no fruiting structures whatsoever such as Mycelia-Sterilia.

Cryptosporiopsis cf. *quercina* is the imperfect stage of *Pezicula cinnamomea*, a fungus commonly associated with hardwood species in Europe. It was isolated as an endophyte from *Tripterigeum wilfordii*, a medicinal plant native to Eurasia (Strobel *et al.*, 1999). On Petri plates, *C. quercina* demonstrates excellent antifungal activity against some important human fungal pathogens including *C. albicans* and *Trichophyton* spp. Cryptocandin, a unique peptide antimycotic related to the echinocandins and the pneumocandins (Walsh, 1992), was isolated and characterized from *C. quercina* (Strobel *et al.*, 1999). This compound contains a number of peculiar hydroxylated amino acids and a novel amino acid, 3-hydroxy-4-hydroxymethyl proline (Fig. 11.2). It is generally true that not one but several bioactive and related compounds are produced by an endophytic microbe. So it is that other antifungal agents related to cryptocandin are also produced by *C. quercina*. Cryptocandin is also active against a number of plant pathogenic fungi including *S. sclerotiorum* and *Botrytis cinerea*. Cryptocandin and its related compounds are currently being considered for use against a number of fungi causing diseases of the skin and nails.

Cryptocin, a unique tetramic acid, is also produced by *C. quercina* (Fig. 11.3) (Li *et al.*, 2000). This unusual compound possesses potent activity against *Pyricularia oryzae*, the causal organism of 'rice blast', one of the worst plant diseases in the world, as well as a number of other plant pathogenic fungi (Li *et al.*, 2000). The compound was generally ineffective against an array of human and plant pathogenic fungi. Nevertheless, with minimum inhibitory concentrations against *P. oryzae* at 0.39 µg/ml, this compound is being examined as a natural chemical control agent for rice blast and is being used as a model to synthesize other antifungal compounds.

Pestelotiopsis microspora is a common rainforest endophyte (Strobel, 2002; Strobel and Daisy, 2003). It turns out that enormous biochemical diversity

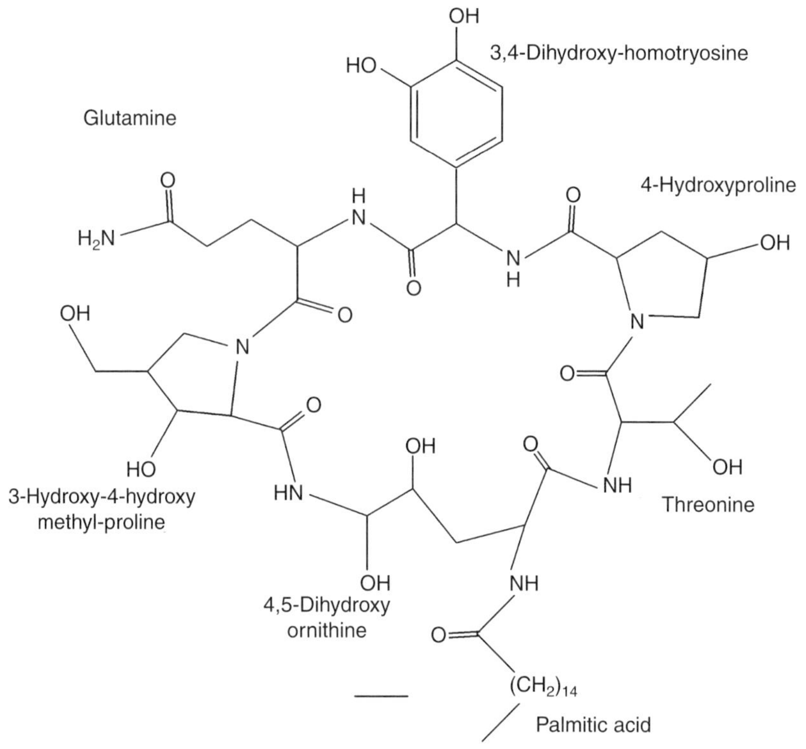

Fig. 11.2 Cryptocandin A, an antifungal lipopeptide obtained from the endophytic fungus *Cryptosporiopsis* cf. *quercina*.

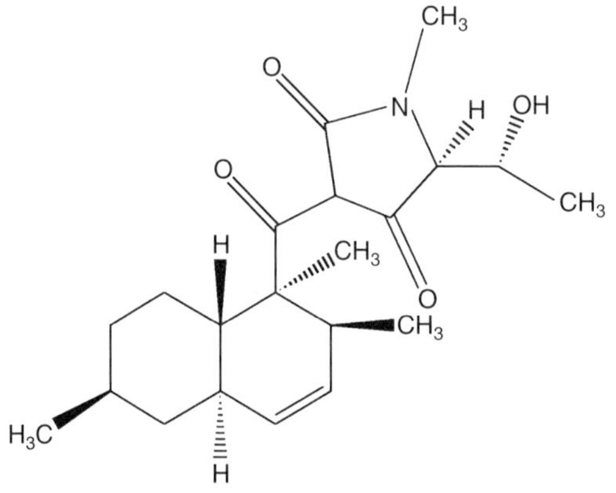

Fig. 11.3 Cryptocin, a tetramic acid antifungal compound found in *Cryptosporiopsis* cf. *quercina*.

does exist in this endophytic fungus and many secondary metabolites are produced by various strains of this widely dispersed organism. One such secondary metabolite is ambuic acid, an antifungal agent, which has been recently described from several isolates of *P. microspora* found as representative isolates in many of the world's rainforests (Fig. 11.4; Li *et al.*, 2001). As an interesting spin-off to endophyte studies, ambuic acid and another endophyte product, terrein, have been used as models to develop new solid-state NMR tensor methods to assist in the characterization of molecular stereochemistry of organic molecules (Harper *et al.*, 2001).

A strain of *P. microspora* was isolated from the endangered tree *Torreya taxifolia* and produces several compounds that have antifungal activity, including pestaloside, an aromatic β-glucoside, and two pyrones, pestalopyrone and hydroxypestalopyrone (Lee *et al.*, 1995a,b). These compounds also possess phytotoxic properties. Other newly isolated secondary products obtained from *P. microspora* (endophytic on *Taxus brevifolia*) include two new caryophyllene sesquiterpenes, pestalotiopsins A and B (Pulici *et al.*, 1996a,b), 2α-hydroxydrimeninol and a highly functionalized humulane. Variation in the amount and types of products found in this fungus depends on both the culture conditions and the original plant source from which it was isolated.

Pestalotiopsis jesteri is a newly described endophytic fungal species from the Sepik river area of Papua New Guinea, and it produces the highly functionalized cyclohexenone epoxides, jesterone and hydroxy-jesterone, which exhibit antifungal activity against a variety of plant pathogenic fungi (Li and Strobel, 2001). Jesterone has subsequently been synthesized with complete retention of biological activity (Fig. 11.5; Hu *et al.*, 2001). Jesterone is one of

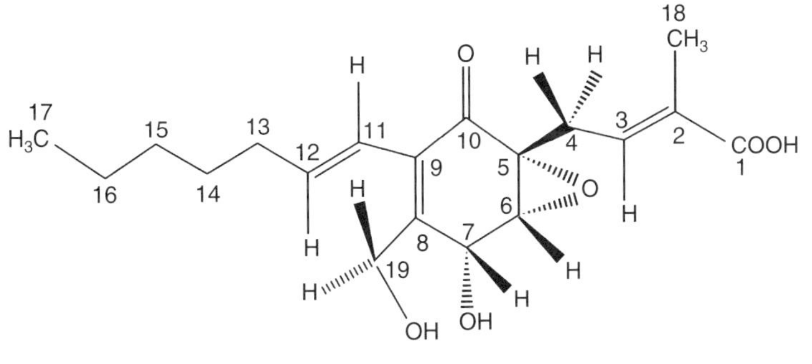

Fig. 11.4. Ambuic acid, a highly functionalized cyclohexenone epoxide produced by a number of isolates of *Pestalotiopsis microspora* found in rainforests around the world.

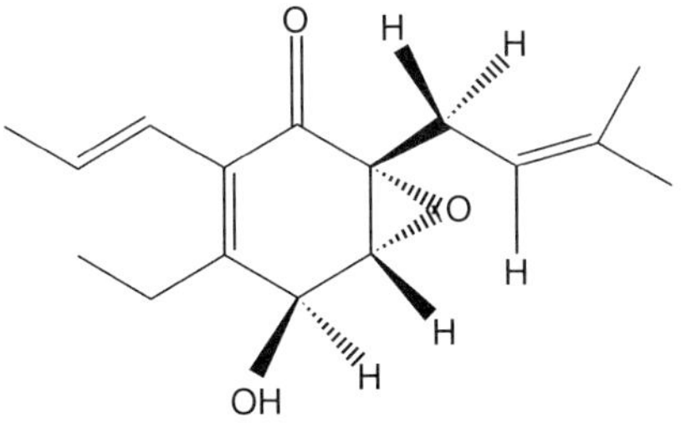

Fig. 11.5. Jesterone, a cyclohexenone epoxide from *Pestaliotiopsis jesteri*, has antioomycete activity.

only a few products from endophytic microbes for which total synthesis has been successfully accomplished.

Phomopsichalasin, a metabolite from an endophytic *Phomopsis* sp., represents the first cytochalasin-type compound with a three-ring system replacing the cytochalasin macrolide ring. This metabolite exhibits antibacterial activity in disc diffusion assays (at a concentration of 4 µg/disc) against *Bacillus subtilis*, *Salmonella gallinarum* and *Staphylococcus aureus*. It also displays a moderate activity against the yeast *Candida tropicalis* (Horn *et al.*, 1995).

An endophytic *Fusarium* sp. isolated from *Selaginella pallescens* collected in the Guanacaste Conservation Area of Costa Rica was screened for antifungal activity. A new pentaketide antifungal agent, CR377, was isolated from the culture broth of the fungus and showed potent activity against *C. albicans* in agar diffusion assays (Brady and Clardy, 2000). Colletotric acid, a metabolite of *Colletotrichum gloeosporioides*, an endophytic fungus isolated from *Artemisia mongolica*, displays antibacterial activity against bacteria as well as against the fungus *Helminthsporium sativum*. Antimicrobial products have been identified from another *Colletotrichum* sp. isolated from *Artemisia annua*, a traditional Chinese herb that is well recognized for its synthesis of artemisinin (an antimalarial drug) and its ability to inhabit many geographically different areas. Not only did the *Colletotrichum* sp. found in *A. annua* produce metabolites with activity against human pathogenic fungi and bacteria, but also metabolites that were fungistatic to plant pathogenic fungi (Lu *et al.*, 2000).

A novel antibacterial agent, guignardic acid, was isolated from the endophytic fungus *Guignardia* sp.; the organism was obtained from the medicinal plant *Spondias mombin* of the tropical plant family Anacardiaceae found in Brazil. The compound was isolated by UV-guided fractionation of the fermentation products of this fungus and is the first member of a novel class of natural compounds containing a dioxolanone moiety formed by the fusion of 2-oxo-3-phenylpropanoic acid and 3-methyl-2-oxobutanoic acid, which are products of the oxidative deamination of phenylalanine and valine, respectively (Fig. 11.6; Rodriguez-Heerklotz *et al.*, 2001).

Another antibacterial compound javanicin is produced by endophytic fungus *Chloridium* sp., under liquid- and solid-media culture conditions

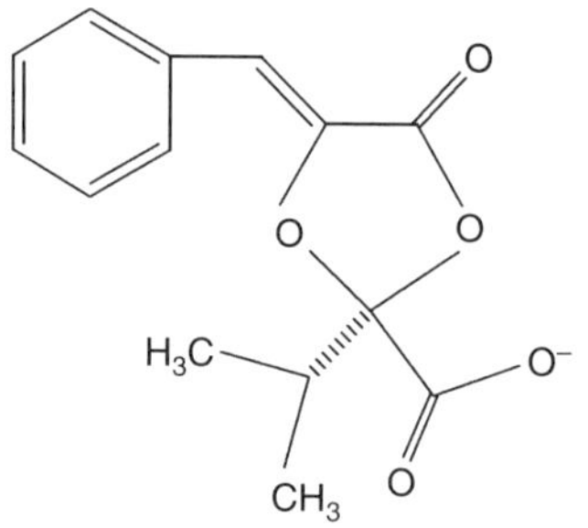

Fig. 11.6. Guignardic acid from *Guignardia* sp. obtained from *Spondias mombin*, an Anacardiaceaeous plant in Brazil.

(Kharwar *et al.*, 2009). This highly functionalized naphthaquinone exhibits strong antibacterial activity against *Pseudomonas* spp., representing pathogens to both humans and plants. The compound was crystallized and the structure was elucidated using X-ray crystallography. The X-ray structure confirms the previously elucidated structure of the compound that was done through standard spectroscopic methods (Fig. 11.7).

Volatile antibiotics from fungal endophytes

Muscodor albus is a newly described endophytic fungus obtained from small limbs of *Cinnamomum zeylanicum* (Worapong *et al.*, 2001). This xylariaceaous (non-spore-producing) fungus effectively inhibits and kills certain other fungi and bacteria by producing a mixture of volatile compounds (Strobel *et al.*, 2001). The majority of these compounds have been identified by gas chromatography / mass spectrometry (GC/MS), synthesized or acquired, and then ultimately formulated into an artificial mixture that mimicked the antibiotic effects of the volatile compounds produced by the fungus (Strobel *et al.*, 2001). Individually, each of the five classes of volatile compounds produced by the fungus had some antimicrobial effects against the test fungi and bacteria but none was lethal. However, collectively they acted synergistically to cause death in a broad range of plant and human pathogenic fungi and bacteria. The most effective class of inhibitory compounds was the esters, of which isoamyl acetate was the most biologically active; however, in order to be lethal it needs to be combined with other volatiles (Strobel *et al.*, 2001). The composition of the medium on which *M. albus* is grown dramatically influences the kind of volatile compounds that are produced (Ezra and Strobel, 2003).

The ecological implications and potential practical benefits of the 'myco-fumigation' effects of *M. albus* are very promising given the fact that soil fumigation utilizing methyl bromide will soon be illegal in the USA. Methyl bromide is not only a hazard to human health but it has been implicated in causing destruction of the world's ozone layer. The potential use of mycofumigation to treat soil, seeds, and plants will soon be a reality as AgraQuest of Davis, California, has EPA approval to release this organism for agricultural uses. The artificial mixture of volatile compounds may also have usefulness

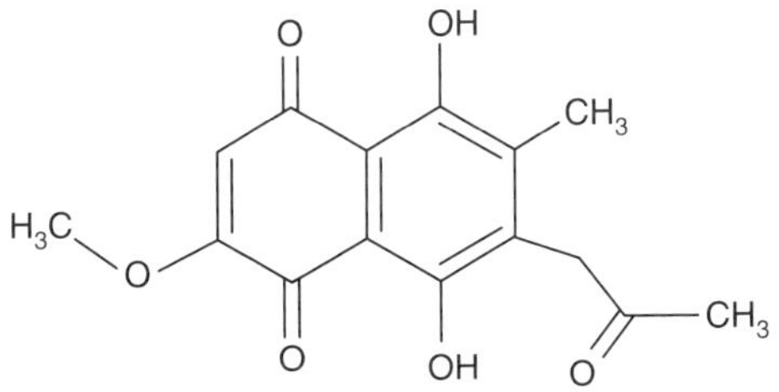

Fig. 11.7. Javanicin, an antibacterial compound isolated from endophytic fungus *Chloridium* sp.

in treating seeds, fruits and plant parts in storage and while being trans-ported. In addition, *M. albus* is already in a limited market for the treatment of human wastes. Its gases have both inhibitory and lethal effects on such faecal-inhabiting organisms as *Escherichia coli* and *Vibrio cholera*. It will be used for this purpose in coming years. Studies are underway that show its promise to fumigate buildings, thus removing the potential for fungi to contaminate building surfaces and cause health risks.

Using *M. albus* as a screening tool, it has now been possible to isolate other endophytic fungi producing volatile antibiotics. The newly described *M. roseus* was twice obtained from tree species found in the Northern Territory of Australia. This fungus is just as effective in causing inhibition and death of test microbes in the laboratory as *M. albus* (Worapong *et al.*, 2002). Other inter-esting *M. albus* isolates have been obtained from several plant species growing in the Northern Territory of Australia and the jungles of the Tesso Nilo area of Sumatra, Indonesia (Ezra *et al.*, 2004; Atmosukarto *et al.*, 2005).

A non-muscodor species (*Gliocladium* sp.) has also been discovered to produce volatile antibiotics. The volatile components of this organism are totally different from those of either *M. albus* or *M. roseus*. In fact, the most abundant volatile inhibitor is [8]-annulene, formerly used as a rocket fuel and discovered here for the first time as a natural product. However, the bioactivity of the volatiles of this *Gliocladium* sp. is not as good or as compre-hensive as that of the *Muscodor* spp. (Stinson *et al.*, 2003). Due to the volatile antibiotic producing properties of these fungi, they could be used against several soil plant pathogens to reduce their inoculum. The latter, *G. roseum*, has already shown its potential through the production of a series of hydro-carbons and hydrocarbon derivatives as mycodiesel (Strobel *et al.*, 2008). Interestingly, another novel species of *Muscodor* sp., *M. crispans*, has been reported to produce a mixture of strong volatile compounds (VOCs) that were effective against a wide range of plant pathogens, including the fungi *Pythium ultimum*, *Phytophthora cinnamomi*, *S. sclerotiorum* and *Mycosphaerella fijiensis* (the black sigatoka pathogen of bananas), and the serious bacterial pathogen of citrus, *Xanthomonas axonopodis* pv. *citri* (Mitchell *et al.*, 2010).

Antiviral compounds from endophytic fungi

Another fascinating use of products from endophytic fungi is the inhibition of viruses. Two novel human cytomegalovirus (hCMV) protease inhibitors, cytonic acids A and B, have been isolated from solid-state fermentation of the endophytic fungus *Cytonaema* sp. Their structures were elucidated as *p*-tridepside isomers using MS and NMR methods (Guo *et al.*, 2000). It is apparent that the potential for the discovery of compounds having antiviral activity from endophytes is in its infancy. The fact, however, that some com-pounds have been found already is promising. The main limitation to com-pound discovery to date is probably related to the absence of common antiviral screening systems in most compound discovery programs. Since no antiviral compound isolated from endophytes has been tested against virulent plant pathogens, an intensive exploration is therefore needed to identify some potential antiviral compounds which could be very effective against variety of pathogenic plant viruses.

Antioxidants from fungal endophytes

Two compounds, pestacin and isopestacin, have been obtained from culture fluids of *Pestalotiopsis microspora*, an endophyte isolated from a combretaceaous plant, *Terminalia morobensis*, growing in the Sepik River drainage system of Papua New Guinea (Strobel *et al.*, 2002; Harper *et al.*, 2003). Both pestacin and isopestacin display antimicrobial as well as antioxidant activity. Isopestacin was attributed with antioxidant activity based on its structural similarity to the flavonoids (Fig. 11.8). Electron spin resonance spectroscopy measurements confirmed this antioxidant activity showing that the compound is able to scavenge superoxide and hydroxyl free radicals in solution (Strobel *et al.*, 2002). Pestacin occurs naturally as a racemic mixture and also possesses potent antioxidant activity, at least one order of magnitude more potent than that of Trolox, a vitamin E derivative (Harper *et al.*, 2003). The antioxidant activity of pestacin arises primarily via cleavage of an unusually reactive C–H bond and, to a lesser extent, through O–H abstraction (Harper *et al.*, 2003). Endophytes also help plants to keep fit through balancing the generation and elimination of reactive oxygen species (Rodriguez and Redman, 2005). The use of chemical antioxidants is being slowly prohibited due to their small carcinogenic effect, and thus the demand for natural

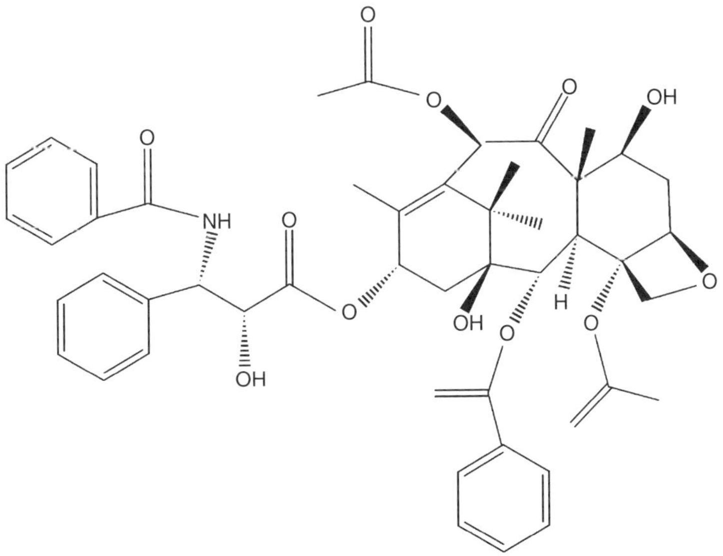

Fig. 11.8. Isopestacin, an antioxidant produced by an endophytic *Pestalotiopsis microspora* strain, isolated from *Terminalia morobensis* growing on the north coast of Papua New Guinea.

antioxidant compounds from natural resources has increased, so that the shelf life of stored crop produce and plant health could be enhanced to cater for a hungry population, and endophytes may be one of the novel natural resources to achieve this goal (Tejesvi *et al.*, 2008).

Immunosuppressive compounds from fungal endophytes

Immunosuppressive drugs are used today to prevent allograft rejection in transplant patients and in the future they could be used to treat autoimmune diseases such as rheumatoid arthritis and insulin-dependent diabetes. The endophytic fungus, *Fusarium subglutinans*, isolated from *Tripterygium wilfordii*, produces the immunosuppressive, but non-cytotoxic diterpene pyrones, subglutinols A and B (Lee *et al.*, 1995). Subglutinols A and B are equipotent in the mixed lymphocyte reaction (MLR) assay and thymocyte proliferation (TP) assay with an IC_{50} of 0.1 μM. The famed immunosuppressant drug cyclosporin A, also a fungal metabolite, was roughly as potent in the MLR assay and 10^4 more potent in the TP assay. However, the lack of toxicity associated with subglutinols A and B suggests that they should be explored in greater detail as potential immunosuppressants (Lee *et al.*, 1995).

11.5 Conclusion

Geographical, floristic and significant seasonal variations that exist in different parts of globe provide conducive/or adverse conditions for the luxuriant growth of microbes in a wide range of different habitats, including living tissues of higher plants where they grow as endophytes. Due to great variation in plant biodiversity and seasonal changes in tropical and subtropical regions, there is a need to collect/isolate various types of promising endophytic fungi, especially from rainforests and mangrove swamps, which may be able to produce an enormous variety of potential bioactive natural compounds. The fungi, as a group, hold enormous potential as sources of antimicrobials. The observations prove that this group of organisms resides inside the healthy plants tissue, as endophytes, without causing any detectable symptoms. Therefore, we strongly feel that there is a need to accelerate and focus the research to exploit the maximum potential of the promising endophytes for natural-product discovery, which could at least facilitate some existing problems of the huge population.

The past history of endophytic research in India especially with fungi is not so encouraging. It seems that workers who have started this research in India are still actively involved in advancing their research manifesto with this 'under-studied' group of microbes, and have not advanced to the fields and forests of the countryside looking for novel microbe–plant associations. Prof. T.S. Suryanarayanan and his group (Chennai) have initiated biodiversity and distribution patterns of fungal endophytes with some medicinal plants in India and have published several papers along this line. They have also isolated some bioactive compounds and melanin from endophytic fungi (Suryanarayanan *et al.*, 2004). To date, the overall situation in India has not drastically improved, as is the case in other countries such as China, Japan and Brazil (Verma *et al.*, 2009). However, several research groups have started paying more attention to various aspects of endophytic fungi. No more than a dozen research groups at different places in India are vigorously involved either with biodiversity or in natural-product discovery from this untapped and alternative resource. It has become obvious to many workers throughout the world that endophytic microbes have enormous potential for solving many of humankind's problems. Thus, with the discovery of new compounds, we can protect our interests in agricultural and medical industries as well as plant health (Omacini *et al.*, 2001; Gunatilaka, 2006; Gimenez *et al.*, 2007). Currently, some focused research on the different aspects of endophytes is being carried out and valuable results have been published (Shankar *et al.*, 2003; Seena and Sridhar, 2004; Amna *et al.*, 2006; Verma *et al.*, 2006, 2007, 2008, 2009; Gond *et al.*, 2007; Tejesvi *et al.*, 2007; Gangadevi and Muthumarry, 2008; Kharwar *et al.*, 2008, 2009).

It is important for all involved in this work to realize the importance of acquiring the necessary permits from governmental, local, and other sources to collect and transport plant materials (especially from abroad) from which endophytes are to be eventually isolated. In addition to this aspect of the work is the added activity of producing the necessary agreements and financial-sharing arrangements with indigenous people or governments in case a product does develop an income stream.

Another concern is that if endophytic fungi may produce a natural product of host origin, a host plant may also acquire the ability to produce some 'mycotoxins' or other secondary metabolites of fungal origin. The need for new and useful compounds to provide assistance and relief in all aspects of the human condition is ever growing. Drug resistance in bacteria, the appearance of new life-threatening viruses, the recurrent problems of diseases in people with organ transplants, a variety of severe diseases affecting plants and the tremendous increase in the incidence of fungal infections in the world's population all underscore our inadequacy to cope with these problems. Environmental degradation, loss of biodiversity, and spoilage of land and water also add to problems facing us, and each of these in turn can have health- and hunger-related consequences.

The researchers who are interested in exploiting endophytes should have access to, or have some expertise in, microbial taxonomy and this includes modern molecular techniques involving sequence analyses of 16S and 18S rDNA. Currently, endophytes are viewed as an outstanding source of bioactive natural products because there are so many of them occupying literally millions of unique biological niches (higher plants) growing in so many unusual environments. Thus, it would appear that a myriad of biotopical factors associated with plants can be important in the selection of a plant for study. It may be the case that these factors govern which microbes are present in the plant as well as the biological activity of the products associated with these organisms.

Certainly, one of the major problems facing the future of endophyte biology and natural-product discovery is the rapidly diminishing rainforests, which hold the greatest possible resource for acquiring novel microorganisms and their products. The total land mass of the world that currently supports rainforests is about equal to the area of the USA (Mittermeier *et al.*, 1999). Each year, an area the size of Vermont or greater is lost to clearing, harvesting, fire, agricultural development, mining or other human-oriented activities (Mittermeier *et al.*, 1999). Presently, it is estimated that only a small fraction (10–20%) of what were the original rainforests existing 1000–2000 years ago are currently present on the earth (Mittermeier *et al.*, 1999). The advent of major negative pressures on them from these human-related activities appears to be eliminating entire mega-life forms at an alarming rate. Few have ever expressed information or opinions about what is happening to the potential loss of microbial diversity as entire plant species disappear. It can only be guessed that this microbial diversity loss is also happening, perhaps with the same frequency as the loss of mega-life forms, especially since certain microorganisms may have developed unique symbiotic relationships with their plant hosts. Thus, when a plant species disappears, so too does its entire suite of associated endophytes. Consequently all of the capabilities that the endophytes might possess to provide natural products with medicinal potential are also lost. Multi-step processes are needed now to secure information and life forms before they continue to be lost. Areas of the planet that represent unique places housing biodiversity need immediate preservation. Countries need to establish information bases of their biodiversity and at the same time begin to make national collections of microorganisms that live in these areas. Endophytes are only one example of a life form source that holds enormous promise to impact many aspects of human existence. The problem of the loss of biodiversity should be one of concern to the entire world.

Acknowledgements

The authors express appreciation to the NSF, USDA, NIH, the R&C Board of the State of Montana and the Montana Agricultural Experiment Station for providing financial support for some of the work reviewed in this report. RNK expresses his sincere thanks to DST, New Delhi, for financial assistance.

References

Amna, T., Khajuria, R.K., Puri, S.C., Verma, V. and Qazi, G.N. (2006) Determination and quantification of Camptothecin in an endophytic fungus by liquid chromatography-positive mode electrospray ionization tandem mass spectrometry. *Current Science* 91, 208–211.

Anke, H., Weber, R.W., Heil, N., Pauls, S., Teinert, M., Pauluat, T., Kuenzel, E., Huth, F., Eckard, P. and Kappe, R. (2003) Antifungal compounds from endophytic fungi. *Interscience Conference on Antimicrobial Agents and Chemotherapy.* 43, Sep 14–17, 2003; abstract no. F-1236.

Atmosukarto, I., Castillo, U., Hess, W.M., Sears, J. and Strobel, G.A. (2005) Isolation and characterization of *Muscodor albus* I-41.3s, a volatile antibiotic producing fungus. *Plant Science* 169, 854–861.

Bacon, C.W. and White, J.F. (2000) *Microbial Endophytes*, Marcel Dekker, New York.

Bashyal, B., Li, J.Y., Strobel, G.A. and Hess, W.M. (1999) *Seimatoantlerium nepalense*, an endophytic taxol producing coelomycete from Himalayan yew (*Taxus wallachiana*). *Mycotaxon*, 72, 33–42.

Bills, G., Dombrowski, A., Pelaez, F. and Polishook, J. (2002) Recent and future discoveries of pharmacologically active metabolites from tropical fungi. In:, Watling R., Frankland J.C., Ainsworth A.M., Issac S., Robinson C.H. and Eda Z. (eds) *Tropical Mycology Micromycetes*, CABI Publishing, New York, 2, 165.

Borrow, A., Brian, P.W., Chester, V.E., Curtis, P.J., Hemming, H.G., Henehan, C., Jefferys, E.G., Lloyd, P.B., Nixon, I.S., Norris, G.L.F. and Radley, M. (1955) Gibberellic acid, a metabolic product of the fungus *Gibberella fujikuroi*: Some observation on its production and isolation. *Journal of the Science of Food and Agriculture* 6, 340–348.

Brady, S.F. and Clardy, J. (2000) CR377, a new pentaketide antifungal agent isolated from an endophytic fungus. *Journal of Natural Product* 63, 1447–1449.

Castillo, U.F., Myers, S., Browne, L., Strobel, G.A., Hess, W.M., Hanks, J. and Reay, D. (2005) Scanning electron microscopy of some endophytic streptomycetes in snakevine. *Journal of Scanning Microscopies* 27, 305–311.

Castillo U.F., Strobel, G.A., Ford, E.J., Hess, W.M., Porter, H., Jensen, J.B., Albert, H., Robison, R., Condron, M.A.M., Teplow, D.B., Stevens, D., Yaver, D. (2002) Munumbicins, wide-spectrum antibiotics produced by *Streptomyces* NRRL 30562, endophytic on *Kennedia nigriscans*. *Microbiology-GSM* 148, 2675–2685.

Concepcion, G.P., Lazaro, J.E. and Hyde, K.D. (2001) Screening for bioactive novel compounds. In: Pointing, S.B. and Hyde, K.D. (eds) *Bio-exploitation of Filamentous Fungi*. Fungal Diversity Press, Hong Kong, 93.

Demain, A.L. (1981) Industrial microbiology. *Science* 214, 987–990.

Dimetry, N.Z. and El-Hawary, F.M.A. (1995) Neem Azal-F as an Inhibitor of Growth and Reproduction in the Cowpea Aphid *Aphis craccivora* Koch. Journal of Applied Entomology 119, 67–71.

Dingle, J and Mcgee, P.A. (2003). Some endophytic fungi reduce the density of pustules of *Puccinia recondita* f. sp. *tritici* in wheat. *Mycological Research*, 107, 310–316.

Duijff, B.J., Poulain, D., Olivain, C., Alabouvette, C. and Lemanceau, C. (1998) Implication of systemic induced resistance in the suppression of fusarium wilt of tomato by Pseudomonas fluorescens WCS417r and by nonpathogenic *Fusarium oxysporum* Fo47. *European Journal of Plant Pathology* 104, 903–910.

Dreyfuss, M.M. and Chapela, I.H. (1994) Potential of fungi in the discovery of novel, low-molecular weight pharmaceuticals. In: Gullo, V.P. (eds) *The Discovery of Natural Products with Therapeutic Potential*. Butterworth-Heinemann, Boston, p. 49.

Ezra, D. and Strobel, G.A. (2003) Substrate affects the bioactivity of volatile antimicrobials emitted by *Muscodor albus*. *Plant Science* 65, 1229–1238.

Ezra, D., Hess, W.M. and Strobel, G.A. (2004) Unique wild type endophytic isolates of *Muscodor albus*, a volatile antibiotic producing fungus. *Microbiology* 150, 4023–4031.

Gahukar, R. T. (2005). Plant-derived products against insect pests and plant diseases of tropical grain legumes. *International pest control* 47, 315–318.

Gangadevi, V. and Muthumarry, J. (2008) Taxol, an anticancer drug produced by an endophytic fungus *Bartalinia robillardoides* Tassi, isolated from a medicinal plant, *Aegle marmelos* Correa ex Roxb. *World Journal of Microbiology and Biotechnology* 24, 717–724.

Gimenez, C., Cabrera, R., Riena, M. and Gonzalez-Coloma, A. (2007) Fungal endophyte and their role in plant protection. *Current Organic Chemistry* 11, 707–720.

Gond, S.K., Verma, V.C., Kumar, A., Kumar, V. and Kharwar, R.N. (2007). Study of endophytic fungal community from different parts of *Aegle marmelos* Correae (Rutaceae) from Varanasi (India). *World Journal of Microbiology and Biotechnology* 23, 1371–1375.

Griffith, J.M., Davis, A.J. and Grant, B.R. (1992) Target sites of fungicides to control oomycetes. In: Koller, W. (ed.) *Target Sites of Fungicide Action*, pp. 69–100.

Gunatilaka, A.A.L. (2006) Natural Products from Plant-Associated Microorganisms: Distribution, Structural Diversity, Bioactivity, and Implications of Their Occurrence. *Journal of Natural Product* 69, 509–526.

Guo. B., Dai, J.R., N.G.S., Huang, Y., Leong, C., Ong, W. and Carte, B.K. (2000) Cytonic Acids A and B: Novel tridepside inhibitors of hCMV protease from the endophytic fungus *Cytonaema* species. *Journal of Natural Products* 63, 602–606.

Harper, J.K., Arif, A.M., Ford, E.J., Strobel, G.A. (2003) Pestacin: a 1, 3–dihydro isobenzofuran from *Pestalotiopsis microspora* possessing antioxidant and antimycotic activities. *Tetrahedron* 59, 2471–2476.

Harper, J.K., Mulgrew, A.E., Li, J.Y., Barich, D.H., Strobel, G.A. and Grant, D.M. (2001) Characterization of stereochemistry and molecular conformation using solid-state NMR tensors. *Journal of American Chemical Society* 123, 9837–9842.

Hawksworth, D.C. and Rossman, A.Y. (1987) Where are the undescribed fungi ? *Phytopathology* 87, 888–891.

Hoffman, A., Khan, W., Worapong, J., Strobel, G., Griffin, D., Arbogast, B., Borofsky, D., Boone, R.B., Ning, L., Zheng, P. and Daley, L. (1998). Bioprospecting for taxol in Angiosperm plant extracts. *Spectroscopy* 13, 22–32.

Horn, W.S., Simmonds, M.S., Schwartz, R.E. and Blaney, W.M. (1995) Phomopsichalasin, a novel antimicrobial agent from an endophytic *Phomopsis* sp. *Tetrahedron* 14, 3969–3973.

Hu, Y., Chaomin, L., Kulkarni, B., Strobel, G.A., Lobkovsky, E., Torczynski, R. and

Porco, J. (2001) Exploring chemical diversity of epoxyquinoid natural products: synthesis and biological activity of jesterone and related molecules. *Organic Letters* 3, 1649–1652.

Huang, Y, Wang, J, Li, G, Zheng, Z, Su, W (2001). Antitumor and antifungal activities in endophytic fungi isolated from pharmaceutical plants *Taxus mairei*, *Cephalataxus fortunei* and *Torreya grandis*. *FEMS Immunology and Medical Microbiology*, 31, 163–167.

Inacio, M.L., Silva, G.H., Teles, H.L., Trevisan, H.C., Cavalheiro, A.J., Bolzani, V.S., Young, M.C.M., Pfenning, L.H. and Araujo, A.R. (2006) Antifungal metabolites from *Colletotrichum gloeosporioides*, an endophytic fungus in *Cryptocarya mandioccana* Nees (Lauraceae). *Biochemical Systematics and Ecology*, 34, 822–824.

Kharwar, R.N., Strobel, G., Verma, V.C. and Ezra, D. (2008) Endophytic Fungal Complexes of *Catharanthus roseus* (L.) G. Don. *Current Science* 95, 228–233.

Kharwar, R.N., Verma, V.C., Kumar, A., Gond, S.K., Harper, J.K., Hess, W.M., Lobkovosky, E., Ma, C., Ren, Y. and Strobel, G.A. (2009) Javanicin, an antibacterial naphthaquinone from an endophytic fungus of neem, *Chloridium* sp. *Current Microbiology*, 58, 233–238.

Kim, H.Y., Choi, G.J., Lee, H.B., Lee, S.W., Lim, H.K., Jang, K.S., Son, S.W., Lee, S.O., Cho, K.Y., Sung, N.D. and Kim, J.C. (2007). Some fungal endophytes from vegetable crops and their anti-oomycete activities against tomato late blight. *Letters in Applied Microbiology*, 44, 332–337.

Lane, G.A., Christensen, M.J. and Miles, C.O. (2000) Coevolution of fungal endophytes with grasses: The significance of secondary metabolites. In: Bacon, C.W. and White, J.F. (eds) *Microbial Endophytes*. Marcel Dekker, Inc., New York.

Lee, J.C., Lobkovsky, E., Pliam, N.B., Strobel, G.A. and Clardy, J. (1995a) Subglutinols A and B: immunosuppressive compounds from the endophytic fungus *Fusarium subglutinans*. *Journal of Organic Chemistry* 60, 7076–7077.

Lee, J.C., Yang, X., Schwartz, M., Strobel, G.A. and Clardy, J. (1995b) The relationship between the rarest tree in North America and an endophytic fungus. *Chemistry and Biology* 2, 721–727.

Li, J.Y., Strobel, G.A., Harper, J.K., Lobkovsky, E. and Clardy, J. (2000) Cryptocin, a potent tetramic acid antimycotic from the endophytic fungus *Cryptosporiopsis cf. quercina*. *Organic Letters* 2, 767–770.

Li, J.Y., Harper, J.K., Grant, D.M., Tombe, B.O., Bashyal, B., Hess, W.M. and Strobel, G.A. (2001) Ambuic acid, a highly functionalized cyclohexenone with antifungal activity from *Pestalotiopsis* spp. and *Monochaetia* sp. *Phytochemistry* 56, 463–467.

Li, J.Y., Sidhu, R.S., Ford, E., Hess, W.M. and Strobel, G.A. (1998) The induction of taxol production in the endophytic fungus – *Periconia* sp. from *Torreya grandifolia*. *Journal of Industrial Microbiology* 20, 259–264.

Li, J.Y. and Strobel, G.A. (2001) Jesterone and hydroxy-jesterone antioomycete cyclohexenenone epoxides from the endophytic fungus- *Pestalotiopsis jesteri*. *Phytochemistry* 57, 261–265.

Liu, C.H., Zou, W.X., Lu, H. and Tan, R.X. (2001) Antifungal activity of *Artemisia annua* endophyte cultures against phytopathogenic fungi. *Journal of Biotechnology* 88, 277–282.

Long, D.E., Smidmansky, E.D., Archer, A.J. and Strobel, G.A. (1998) *In vivo* addition of telomeric repeats to foreign DNA generates chromosomal DNAs in the taxol-producing fungus *Pestalotiopsis microspora*. *Fungal Genetics and Biology* 24, 335–344.

Lu, H., Zou, W.X., Meng, J.C., Hu, J. and Tan, R.X. (2000) New bioactive metabolites produced by *Colletotrichum* sp., an endophytic fungus in *Artemisia annua*. *Plant Science* 151, 67–73.

Mejia, L.C., Rojas E.I., Maynard Z., Arnold E.A., Hebbar P., Samuels G.J., Robbins N., and Herre A.E. (2008) Endophytic fungi as biocontrol agents of *Theobroma cacao* pathogens. *Biocontrol* 46, 4–14.

Mitchell, A.M., Strobel, G.A., Moore, E., Robinson, R. and Sears, J. (2010) Volatile antimicrobials from *Muscodor crispans* a novel endophytic fungus. *Microbiology-GSM* 156, 270–277.

Mittermeier, R.A., Myers, N., Gil, P.R. and Mittermeier, C.G. (1999) Hotspots: Earth's Biologically Richest and Most Endangered Ecoregions, CEMEX Conservation International, Washington, USA.

Narisawa, K., Ohki T. and Hashiba T. (2000) Suppression of clubroot and *Verticillium* yellows in Chinese cabbage in the field by the endophytic fungus, *Heteroconium chaetospira*. *Plant Pathology* 49, 141–146.

Newman, D.J., Cragg, G.M. and Snader, K.M. (2003) Natural products as sources of new drugs over the period 1981–2002. *Journal of Natural Product* 66, 1022–1037.

NIAID Global Health Research Plan for HIV/AIDS, Malaria and Tuberculosis, U.S. (2001) Department of Health and Human Services, Bethesda, USA.

Omacini, M., Chaneton, E.J. and Ghersa, C.M. (2001) Symbiotic fungal endophytes control insect host-parasite interaction. *Nature* 409, 78–81.

Park, J.H., Choi G.J., Lee H.B., Kim K.M., Jung H.S., Lee S.W., Jang K.S. and Cho K.Y. (2005). Griseofulvin from *Xylaria* sp. strain F0010, an endophytic fungus of *Abies holophylla* and its antifungal activity against plant pathogenic fungi. *Journal of Microbiology and Biotechnology* 15, 112–117.

Petrini, O. (1991) Fungal endophytes of tree leaves. In: Andrews, J.A. and Hirano, S.S. (eds) *Microbial Ecology of Leaves*. Springer-Verlag. New York, 179–197.

Promputtha, I., Lumyong, S., Dhanasekaran, V., Mckenzie, E.H.C., Hyde, K.D. and Jewoon, R. (2007) A phylogenetic Evaluation of Whether Endophytes Become Saprotrophs at Host Senescence. *Microbial Ecology* 53, 579–590.

Pulici, M., Sugawara, F., Koshino, H., Uzawa, J., Yoshida, S., Lobkovsky, E. and Clardy, J. (1996a) Pestalotiopsin-A and pestalotiopsin-B- new caryophyllenes from an endophytic fungus of *Taxus brevifolia*. *Journal of Organic Chemistry* 61, 2122–2124.

Pulici, M., Sugawara, F., Koshino, H., Uzawa, J., Yoshida, S., Lobkovsky, E. and Clardy, J.

(1996b) Metabolites of endophytic fungi of *Taxus brevifolia*-the first highly functionalized humulane of fungal origin. *Journal of Chemical Research* (S), 378–379.

Raviraja, N.S., Maria G.L. and Sridhar K.R. (2006). Antimicrobial evaluation of endophytic fungi inhabiting medicinal plants of the Western Ghats of India. *Engineering in Life Sciences* 6, 515–520.

Redlin, S.C. and Carris, L.M. (1996) Endophytic Fungi in Grasses and Woody Plants, APS Press, St. Paul, USA.

Redell, P. and Gordon, V. (2000) Lessons from nature: can ecology provide new leads in the search for novel bioactive chemicals from rainforests? In: Wrigley S.K., Hayes MA., Thomas, R., Chrystal, E.J.T. and Nicholson, N. (eds) *Biodiversity: New Leads for Pharmaceutical and Agrochemical Industries,*The Royal Society of Chemistry, Cambridge, UK, p. 205.

Rodriguez-Heerklotz, K.F., Drandarov, K., Heerklotz, J., Hesse, M. and Werner, C. (2001) Guignardic acid, a novel type of secondary metabolite produced by the endophytic fungus *Guignardia* sp: Isolation, structure elucidation, and asymmetric synthesis. *Helvetica Chimica Acta* 84, 3766–3772.

Rodriguez, R.J., Henson, J., Van Volkenburgh, E., Hoy, M., Wright, L., Beckwith, F., Kim, Y., Redman, R.S. (2008) Stress tolerance in plants via habitat-adapted symbiosis. *ISME-Nature* 2, 404–416.

Rodriguez, M.A., Cabrera G. and Godeas A. (2006) Cyclosporine A from a nonpathogenic *Fusarium oxysporum* suppressing *Sclerotinia sclerotiorum*. *Journal of Applied Microbiology*, 100, 575–586.

Schulz, B. and Christine, B. (2005) The Endophytic Continum. *Mycological Research* 109, 661–686.

Schulz, B. (2001) Bioactive fungal metabolites – impact and exploitation, In: *International Symposium Proceedings, British Mycological Society,* University of Wales, Swansea, UK, p. 20.

Schulz, B., Christine, B., Draeger, S., Romert, AK and Krohn, K. (2002) Endophytic fungi: a source of novel biologically active secondary metabolites, *Mycological Research* 106, 996–1002.

Seena, S. and Sridhar, K.R. (2004) Endophytic fungal diversity of 2 sand dune wild legumes from southwest coast of India. *Canadian Journal of Microbiology* 50, 1015–1018.

Shankar, S.S., Ahmad, A., Pasricha, R. and Sastry, M. (2003) Birduction of chloroauret ions by geranium leaves and its endophytic fungus yields gold nanoparticles of different shapes. *Journal of Meterial Chemistry* 13, 1822–1826.

Stierle, A., Strobel, G.A. and Stierle, D. (1993) Taxol and taxane production by *Taxomyces andreanae*. *Science* 260, 214–216.

Stinson, M., Ezra, D. and Strobel, G.A. (2003) An endophytic *Gliocladium* sp. of *Eucryphia cordifolia* producing selective volatile antimicrobial compounds. *Plant Science* 165, 913–922.

Strobel, G.A. (2002) Microbial gifts from rain forests. *Canadian Journal of Plant Pathology* 24, 14–20.

Strobel, G.A., Dirksie, E., Sears, J. and Markworth, C. (2001) Volatile antimicrobials from *Muscodor albus* a novel endophytic fungus. *Microbiology* 147, 2943–2950.

Strobel, G.A., Miller, R.V., Martinez-Miller, C., Condron, M.D., Teplow, B. and Hess, W.M. (1999) Cryptocandin, a potent antimycotic from the endophytic fungus *Cryptosporiopsis* cf. *quercina*. *Microbiology* 145, 1919–1926.

Strobel, G.A., Ford, E., Worapong, J., Harper, J.K., Arif, A.M., Grant, D., Fung, P.C.W. and Chan, K. (2002) Ispoestacin, an isobenzofuranone from *Pestalotiopsis microspora*, possessing antifungal and antioxidant activities. *Phytochemistry* 60, 179–184.

Strobel, G.A. (2002) Rainforest endophytes and bioactive products. *Critical Review in Biotechnology* 22, 315–333.

Strobel, G.A. and Daisy, B. (2003) Bioprospecting for microbial endophytes and their natural products. *Microbiology and Molecular Biology Review* 67, 491–502.

Strobel, G.A., Daisy, B., Castillo, U.F. and Harper, J. (2004) Natural products from endophytic microorganisms. *Journal Natural Product* 67, 257–268.

Strobel, G.A., Stierle, A., Stierle, D. and Hess, W.M. (1993) *Taxomyces andreanae* a proposed new taxon for a bulbilliferous hyphomycete associated with Pacific yew. *Mycotaxon* 47, 71–78.

Strobel, G.A., Knighton, B., Kluk, K., Ren, Y., Livinghouse, T., Griffin, M. and Sears, J. (2008) The production of myco-diesel hydrocarbons and their derivatives by the endophytic fungus *Gliocladium roseum* (NRRL 50072). *Microbiology*-GSM 154, 3319–3328.

Suffness, M. (1995) (ed) *Taxol®: Science and Applications*, CRC Press, Boca Raton, USA.

Suryanarayanan, T.S., Ravishanker, J.P., Venkateshan, G. and Murali, T.S. (2004) Characterisation of the melanin pigment of a cosmopolitan fungal endophyte. *Mycological Research* 180, 974–979.

Tan, R.X. and Zou, W.X. (2001) Endophytes: a rich source of functional metabolites. *Natural Product Report* 18, 448–459.

Tejesvi, M.V., Kini, K.R., Prakash, H.S., Ven Subbiah, and Shetty, H.S. (2007) Genetic diversity and antifungal activities of *Pestalotiopsis* isolated as endophyte from medicinal plant. *Fungal Diversity* 24, 37–54.

Tejesvi, M.V., Kini, K.R., Prakash, H.S., Ven Subbiah, and Shetty, H.S. (2008). Antioxidant, antihypertensive and antibacterial properties of endophytic Pestalotiopsis species from medicinal plants. *Canadian Journal of Microbiology* 54, 769–780.

Verma, V.C. and Kharwar, R.N. (2006) Efficacy of neem leaf extract against it's own fungal endophyte *Curvularia lunata*. *Journal of Agricultural Technology* 2, 329–337.

Verma, V.C., Gond, S.K., Kumar, A., Kharwar, R.N. and Strobel, G.A. (2007) Endophytic mycoflora from leaf stem and bark tissues

of *Azadirachta indica* A. Juss from Varanasi India, *Microbial Ecology* 54, 119–125.

Verma, V.C., Kharwar, R.N. and Strobel, G.A. (2009). Chemical and functional diversity of natural products from plant associated endophytic fungi. *Natural Product Communications* 4, 1511–1532.

Verma, V.C., Gond, S.K, Misra, A., Kumar, A. and Kharwar, R.N. (2008) Selection of Natural Strains of Fungal Endophytes from *Azadirachta indica* A. Juss, with Anti-Microbial Activity Against Dermatophytes. *Current Bioactive Compounds* 4, 36–42.

Walsh, T.J., Lee, J.W., Roilides, E., Francis, P., Bacher,J., Lyman, C.A., Pizzo, P.A. (1992) Experimental antifungal chemotherapy in granulocytopenic animal-models of disseminated candidiasis – approaches to understanding investigational antifungal compounds for patients with neoplastic diseases. *Clinical Infectious Diseases* 14, S139–147.

Wani, M.C., Taylor, H.L., Wall, M.E., Goggon, P. and McPhail, A.T. (1971) Plant antitumor agents, VI. The isolation and structure of taxol, a novel antileukemic and antitumor agent from *Taxus brevifolia*. *Journal of American Chemical Society* 93, 2325–2327.

Worapong, J., Strobel, G.A., Daisy, B., Castillo, U., Baird, G., and Hess, W.M. (2002) *Muscodor roseus* anna. nov. an endophyte from *Grevillea pteridifolia*. *Mycotaxon* 81, 463–475.

Worapong, J., Strobel, G.A., Ford, E.J., Li, J.Y., Baird, G. and Hess, W.M. (2001) *Muscodor albus* gen. et sp. nov., an endophyte from *Cinnamomum zeylanicum*. *Mycotaxon* 79, 67–79.

12 Suppressive Effects of Compost Tea on Phytopathogens

MILA SANTOS, FERNANDO DIÁNEZ
AND FRANCISCO CARRETERO

Plant Production Department, University of Almería, Almería, Spain

Abstract

Suppression of soilborne and airborne diseases of horticultural crops by compost has been attributed to the activities of antagonistic microorganisms. A great diversity of biological control agents naturally colonize compost. The knowledge of mechanisms for biological control through the action of compost or its water extracts (compost tea) is necessary in order to increase the efficiency of the suppressing power. Modes of action of biocontrol agents include: inhibition of the pathogen by antimicrobial compounds (antibiosis), competition for iron through production of siderophores, competition for colonization sites and nutrients supplied by seeds and roots, induction of plant resistance mechanisms, inactivation of pathogen germination factors present in seed or root exudates, degradation of pathogenicity factors of the pathogen such as toxins, parasitism that may involve production of extracellular cell wall-degrading enzymes, for example, chitinase and β-1,3 glucanase that can cause lysis of pathogen cell walls. None of the mechanisms are necessarily mutually exclusive and frequently several modes of action are exhibited by a single biocontrol agent. Indeed, for some biocontrol agents, different mechanisms or combinations of mechanisms may be involved in the suppression of different plant diseases. The healthy development of plants, as well as the biological control of soilborne fungi, originates from many microorganisms, partly due to the production of siderophores under iron-restricting conditions. The siderophores are natural chelators that keep iron available for plants in soil. Iron also plays a major role in nutrient competition among pathogens and beneficial microorganisms in infection sites.

12.1 Introduction

Suppressive soils or suppressive substrates are characterized by a very low level of disease development even though a virulent pathogen and susceptible host are present. Biotic and abiotic elements of the soil environment contribute to suppressiveness; however, most defined systems have identified biological elements as primary factors in disease suppression. Many soils possess

© CAB International 2011. *Natural Products in Plant Pest Management*
(ed. N.K. Dubey)

similarities with regard to microorganisms involved in disease suppression, while other attributes are unique to specific pathogen-suppressive soil systems. Modes of action of biocontrol agents include: inhibition of the pathogen by antimicrobial compounds (antibiosis); competition for iron through the production of siderophores; competition for colonization sites and nutrients supplied by seeds and roots; induction of plant-resistance mechanisms; inactivation of pathogen germination factors present in seed or root exudates; degradation of pathogenicity factors of the pathogen such as toxins; parasitism that may involve production of extracellular cell-wall-degrading enzymes, for example, chitinase and β-1,3 glucanase that can cause lysis to pathogen cell walls (Keel and Défago, 1997; Whipps, 1997). None of the mechanisms are necessarily mutually exclusive and frequently several mode of action are exhibited by a single biocontrol agent. Indeed, for some biocontrol agents, different mechanisms or combinations of mechanisms may be involved in the suppression of different plant diseases (Whipps, 2001). So the organism operatives in pathogen suppression do so via diverse mechanisms including competition for nutrients, antibiosis and induction of host resistance. Non-pathogenic *Fusarium* spp. and fluorescent *Pseudomonas* spp. play a crucial role in naturally occurring soils that are suppressive to Fusarium wilt. Suppression of take-all of wheat (*Triticum aestivum*), caused by *Gaeumannomyces graminis* var. *tritici*, is induced in soil after continuous wheat monoculture and is attributed, in part, to selection of fluorescent *Pseudomonads* with the capacity to produce the antibiotic 2,4-diacetylphloroglucinol. Cultivation of orchard soils with specific wheat varieties induces suppressiveness to *Rhizoctonia* root rot of apple (*Malus domestica*) caused by *Rhizoctonia solani* AG 5. Wheat cultivars that stimulate disease suppression enhance populations of specific fluorescent pseudomonad genotypes with antagonistic activity toward this pathogen. Methods that transform resident microbial communities in a manner that induces natural soil suppressiveness have the potential as components of environmentally sustainable systems for the management of soilborne plant pathogens (Mazzola, 2002).

During the past decades, compost prepared from a variety of organic waste has been utilized successfully in container media for suppressing soil-borne diseases (Table 12.1). Examples are composts prepared from tree barks, liquorice (*Glycyrrhina glabra*) root waste, grape pomace, separated cattle manure, and municipal sewage sludge (Kuter *et al.*, 1998). Suppressive soils or substrates are characterized by a very low level of disease development even though a virulent pathogen and susceptible host are present. Diseases that have been shown to be effectively suppressed by compost use include those caused by *Fusarium, Phytophthora, Pythium* and *R. solani*. Disease suppression in compost has been attributed mainly to elevated levels of microbial activity (Chen *et al.*, 1988b, Santos *et al.*, 2008). There are several mechanisms involved in the suppression: the competition among microbial populations for available carbon or nitrogen, iron competition, mycoparasitism, production of inhibitors or hydrolytic enzymes by microorganisms and interactions with some saprophytes. Since 1991, several reports have shown that some rhizosphere microorganisms can induce systemic resistance in

Table 12.1.   Different composted material suppressive to soilborne disease.

Composted material	Pathogens	References
Tree bark	*Phytophthora cinnamomi*	Hoitink *et al.*, 1977; Sivasithamparam, 1981; Spencer and Benson, 1981, 1982; Blaker and MacDonald, 1983; Hardy and Sivasithamparam, 1991
	Phytophthora citricola	Spencer and Benson, 1981; Hardy and Sivasithamparam, 1991
	Phytophthora drechleri	Hardy and Sivasithamparam, 1991
	Phytophthora nicotianae	Hardy and Sivasithamparam, 1991
	Pythium ultimum	Daft *et al.*, 1979; Chen *et al.*, 1987; Chen *et al.*, 1988 a,b
	Rhizoctonia solani	Daft *et al.*, 1979; Stephens *et al.*, 1981; Nelson and Hoitink,1983
	Fusarium oxysporum	Chef *et al.*, 1983; Cebolla and Pera, 1983; Trillas-Gay *et al.*, 1986
	Nematodes	Malek and Gartner, 1975; Mc Sorley and Gallaher, 1995
Pine bark	*Phytophthora* spp.	Sivasithamparam, 1981; Spencer and Benson, 1981, 1982
	Pythium spp.	Gugino *et al.*, 1973; Zhang *et al.,* 1996
	Fusarium oxysporum	Chef *et al.*, 1983; Pera and Calvet, 1989
Grape marc	*Pythium aphanidermatum*	Mandelbaum *et al.*, 1988
	Rhizoctonia solani	Gorodecki and Hadar, 1990
	Sclerotium rolfsii	Gorodecki and Hadar, 1990; Hadar and Gorodecki, 1991
Olive marc	*Fusarium oxysporum* f. sp. *dianthi*	Pera and Calvet, 1989
Municipal sewage sludge	*Pythium* spp.	Lumsden *et al.*, 1983
Municipal solid residues	*Rhizoctonia* spp.	Mathot, 1987
Cattle manure	*Pythium aphanidermatum*	Mandelbaun *et al.*, 1988
	Phytophthora nicotianae	Szczech *et al.*, 1993
	Rhizoctonia solani	Gorodecki and Hadar, 1990
	Sclerotium rolfsii	Gorodecki and Hadar, 1990
	Fusarium oxysporum	Garibaldi, 1988; Szczech *et al.*, 1993
Beer industry sludge	*Pythium graminicola*	Craft and Nelson, 1996
Liquorice roots	*Pythium aphanidermatum*	Hadar and Mandelbaun, 1986

plants to root and foliar disease (Han *et al.*, 2000). This systemic resistance was found in compost by Zhang *et al.* (1996).

Most of the composts are naturally suppressive against root rot disease caused by *Phytophthora* and *Pythium* (Santos *et al.*, 2008), nearly 20% are naturally suppressive against damping off caused by *Rhizoctonia* (Hoitink and

Boehm, 1999), less than 10% of the compost induced systemic resistance in plants (Zhang *et al.*, 1996, 1998; Hoitink and Boehm, 1999).

There are two primary mechanisms by which the colonies of biocontrol organisms in compost combat disease, general suppression and specific suppression.

General suppression occurs when a high-microbial activity environment is created in which the germination of pathogen propagules is inhibited. General suppression occurs when many different organisms compete with pathogens for nutrients (as root or seed exudates) and/or produce general antibiotics that reduce pathogen survival and growth. In compost there is a slow release of nutrients which supports beneficial activity of this microbiotic. Biocontrol agents that colonize composts include bacteria such as *Bacillus, Enterobacter, Flavobacterium balusstinum*, and *Pseudomonas*; actinomycetes such as *Streptomyces*; and fungi such as *Trichoderma* and *Gliocladium* (Hoitink *et al.*, 1991b).

General suppression owes its activity to the total microbial biomass in the soil or substrate and is not transferable between them. Whether soil organic matter can support biological control depends on its decomposition level and the types of biocontrol agents present on the substrate (Hoitink and Boehm, 1999). The carrying capacity of organic matter in the substrate limits suppressiveness to pathogens that depend on exogenous sources of nutrients (root exudates) for germination and infection. Excessively stabilized organic matter such as dark peat has a limited ability to sustain activity of the general microbial biomass in soil (Hoitink *et al.*, 1993). Dark decomposed sphagnum peat is consistently conductive to *Pythium* root rot, whereas light sphagnum peat harvested from the surface of peat bogs is less decomposed and has a higher microbial activity (Hoitink *et al.*, 1991a). Light peat moss is suppressive against *Pythium* for a short time (6–7 weeks).

Specific suppression involves the action of one or a few specific microbial agents in suppressing a specific pathogen (Hoitink, 1993). They exert hyperparasitism on the pathogen or induce systemic resistance in the plant to specific pathogens. Specific suppression owes its activity to the effects of individual or select groups of microorganisms and is transferable. This can be achieved by inoculating the compost with the desired microbial agent (Hoitink, 1993).

The suppression effect on pathogens is linked to the type of pathogens. Those pathogens that have a small propagule size, such as *Pythium* and *Phytophthora* species, are susceptible to general suppression ('nutrient-dependent' pathogens). They have small nutrient reserves and need to rely on an external carbon source and other nutrients. Pathogens with a large propagule size, such as *Sclerotium rolfsii* and *R. solani*, are susceptible to specific suppression. Structures like sclerotia are less susceptible to microbial competition but specific hyperparasites such as *Trichoderma* species will colonize the sclerotia.

Composting conditions as well as the materials that are composted are critical, with the type of the material impacting on the sort of active microflora. Thus the composting of lignocellulosic wastes will induce a specific suppression of *Rhizoctonia* by *Trichoderma* species, while *Penicillium* fungi predominate in grape pomace (high sugar, low cellulose) in suppressing *S. rolfsii*. In this medium, *Trichoderma* spp. were absent and would have been ineffective even if applied. *Rhizoctonia* (a common nursery pathogen) is also a very active saprophyte in fresh wastes but not in low-cellulose mature composts. *Trichoderma*, by contrast, colonizes both. However, the hyperparasitic activity and antibiotic production by *Trichoderma* is repressed in fresh compost because of the high glucose level. Saline composts (for instance overfertilized media) enhance *Pythium* and *Phytophthora*, whereas composts with low C/N ratios that release nitrogen, especially as ammonia, promotes the growth of *Fusarium* on host species. The appropriate control of the potting medium with a pH greater than 5.0 and moisture content (oven dried) maintained between 40 and 50% is also important (Hoitink *et al.*, 1997).

12.2 Competition Among Microbial Populations

Microbiostasis

There is a competition for nutrients in exudates such as sugars and amino acids which leak out of seeds during germination or out of root tips as plants grow through the soil. Pathogens that grow or swim to these sources of nutrients must compete with the beneficial microflora at the infection site on the surface of the seed or root (Hoitink and Changa, 2004). This type of competition plays a major role in general suppression and with 'nutrient-dependent' pathogens such as *Pythium* and *Phytophthora* species. The competition includes microbial competition for nutrients and competition for infection sites and root colonization. Limited nutrient availability to the germinating spores or invading hyphae is the most common explanation given for the microbial cause of fungistasis. On the other hand, fungistasis has been attributed to the presence of antifungal compounds of microbiological origin. The production of antifungal compounds is correlated with available carbon, which is not surprising, as microbial production of antibiotics can be induced by interspecific competition for substrates (Ellis *et al.*, 2000; Slattery *et al.*, 2001). The sensitivity of plant pathogenic fungi to fungistasis is thought to protect them from germinating and initiating growth under unfavourable conditions (e.g., the absence of a host plant) (Garrett, 1970; Lockwood, 1977). However, this protection comes at a cost, since the viability of resting structures decreases during prolonged incubation in soils (Lockwood, 1986). The negative effects of fungistasis on the inoculum density of plant pathogenic fungi has been suggested as a mechanism to explain the commonly found correlation between fungistasis and disease suppressiveness (Lockwood, 1977; Hornby, 1983; Larkin *et al.*, 1996; Knudsen *et al.*, 1999).

Iron competition also plays a major role in nutrient competition among pathogens and beneficial microorganisms in infection sites (Hoitink and Changa, 2004). Ionic forms of iron, especially iron (III), its most common state, are very insoluble, especially when plants are grown on calcareous soils or soils with high pH. Iron is only available to organisms at concentrations at or below 10^{-18} M in soil solutions at neutral pH (Handelsman and Stabb, 1996). Iron salts that have low solubility such as iron oxides, carbonates, phosphates, hydroxides and some forms of insoluble chelates are created in certain soil types that make iron not readily available to plants. To circumvent the solubility problem, many microorganisms synthesize and utilize very specific low molecular weight (500–1000 Da) iron chelators called siderophores (that can bind iron and are sent out by the cells to absorb iron, as well as being used inside the cell to store it). When grown under iron-deficient conditions, many microbes will synthesize and excrete siderophores in excess of their own dry cell weight to kidnap and solubilize iron and made it unavailable to the pathogens. The typical system involves a siderophore, which is an iron-binding ligand, and an uptake protein, which transports the siderophore into the cell. The fluorescent pseudomonads produce a class of siderophore known as the pseudobactins, which are structurally complex iron-binding molecules. Analyses of mutants lacking the ability to produce siderophores suggest that they contribute to the suppression of certain fungal and oomycete diseases (Duijff *et al.*, 1994; Buysens *et al.*, 1996). An interesting aspect of siderophore biology is that diverse organisms can use the same type of siderophore. Microorganisms may use each other's siderophores if they contain the appropriate uptake protein (Koster *et al.*, 1993; Raaijmakers *et al.*, 1995), and plants can even acquire iron from certain pseudobactins (Duijff *et al.*, 1994).

Microorganisms in compost produce siderophores that keep iron in an available form for plants in the soil, even at high pH. Compost also produces water-soluble humic substances, including fulvic acids, which keep iron, zinc, manganese and other trace elements in solution (Chen and Inbar, 1993).

Antibiosis

Antibiosis is defined as antagonism mediated by specific or non-specific metabolites of microbial origin, by lytic agents, enzymes, volatile compounds or other toxic substances (Jackson, 1965; Fravel, 1988). Antibiotic production appears to be important to the survival of microorganisms through the elimination of microbial competition for food sources, which are usually very limited in soil (Ellis *et al.*, 2000; Slattery *et al.*, 2001). Antibiotic production is very common among soil-dwelling bacteria and fungi. Inhibition in the Petri dish may be the result of antibiosis, but it is not easy to show that this antibiosis is actually responsible for disease suppression. First, the antibiotic must be extracted, purified and identified chemically. Then it is necessary to show that the microorganism grows in the microhabitat of the pathogen, and that the antibiotic is produced in the right place, at the right time, and in sufficient amounts to control disease. It is also necessary to demonstrate that the pathogen is sensitive to the antibiotic. Genetic analyses have been particularly informative in determining the role of antibiotics in biocontrol, in part because mutants can be screened easily *in vitro* for changes in antibiotic accumulation, providing the means to conduct thorough genetic analyses (Handelsman and Stabb, 1996).

Antibiotic production by strains of the bacteria *Pseudomonas* and *Bacillus* has been shown to be important to the successful biocontrol of several crop diseases. The antibiotics zwittermicin A and kanosamine, produced by the biocontrol agent *Bacillus cereus* UW85, appear to be important in the biocontrol of oomycetes such as *Phytophthora* (Silo-Suh *et al.*, 1994; Milner *et al.*, 1996). The antibiotic phenazine derivatives, produced by *Pseudomonas fluorescens* strain 2-79 and *P. aureofaciens* strain 30-48, control take-all of wheat (Weller and Cook, 1983; Brisbane and Rovira, 1988).

Trichoderma and *Gliocladium* are closely related fungal biocontrol agents. Each one produces antimicrobial compounds and suppresses disease by diverse mechanisms, including the production of the structurally complex antibiotics gliovirin and gliotoxin (Howell *et al.*, 1993). Mutants with increased or decreased antibiotic production show a corresponding effect on biocontrol (Howell and Stipanovic, 1983; Handelsman and Stabb, 1996).

Hyperparasitism

Hyperparasitism is parasitism on another parasite. The mycelium and resting spores (oospores), hyphae or sclerotia of several pathogenic soil fungi such as *Pythium*, *Phytophthora*, *Verticillium*, *Rhizoctonia*, *Sclerotinia* and *Botrytis* (Fig. 12.1) are invaded and parasitized (mycoparasitism) or are lysed (mycolysis) by several non-pathogenic microbes.

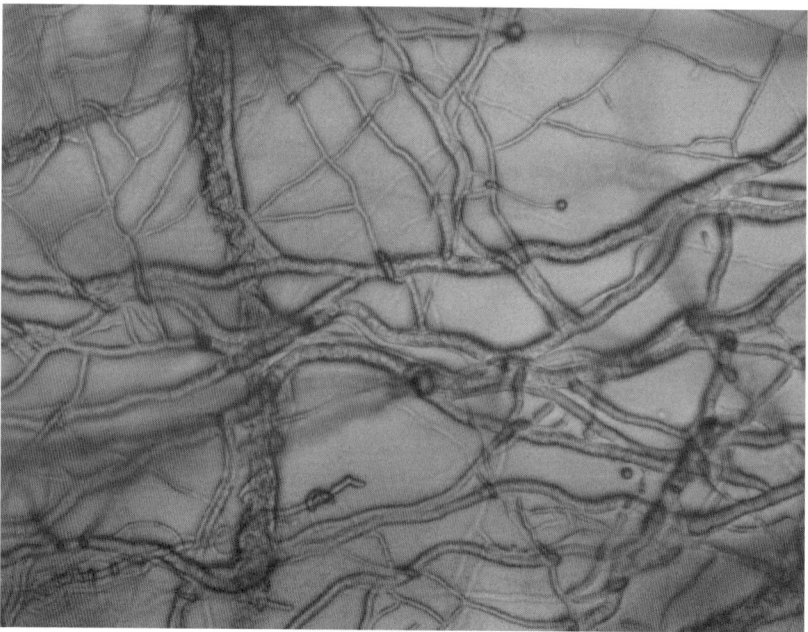

Fig. 12.1. Microscopic photography (× 400) of *Trichoderma saturnisporum* catching *Botrytis cinerea.*

Mycoparasitism is parasitism of a pathogenic fungus by another fungus. These events require specific interactions between the parasite and fungal host. The process involves direct contact between the fungi, resulting in death of the plant pathogen and nutrient absorption by the parasite. *Trichoderma* spp. parasitize fungal plant pathogens. The parasite extends hyphal branches toward the target host, coils around and attaches to it with appressorium-like bodies, and punctures its mycelium (Chet *et al.*, 1981; Goldman *et al.*, 1994). Mycoparasites produce cell-wall-degrading enzymes, which allow them to bore holes into other fungi and extract nutrients for their own growth. But many so-called mycoparasites also produce antibiotics, which may first weaken the fungi they parasitize. So the digestion of host cell walls is accomplished by a battery of excreted enzymes, including proteases, chitinases and glucanases. These enzymes often have antifungal activity individually and are synergistic in mixtures or with antibiotics (Di Pietro *et al.*, 1993; Lorito *et al.*, 1993, 1994; Handelsman and Stabb, 1996). By manipulating their activity through the construction of 'overproducing' mutants, enzyme-negative mutants or even transgenic plants expressing the enzyme, a role for their production in biocontrol has been implied (Whipps, 2001).

The best documented examples of hyperparasitism involve the myco-parasitic *Trichoderma* spp. on *R. solani*. Highly competitive as a saprophyte, *R. solani*, can utilize cellulose and colonize fresh bark but cannot colonize the low-in-cellulose mature bark compost. However, isolates of *Trichoderma* that function as biological agents for *R. solani* are capable of colonizing mature compost. Biological control does not occur in fresh, undecomposed organic matter because both fungi grow as saprophytes and *R. solani* remains capable of causing disease. In mature compost, on the other hand, sclerotia of *R. solani* are killed by the hyperparasites and biological control prevails (Hoitink and Fahy, 1986; Chung and Hoitink, 1990). One way to ensure specific suppression is to inoculate the compost with the appropriate beneficial microorganisms, fungus like *Trichoderma* sp., *Talaromyces flavus*, non-pathogenic *Fusarium oxysporum*, or bacterial like *Bacillus subtilis* and *Streptomyces griseoviride*, which leads to increased levels of disease suppressiveness.

Induced resistance

The natural resistance of plants to pathogens is based on the combined effects of preformed barriers and induced mechanisms. In both cases, plants use physical and antimicrobial defences against the invaders. In contrast to constitutive resistance, induced resistance relies on recognition of an invader and subsequent signal transduction events leading to the activation of defences (Mauch-Mani and Métraux, 1998). Plants possess active defence mechanisms against pathogen attack; some biotic and abiotic stimuli increase their tolerance to infection by a pathogen, by activation of these active defence mechanisms. This phenomenon is known as induced resistance. Induced resistance was defined by Kloepper *et al.* (1992) as 'the process of active resistance dependent on the host plant's physical or chemical barriers, activated by biotic or abiotic agents'. Induced defence responses are regulated by a network of interconnecting signal transduction pathways in which the hormonal signals salicylic acid (SA), jasmonic acid (JA) and ethylene (ET) play a major role (Pieterse and Van Loon, 1999; Glazebrook, 2001), and other hormones such as brassinosteroids and abscisic acid can also be involved (Nakashita *et al.*, 2003; Ton and Mauch-Mani, 2004). The phenotypic effects of activated defences are quite similar when the stimuli are biotic or abiotic agents, but the biochemical and mechanistic changes appear to be subtly different (Van Loon *et al.*, 1998; Whipps, 2001). This has resulted in the term induced systemic resistance (ISR) for bacterially induced resistance and systemic acquired resistance (SAR) for that induced by abiotic agents or microorganisms that cause localized damage (Pieterse *et al.*, 1996).

SAR confers long-lasting protection against a broad spectrum of micro-organisms. SAR requires the signal molecule SA and is associated with the accumulation of pathogenesis-related (PR) proteins such as chitinase, β-1-3-glucanases or proteinase inhibitor, which are thought to contribute to resistance (Durrant and Dong, 2004). One of the major differences between SAR and ISR is that PR proteins are not universally associated with bacterially induced resistance (ISR) (Hoffland *et al.*, 1995) and SA is not always involved in the expression of ISR.

Several reports suggest that compost and compost-amended soil may alter the resistance of a plant to disease. This was observed in airborne diseases such as powdery mildew of wheat and barley (Tränkner, 1992), early blight and bacterial spot of tomato (*Lycopersicum esculentum*) (Roe *et al.*, 1993) or *Anthracnose* and *Pythium* root rot in cucumber (*Cucumis sativus*) (Zhang *et al.*, 1996). A select few beneficial microorganisms of compost can induce mechanisms of ISR. These microorganisms activate biochemical pathways in plants leading to ISR to root as well as some foliar diseases. This mechanism helps explain the often heard statement that plants raised on healthy organic soils are more able to resist diseases (Zhang *et al.*, 1998). Plants produced in compost-amended mixes harbouring biocontrol agents that induce systemic resistance have higher concentrations of proteins related to host defence mechanisms (Zhang *et al.*, 1998). The activation of plant defence produces changes in cell-wall composition, *de novo* production of PR proteins such as chitinases and glucanases, and synthesis of phytoalexins, although further defensive compounds are likely to exist but remain to be identified (Heil and Bostock, 2002).

Zhang *et al.* (1996, 1998) showed that composted pine-bark-amended potting mix provide *Pythium* root rot and *Anthracnose* control in cucumber by inducing systemic resistance (reported as SAR) utilizing split-root techniques. Experiments showed that when only some of the roots of a plant are in compost-amended soil, while the other roots are in diseased soil, the entire plant can still acquire resistance to the disease. On those plants, the systemic protection induced by compost was accompanied by increased peroxidase activity in leaf tissue. Acid peroxidase was previous reported as a putative molecular marker of SAR in cucumber.

ISR plays a role in the suppression of plant pathogens that colonize aerial and soil parts (Pharand *et al.*, 2002). Krause *et al.* (2003) demonstrated that less than 2% of 80 different batches of compost tested induced systemic resistance in radish (*Raphanus sativus*) against bacterial leaf spot. The effect was due to the activity of specific biocontrol agents in the batches of compost that suppressed bacterial leaf spot. They identified *Trichoderma hamatum* 382 (Bonord.) Bainer (T_{382}) as the most active inducer of ISR in radish (Khan *et al.*, 2003). Another species of the genera *Trichoderma* (*Trichoderma harzianum* Rifai T-203) was reported as an inducer of ISR in pepper (*Capsicum annum*) seedlings against *Phytophthora capsici* when the seeds were previously treated with this biological agent (Ahmed *et al.*, 2000). Khan *et al.* (2003) report how *Phytophthora* root and crown rot of cucumber caused by *P. capsici* was suppressed significantly in cucumber transplants produced in a composted cow-manure-amended mix compared with those in a dark-sphagnum-peat mix. In split-root bioassays, *Trichoderma hamatum* 382 (T_{382}) inoculated into the compost-amended potting mix significantly reduced the severity of *Phytophthora* root and crown rot of cucumber caused by *P. capsici* on paired roots in the peat mix. This effect did not differ significantly from that provided by a drench with benzothiadiazole (BTH) or mefenoxam (Subdue MAXX).

12.3 Suppression of Plant Pathogens by Compost Tea

The disease-suppressive characteristic of organic tea was reported as early as 1973 by Hunt *et al.* immobilizing the sting nematode (*Belonolaimus longicaudatus*). Compost tea or compost water extract can be defined as a watery extract of compost produced through a deliberate process. The goal is to enhance populations of beneficial microbes that can then exert a biological control over pathogens. The terminology used is not clear because there are several different methods used for making compost teas, those where tea is aerated during the production of the tea, and where no aeration is used and compost is just passively steeped with little agitation. The review of Scheuerell and Mahaffee (2002a) clarified the numerous terms that have been used to describe composts' fermentation; many are synonymous or easy confused with other concepts. Terms used include compost tea, aerated compost tea, organic tea, compost extracts, watery fermented compost extract, amended extracts, steepages and slurries. Compost tea is produced by mixing compost with water and culturing for a defined period, either actively aerating (aerated compost tea; ACT) or not (non-aerated compost tea; NCT) and with or without additives that are intended to increase microbial population densities during production (Scheuerell and Mahaffee, 2002a). These teas have been shown to act as a natural fungicide; they contain populations of various biofungicidal microbes and organic chelators Teas can be used as a foliar spray to inhibit late blight caused by *Phytophthora infestans* on tomatoes and potatoes, the suppressive effect of organic teas are of a living microbial nature and the sterilized or micron-filtered tea had little ability to impact on pathogens (Weltzien and Ketterer, 1986; Weltzien, 1989).

There are several reports on the control of plant pathogens or plant diseases with organic teas including airborne and soilborne diseases; compost teas coat plant surfaces (foliar application) or roots (liquid drench application) with living microorganisms and provide food for beneficial microbes. *Plasmopara viticola* or downy mildew of grape leaves (Weltzien and Ketterer, 1986), *Botrytis cinerea* or grey mould (Elad and Shtienberg, 1994; Diánez, 2005; Koné *et al.*, 2010), *Phytophthora cinnamomi* (Hoitink *et al.*, 1977), *Fusarium oxysporum* f.sp *pisi* or *Fusarium* wilt of peas (Khalifa, 1965), *Fusarium oxysporum* f.sp *cucumerinum* or *Fusarium* wilt of cucumber (Ma *et al.*, 1999) *Pythium ultimum* or damping-off in pea (*Pisum sativum*) (Tränkner, 1992).

   Multiple modes of activity are involved in suppressing plant disease with NCT, whereas no studies have determined the mechanism involved with ACT (Scheuerell and Mahaffee, 2002a). The microbes in compost teas can suppress diseases in several ways: induced resistance, antibiosis and competition (Brinton, 1995; Scheuerell and Mahaffee, 2002a), and direct destruction of pathogens structures (Ma *et al.*, 2001).

   The microbiotic of NCT (Weltzien, 1991) and ACT (Ingham, 2003) had been described as being dominated by bacteria. It is important to know how the manipulation of the compost tea production process enriches and/or selects for individual microbe populations. Scheuerell and Mahaffee (2002b) studied the use of ACT and NCT produced with and without nutrient additives, to drench peat-perlite growing media that was inoculated with *Pythium ultimum* and planted with cucumber seeds. They used different nutrient additives, fungal nutrients (soluble kelp, humic acids and rock dust) and bacterial nutrients (molasses-based nutrients solution). The most consistent compost tea formula for suppression damping-off in cucumber was ACT produced with the fungal nutrients, whereas the disease was not suppressed with ACT produced with the bacterial nutrients and without nutrients.

12.4 Grape Marc Compost

Grape marc is the waste from wine production; once the juice has been extracted, the skin 7% (w/w), stalks 5% (w/w) and seeds 4% (w/w) are all redundant. In total, 15–20% of wine production is waste, comprising thousands of tonnes. The amount of marc that is generated from fruit is known to vary for a number of reasons, including whether the grapes have been irrigated and the type of equipment used to press the grapes (Jordan, 2002). Spanish production of grapes for wine production is near 5 million tonnes; up to 700,000 tonnes of marc is estimated to have been generated throughout Spain. The marc, if not treated effectively, can cause a number of environmental hazards ranging from surface and groundwater pollution to foul smells. In the European Union all by-products of wine production (grape marc and wine lees) are obliged to be distilled; the legal basis for distillation measures in the EU is given in chapter II of title III, Art.27 of regulation (EC) N°1493/1999. In European countries grape marc is first distilled to recover alcohol, followed by washing for tartrate recovery, then seed separation and finally the remainder is either burnt for energy recovery or composted along with sludge from distillation and tartrate recovery to produce fertilizers for agricultural use (Johnston, 2001).

In general, grape-marc compost (GMC) has low nutrient status and conductivity; this compost has low water-holding capacities. It has a high content of lignin and cellulose and a low content of water-soluble carbohydrates.

There are a few reports about the disease-suppressive effects of GMC. Oka and Yermiyahu (2002) tested in pot and *in vitro* experiments the suppressive effects of GMC, on the root-knot nematode *Meloidogyne javanica*. Very few root-galls were found on tomato roots grown in soil containing 50% grape marc compost. Significant reductions in galling index were also found in tomato plants grown in soils containing smaller concentrations of this compost. The water extract of GMC showed weak nematicidal activity to the juveniles and eggs.

GMC and its water extracts have been reported to suppress fungal diseases such as *B. cinerea* on tomato and pepper (Elad and Shtienberg, 1994).

GMC suppressed the soilborne disease caused by *R. solani* and *S. rolfsii* (Gorodecki and Hadar, 1990; Hadar and Gorodecki, 1991). Hadar and Gorodecki (1991) found an inhibitory effect of GMC on sclerotial germination and viability, and associated this effect with high numbers of *Penicillium* isolated from sclerotia. *Penicillium* and *Aspergillus* spp. have been reported to colonize GMC; *Trichoderma* spp. hyperparasites of *R. solani* were not recovered from this compost (Gorodecki and Hadar, 1990; Hadar and Gorodecki, 1991). The age of the composted grape marc had a major effect on suppression; immature GMC (3 months of composting) failed to inhibit sclerotial germination (Hadar *et al.*, 1992). Suppression of *Pythium aphanidermatum* was reported on grape marc compost, the beneficial effect of the compost was negated when the medium was autoclaved, and restored when compost that had not been autoclaved was mixed with the sterile one (Hadar *et al.*, 1992).

Compost can provide natural biological control of diseases of roots as well as the foliage of plants. Its water extracts (compost tea) has been proposed as a substitute for synthetic fungicides (Zhang *et al.*, 1998). Most of the papers published on the control of pathogens by means of compost tea have studied pathogens from the aerial part of the plants, the number of trials that use NTC being higher. Research into the control of soilborne pathogens by means of compost tea has been lower, although this practice is common in ecological agriculture (Scheuerell and Mahaffee, 2002a).

Our research focuses on soilborne disease suppression by GMC. The grape marc compost was produced in the University of Sevilla. It was composted in 40 m^3 windrows and turned each week. Composting took 5 months. During composting, the grape marc pile was fertilized with ammonium nitrate, superphosphate, iron sulfate and magnesium sulfate. Microbiological analysis of our grape marc was performed using the dilution plate technique (Wakelin *et al.*, 1998) on different agar growth media: water agar (WA) pH11 (actynomicetes), tryptose soy agar (TSA) 1/10 (bacteria), glucose peptone medium (GP) (yeast) and malt extract agar (MEA) (fungi). All the plates were incubated at 25°C, and WA and TSA plates were also incubated at 40°C, for detection of thermophylic bacteria and actynomicetes. Two analyses of

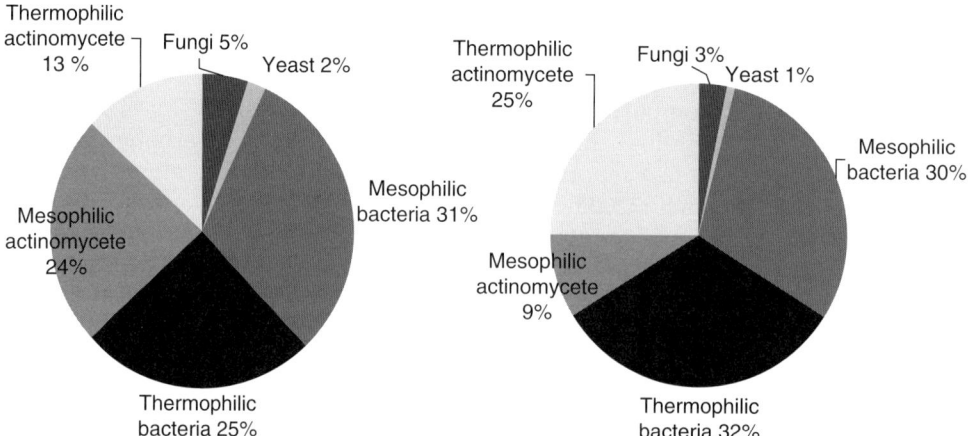

Fig. 12.2. Distribution of the different morphologies found in the first (left) and second (right) analysis of grape marc compost microflora.

the same compost were done, 6 months apart. Microbiological analysis of GMC showed a high number of microbial morphologies in the compost. In the first analysis, 192 different morphologies were found, and 240 in the second, 6 months later. Most of the morphologies present in the first and second analyses were bacteria, with the average percentages being 31% mesophylic bacteria, 28% thermophylic, 16% mesophylic actinomycetes and 20% thermophylic actinomycetes, and only a few moulds 4% and yeasts 1%; the distribution of morphologies found in both assays is shown in Fig. 12.2.

Most microorganisms, both aerobic and anaerobic, respond to conditions that restrict iron ions due to the production of siderophores, whereby ions are kidnapped or assayed thus preventing their availability. Siderophores can act as growth factors and some as powerful antibiotics (Neilands, 1981). Many studies have described suppressiveness in agricultural soil due to siderophores. Thus, the addition of different species of *Pseudomonas* to a conductive soil infected with *Fusarium oxysporum* f. sp. *lini* makes this soil suppressive, preventing the development of the disease. The addition of Fe-EDTA reverses this situation (Kloepper *et al.*, 1980). This principle also applies to *Gaeumannomyces graminis* var. *tritici*, and several species of *Pythium* (Becker and Cook, 1984; Weller *et al.*, 1986).

The presence of siderophores produced by the microorganisms present in the grape marc aerated compost tea and their involvement in the development of eight phytopathogenic fungi in the soil and one mycopathogenic fungus was studied.

The results obtained from the *in vitro* analysis of the inhibiting effect of ACT on fungal development are shown in Fig. 12.3. The effect of the incubation time of ACT on the efficiency of the *in vitro* fungal suppression of the nine fungi tested highlighted an increase in inhibition percentages as the time of incubation of the compost increased (1, 7 or 14 days) being 80–100% for extract F (Filtered), in extractions after 1 and 7 days, and 100% after

14 days, for all the cases examined. In the case of extract C (microfiltered), the results vary depending on the fungus, with the same inhibition tendency percentage increasing as the incubation time and the extract concentration added to the medium increase, reaching 100% inhibition after 14 days of trial for most of the fungi, except for the two races of *Fusarium oxysporum* f. sp. *lycopersici* (Fig. 12.3). From the results obtained, we need to highlight the abrupt change in inhibition percentages for the fungi *R. solani* and *Pythium aphanidermatum*, using the microfiltered tea after 7 and 14 days' incubation. This effect is the result of measuring the radial growth of the fungus, but it

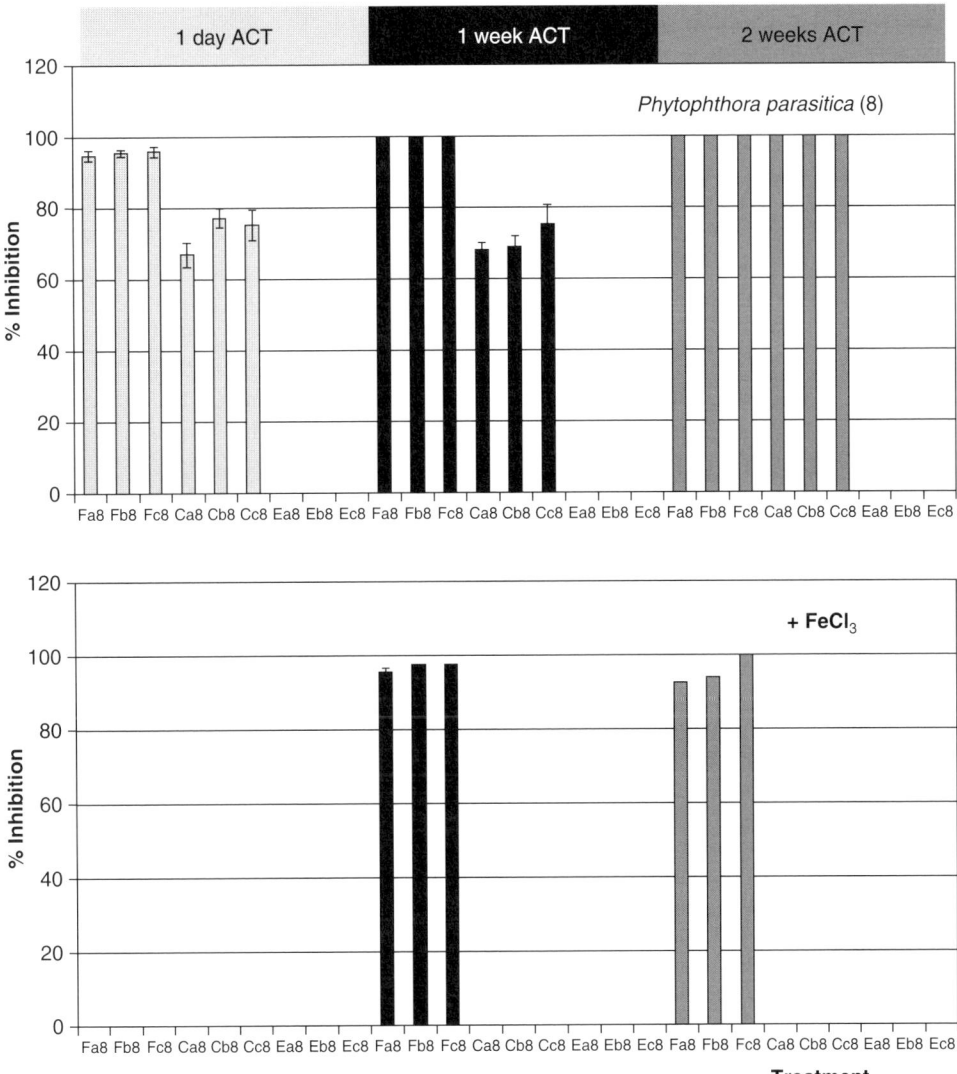

*Vertical T bars represent standard errors

Fig. 12.3. *In vitro* analysis of the inhibiting effect of ACT on fungal development.

does not reflect the effect on the density of the mycelium, which can only be observed under a microscope with a clear inhibition by the compost extracts, although this cannot be expressed in the graph.

The sterilization of compost water extracts annuls the suppressing effect on fungal growth shown by non-thermally treated extracts, except for *Verticillium dahliae* and *V. fungicola*, for which inhibition values reach 60%, in compost tea incubated for 1 day at concentration levels of 10 and 15%.

The detection of siderophores on GMC tea, which can affect the development of the fungal mycelium, occurred equally in the extract obtained simply by filtering (F), in those that were microfiltered (C) and in those that were sterilized (E), with the addition of $FeCl_3$ to the medium. In these, the biological component has not been eliminated, so besides the potential inhibition that siderophores can cause, we must bear in mind other antagonistic effects such as competition for nutrients, space, and so on.

These results suggest that the microorganisms present in GMC produce siderophores, which grow outside the cells and kidnap iron, stopping its availability, thus preventing the *in vitro* development of the phytopathogens studied.

This study confirms the *in vitro* inhibition of the growth of eight pathogens and one mycopathogen, as well as the important role of siderophores in this suppression. In our previous studies, the inhibiting power of GMC had been verified against the nine fungi tested. As can be expected, the suppressiveness shown by GMC extracts is a combination of various factors, such as competition for nutrients, antibiosis, and production of lytic enzymes outside the cells and of low molecular weight molecules that are capable of degrading the fungus wall.

The difference between the use of soil-applied composts and watery compost extracts perhaps is best summarized as that the teas give immediate but very short-term control of surface-spreading pathogens, while soil compost acts more slowly over a longer period of time and requires much larger amounts (York and Brinton, 1996).

12.5 Conclusion

Diseases that have been shown to be effectively suppressed by compost use include those caused by *Fusarium*, *Phytophthora*, *Pythium* and *R. solani*. Several reports suggest that compost and compost-amended soil may alter the resistance of plants to disease. The microbes in compost teas can suppress diseases in several ways: induced resistance, antibiosis and competition, and direct destruction of pathogens structures. The effect is due to the activity of specific biocontrol agents in the batches of compost that suppress the disease. There are several reports on the control of plant pathogens or plant diseases with organic teas including airborne and soilborne diseases. GMC and its water extracts have been reported to suppress fungal diseases. Compost can provide natural biological control of diseases of roots as well as the foliage of plants. Its water extracts (compost tea) have been proposed as substitutes for synthetic fungicides. There are several mechanisms involved in the suppression, including competition among the microbial population for available carbon or nitrogen, iron competition, mycoparasitism, production of inhibitors or hydrolytic enzymes by microorganisms and interactions with some saprophytes. The microorganisms present in GMC produce siderophores, which grow outside the cells and kidnap iron, stopping its availability, thus preventing the development of the phytopathogens. The knowledge of mechanisms for biological control through the action of compost or its water extracts (compost tea) is necessary in order to increase the efficiency of the suppressing power.

References

Ahmed, S.A., Pérez, C. and Candela, M.E. (2000) Evaluation of Induction of Systemic Resistance in pepper plants (*Capsicum annum*) to *Phytophthora capsici* using *Trichoderma harzianum* and its relation with Capsidiol accumulation. *European Journal of Plant Pathology* 106, 817–824.

Becker, J.O. and Cook, R.J. (1984) *Pythium* control by siderophore-producing bacteria on roots of wheat. *Phytopathology* 74, 806.

Blaker, N.S. and MacDonald, J.D. (1983) Influence of container medium pH on sporangium formation, zoospore release, and infection of rhododendron by *Phytophthora cinamomi. Plant Disease* 67, 259–263.

Brinton, W.F. (1995) The control of plant pathogenic fungi by use of compost teas. *Biodynamics,* January/February, 12–15.

Brisbane, P.G. and Rovira, A.D. (1988) Mechanisms of inhibition of *Gaeumannomyces graminis* var. *tritici* by fluorescent *Pseudomonads. Plant Pathology* 37, 104–111.

Buysens, S., Heungens, K., Poppe, J., Hofte, M. (1996) Involvement of pyochelin and pyoverdin in suppression of *Pythium*-induced damping-off of tomato by *Pseudomonas aeruginosa* 7NSK2. *Applied and Environmental Microbiology* 62, 865–871.

Cebolla, V. and Pera, J. (1983) Suppressive effects of certain soil and substrates against *Fusarium* wilt of carnation. *Acta Horticulturae* 150, 113–119.

Chef, D.G., Hoitink, H.A.J. and Madden, L.V. (1983) Effects of organic components in container media on suppression of *Fusarium* wilt of chrysanthemum and flax. *Phytopathology* 73, 279–281.

Chen, Y. and Inbar, Y. (1993) Chemical and spectroscopical analyses of organic matter transformations during composting in relation to compost maturity. In: Hoitinh, H.A.J. and Keener, H.M. (eds.) *Science and Engineering of Composting: Design, Environment, Microbiological and Utilization Aspects*. Renaissance Publications, Worthington, Ohio, pp. 551–600.

Chen, W., Hoitink, H.A.J. and Schmitthenner, A.F. (1987) Factors affecting suppression of *Pythium* damping-off in container media amended with composts. *Phytopathology* 77, 755–760.

Chen, W., Hoitink, H.A.J., Schmitthenner, A.F. and Tuovinen, O.H. (1988a) The role of microbial activity in suppression of damping-off caused by *Pythium ultimum. Phytopathology* 78, 314–322.

Chen, W., Hoitink, H.A.J. and Madden, L.V. (1988b) Microbial activity and biomass in container media for predicting suppressiveness to damping-off caused by *Pythium ultimum. Phytopathology* 78, 1447–1450.

Chet, I., Harman, G.E. and Baker, R. (1981) *Trichoderma hamatum*: its hyphal interactions with *Rhizoctonia solani* and *Pythium* spp. *Microbial Ecology* 7, 29–38.

Chung, Y.R.A. and Hoitink, A.J. (1990) Interactions between thermophilic fungi and *Trichoderma hamatum* in suppression of *Rhizoctonia* damping-off in a bark compost-amended container medium. *Phytopatholology* 80, 73–77.

Craft, C.M. and Nelson, E.B. (1996) Microbial properties of composts that suppress damping-off and root rot of creeping bentgrass caused by *Pythium graminicola*. *Applied and Environmental Microbiology* 62, 1550–1557.

Daft, G.C., Poole, H.A. and Hoitink, H.A.J. (1979) Composted hardwood bark: A substitute for steam sterilization and fungicide drenches for control of poinsettia crown root rot. *HortScience* 14, 185–187.

Diánez, F., Santos, M. de Cara, M. and Tello, J.C. (2006) Presence of siderophores on grape marc aerated compost tea. *Geomicrobiology Journal* 23, 323–331.

Di Pietro, A., Lorito, M., Hayes, C.K., Broadway, R.M. and Harman, G.E. (1993) Endochitinase from *Gliocladium virens:* Isolation, characterization, and synergistic antifungal activity in combination with gliotoxin. *Phytopathology* 83, 308–313.

Duijff, B.J., Bakker, P.A.H.M. and Schlppers, B. (1994) Suppression of *Fusarium* wilt of carnation by *Pseudomonas putida* WCS358 at different levels of disease incidence and iron availability. *Biocontrol Science and technology* 4, 279–288.

Durrant, W.E. and Dong, X. (2004) Systemic acquired resistance. *Annual Review of Phytopathology* 42, 185–209.

Elad, Y. and Shtienberg, D. (1994) Effects of compost water extracts on grey mould, *Botrytis cinerea. Crop Protection*13, 109–114.

Ellis, R.J., Timms-Wilson, T.M. and Bailey, M.J. (2000) Identification of conserved traits in fluorescent *Pseudomonads* with antifungal activity. *Enviromental Microbiology* 2, 274–284.

Fravel, D.R. (1988) Role of antibiosis in the biocontrol of plant diseases. *Annual Review of Phytopathology* 26, 75–91.

Garibaldi, A. (1988) Research on substrates suppressive to *Fusarium oxysporum* and *Rhizoctonia solani. Acta Horticulturae* 221, 271–277.

Garrett, S.D. (1970) Pathogenic root-infecting fungi. (ed) Cambridge University Press, Cambridge, UK.

Glazebrook, J. (2001) Genes controlling expression of defense responses in *Arabidopsis* status. *Current Opinion in Plant Biology* 4, 301–308.

Goldman, G.H., Hayes, C. and Harman, G.E. (1994) Molecular and cellular biology of biocontrol by *Trichoderma* spp. *Trends in Biotechnology* 12, 478–482.

Gorodecki, B. and Hadar, Y. (1990) Suppression of *Rhizoctonia solani* and *Sclerotium rolfsii* in container media containing composted cattle manure and composted grape marc. *Crop Protection* 9, 271–274.

Gugino, J.L., Pokorny, F.A. and Hendrix, F.F.J. (1973) Population dynamic of *Pythium irregulare* Buis. in container plant production as influenced by physical structure of media. *Plant and Soil* 39, 591–602.

Hadar,Y. and Gorodecki, B. (1991) Suppression of germination of sclerotia of *Sclerotium rolfsii* in compost. *Soil Biology and Biochemistry* 23, 303–306.

Hadar, Y. and Mandelbaum, R. (1986) Suppression of *Pythium aphanidermatum* damping-off in container media containing composted liquorice roots. *Crop Protection* 5, 88–92.

Hadar, Y., Mandelbaum, R. and Gorodecki, B. (1992) Biological control of soilborne plant pathogens by suppressive composts. In: E.C. Tjamos, G.C. Papavizas and R.J. Cook (eds) *Biol. Control of Plant Diseases: Progress and Challenges for the Future.* NATO ASI Series No. 230. Plenum Press, New York, pp. 79–83.

Han, D.Y., Coplin, D.L., Bauer, W.D. and Hoitink, H.A.J. (2000) A rapid bioassay for screening rhizosphere microorganisms for their ability to induce systemic resistance. *Phytopathology* 90, 327–332.

Handelsman, J. and Stabb, E.V. (1996) Biocontrol of soilborne plant pathogens. *The Plant Cell* 8, 1855–1869.

Hardy, G.E.S.J. and Sivasithamparan, K. (1991) How container media and matric potential affect the production of sporangial, oospores and chlamydospores by three *Phytophthora* species. *Soil Biology and Biochemistry* 23, 31–39.

Heil, M. and Bostock, R.M. (2002) Induced systemic resistance (ISR) against pathogens in the context of induced plant defenses. *Annals of Botany* 89, 503–512.

Hoffland, E., Pieterse, C.M.J., Bik, L. and Van Pelt, J.A. (1995) Induced systemic resistance in radish is not associated with accumulation of pathogenesis-related proteins. *Physiological and Molecular Plant Pathology* 46, 309–320.

Hoitink, H.A.J., Vandoren, D.M.J. and Schnitthenner, A.F. (1977) Suppression of *Phytophthora cinnamomi* in a composted hardwood bark potting medium. *Phytopathology* 67, 561–565.

Hoitink, H.A.J. (1993) Compost can suppress soil-borne disease in container media. *American Nurseryman* 1993, 91–94.

Hoitink, H.A.J. and Boehm, M.J. (1999) Biocontrol within the context of soil microbial communities: a substrate-dependent phenomenon. *Annual Review of Phytopathology* 37, 427–446.

Hoitink, H.A.J., Boehm, M.J. and Hadar, Y. (1993) Mechanisms of suppression of soilborne plant pathogens in compost-amended substrates. In: Hoitink H.A.J. and Keener, H.M. (eds) *Science and Engineering of Composting: Design, Environmental, Microbiological and Utilization Aspects.* Renaissance Publications, Worthington, USA, pp. 601–621.

Hoitink, H.A.J. and Changa, C.M. (2004) Production and utilization guidelines for disease suppressive composts. *Acta Horticultura* 635, 87–92.

Hoitink, H.A.J. and Fahy, P.C. (1986) Basis for the control of soilborne plant pathogens with composts. *Annual Review of Phytopathology* 24, 93–114.

Hoitink, H.A.J., Inbar, Y. and Boehm, M.J. (1991a) Status of compost amended potting mixes naturally suppressive to soilborne diseases of floricultural crops. *Plant Disease* 75, 869–873.

Hoitink, H.A.J., Inbar, Y. and Boehm, M.J. (1991b) Compost can suppress soil-borne diseases in container media. *American Nurseryman* 15, 91–94.

Hoitink, H.A.J., Stone, A.G. and Han, D.Y. (1997) Suppression of plant diseases by composts. *HortScience* 32, 184–187.

Hoitink, H.A.J., Van Doren, H.M. and Schmitthenner, A.F. (1977) Suppression of *Phytophthora cinnamomi* in a composted hardwood bark potting medium. *Phytopathology* 67, 561–565.

Hornby, D. (1983) Suppressive soils. *Annual Review of Phytopathology* 21, 65–85.

Howell, C.R. and Stipanovic, R.D. (1983) Gliovirin, a new antibiotic from *Gliocladium virens*, and its role in the biological control of *Pythium ultimum*: Fungi used to control plant diseases incited by other fungi. *Canadian Journal of Microbiology* 29, 321–324.

Howell, C.R., Stipanovic, R.D. and Lumsden, R.D. (1993) Antibiotic production by strains of Gliocladium virens and its relation to the biocontrol of cotton seedling diseases. *Biocontrol Science and Technology* 3, 435–441.

Hunt, P., Smart, C. and Eno, C. (1973) Sting Nematode, *Belonolaimus longicaudatus*, immobility induced by extracts of composted municipal refuse. *Journal of Nematology* 5, 60–63.

Ingham, E.R. (2003) *The Compost Tea Brewing Manual.* 3rd Edition. Soil Food Web, Inc., Corvallis, USA.

Jackson, R. (1965) Antibiosis and fungistasis of soil microorganisms. In: Baker, R.W. and Snyder, W.C. (eds) *Ecology of Soil-Borne Plant Pathogens.* University of California Press, Berkeley, USA.

Johnston, T. (2001) Review of Opportunities for the re-use of winery industry solid wastes. SA Waste Management Committee <http://www.winesa.asn.au/environment/pdf/SAWMC-Marc.pdf>

Jordan, R. (2002) Grape marc utilization – cold pressed grapeseed oil and meal. The Co-operative Research Centre for International Food Manufacture and Packaging Science. <http://www.ecorecycle.vic.gov.au/asset/1/upload/Grape_marc_Utilisation_Grape_Marc_to_Grape_Seed_Oil_(2002).pdf>

Keel, C. and Défago, G. (1997) Interactions between beneficial soil bacteria and root pathogens: mechanisms and ecological impact. In: Gange, A.C., Brown, V.K., (eds) *Multitrophic interactions in terrestrial system.* Blackwell Science, Oxford, pp. 27–47.

Khalifa, O. (1965) Biological control of *Fusarium* wilt of peas by organic soil amendments. *Annals of Applied Biology* 56, 129–137.

Khan, J., Ooka, J.J., Miller, S.A., Madden, L.V. and Hoitink, H.A.J. (2003) Systemic Resistance Induced by *Trichoderma hamatum*

382 in cucumber against Phytophthora Crown Rot and Leaf Blight. *Plant Disease* 88, 280–286.

Kloepper, J.W., Leona, J., Teintze, M. and Schroth, M.N. (1980) *Pseudomonas* siderophores: a mechanisms explaining disease-suppressive soils. *Current Microbiology* 4, 317–320.

Kloepper, J., Tuzun, S. and Kuc, J. (1992) Proposed definitions related to induced disease resistance. *Biocontrol Science and Technology* 2, 347–349.

Knudsen, I.M.B., Debosz, K., Hockenhull, J., Jensen, D.F. and Elmholt, S. (1999) Suppressiveness of organically and conventionally managed soils towards brown foot rot of barley. *Applied Soil Ecology* 12, 61–72.

Koné, S.B., Dionne, A., Tweddell, R.J., Antoun, H. and Avis, T.J. (2010) Suppressive effect of non-aerated compost teas on foliar fungal pathogens of tomato. *Biological Control* 52, 167–173.

Koster, M., Van de Vossenberg, J., Leong, J. and Weisbeek, P.J. (1993) Identification and characterization of the *pup*B gene encoding an inducible ferric-pseudobactin receptor of *Pseudomonas putida* WCS358. *Molecular Microbiology* 8, 591–601.

Krause, M.S., De Ceuster, T.J.J., Tiquia, F.C., Michel, J.R.F.C., Madden, L.V. and Hoitink, H.A.J. (2003) Isolation and characterization of rhizobacteria from compost that suppress the severity of bacterial spot of radish. *Phytopathology* 93, 1292–1300.

Kuter, G.A., Hoitink, H.A.J. and Chen, W. (1998) Effects of municipal sludge compost curing time on suppression of *Pythium* and *Rhizoctonia* diseases of ornamental plants. *Plant Disease* 72, 731–756.

Larkin, R.P., Hopkins, D.L. and Martin, F.N. (1996) Suppression of *Fusarium* wilt of watermelon by nonpathogenic *Fusarium oxysporum* and other microorganisms recovered from a disease-suppressive soil. *Phytopathology* 86, 812–819.

Lockwood, J.L. (1977) Fungistasis in soils. *Biological Review* 52, 1–43.

Lockwood, J.L. (1986) Soiborne plant pathogens: Concepts and connections. *Phytopathology* 76, 20–27.

Lorito, M., Harman, G.E., Hayes, C.K., Broadway, R.M., Tronsmo, A., Woo, S.L. and Di Pietro, A. (1993) Chitinolytic enzymes produced by *Trichoderma harzianum*: Antifungal activity of purified endochitinase and chitobiosidase. *Phytopathology* 83, 302–307.

Lorito, M., Peterbauer, C., Hayes, C.K. and Harman, G.E. (1994) Synergistic interaction between fungal cell wall degrading enzymes and different antifungal compounds enhances inhibition of spore germination. *Microbiology* 140, 623–629.

Lumsden, R.D., Lewis, J.A. and Milner, P.D. (1983) Effect of composted sewage sludge on several soilborne pathogens and diseases. *Phytopathology* 73, 1543–1548.

Ma, L.P., Gao, F. and Qiao, X.W. (1999) Efficacy of compost extracts to cucumber wilt (*Fusarium oxysporum* f. sp. *cucumerinum*) and its mechanisms. *Acta Phytopathologica Sinica* 29, 270–274.

Ma, L.P., Qiao, X.W., Gao, F. and Hao, B.Q. (2001) Control of sweet pepper *Fusarium* wilt with compost extracts and its mechanisms. *Chinese Journal of Applied and Environmental Biology* 7, 84–87.

Malek, R.G. and Gartner, J.B. (1975) Hardwood bark as a soil amendment for suppression of plant parasite nematodes on container grown plants. *Hort Science* 10, 261–274.

Mandelbaum, R., Hadar, Y., Chen, Y. (1988) Composting of agricultural wastes for their use as container media. II. Effect of heat treatment on suppression of *Pythium aphidermatum*, and microbial activities in substrates containing compost. *Biological Wastes* 26, 261–274.

Mauch-Mani, B., Métraux, J.P. (1998) Salicylic acid and systemic acquired resistance to pathogen attack. *Annals of Botany* 82, 535–540.

Mathot, P. (1987) Emploi de composts dans la lutte contre *Rhizoctonia* sp. Mededeligen van de Faculteit. Landbouwwetendschappen Rijksunivesiteit. *Gent* 52, 1127–1132.

Mazzola, M. (2002) Mechanisms of natural soil suppressiveness to soilborne diseases. *Antonie Van Leeuwenhoek* 81, 557–564.

McSorley, R. and Gallaher, R.N. (1995) Effect of yard waste compost on plant parasitic nematode densities in vegetable crops. *Journal of Nematology* 27, 545–549.

Milner, J.L., Silo-Suh, L.A., Lee, J.C., He, H., Clardy, J., Handelsman, J. (1996) Production of kanosamine by *Bacillus cereus* UW85. *Applied and Environmental Microbiology* 62, 3061–3065.

Nakashita, H., Yasuda, M., Nitta, T., Asami, T., Fujioka, S., Arai, Y., Sekimata, K., Takatsuto, S., Yamaguchi, I. and Yoshida, S. (2003) Brassino steroid functions in a broad range of disease resistance in tobacco and rice. *The Plant Journal* 33, 887–898.

Neilands, J.B. (1981) Microbial iron compounds. *Annual Review of Biochemistry* 50, 715–731.

Nelson, E.B. and Hoitink, H.A.J. (1983) The role of microorganisms in the suppression of *Rhizoctonia solani* in container media amended with composted hardwood bark. *Phytopathology* 73, 274–278.

Oka, Y., Yermiyahu, U. (2002) Nematode-suppressive effects of composts against *Meloidogyne javanica* on tomato. *Proceedings of Fourth International Congress of Nematology*, 8–13 June, Tenerife, Spain, p. 448.

Pera, J. and Calvet, C. (1989) Suppression of *Fusarium* wilt of carnation in a composted pine bark and a composted olive pumice. *Plant Disease* 73, 699–700.

Pharand, B., Carisse, O., Benhamou, N. (2002) Cytological aspects of compost-mediated induced resistance against *Fusarium* crown and root rot in tomato. *Phytopathology* 92, 424–438.

Pieterse, C.M.J. and Van Loon, L.C. (1999) Salicylic acid-independent plant defense pathways. *Trends in Plant Science* 4, 52–58.

Pieterse, C.M.J., Van Wees, S.C.M., Hoffland, E., Van Pelt, J. and Van Loon, L.C. (1996) Systemic resistance in *Arabidopsis* induced by biocontrol bacteria is independent of salicylic acid accumulation and pathogenesis-related gene expression. *The Plant Cell* 8, 1225–1237.

Raaijmakers, J.M., Vander, S. I., Koster, M., Bakker, P.A.H.M., Welsbeek, P.J. and Schlppers, B. (1995) Utilization of heterologous siderophores and rhizosphere competence of fluorescent Pseudomonas spp. *Canadian Journal of Microbiology* 41, 126–135.

Regulation (EC) No 1493/1999 on the Common Organisation of the Market in wine, 17 May 1999.

Roe, N.E., Stoffella, P.J. and Bryan, H.H. (1993) Utilization of composts and other organic mulches on commercial vegetable crops. *Compost Science and Utilization* 1, 73–84.

Santos M., Diánez F., González del Valle, M. and Tello, J.C. (2008) Grape marc compost: microbial studies and supression of soilborne mycosis in vegetable seedlings. *World Journal Microbiology and Biotechnology* 24, 1493–1505.

Scheuerell, S., Mahaffee, W.F. (2002a) Compost tea: principles and prospects for plant disease control. *Compost Science and Utilization* 10, 313–338.

Scheuerell, S. and Mahaffee, W.F. (2002b) Compost Tea as a container medium drench for suppressing seedling damping-off caused by *Pythium ultimum*. *Phytopathology* 94, 1156–1163.

Silo-Suh, L.A., Lethbridge, B.J., Raffel, S.J., He, H., Clardy, J. and Handelsman, J. (1994) Biological activities of two fungistatic antibiotics produced by *Bacillus cereus* UW85. *Applied and Environmental Microbiology* 60, 2023–2030.

Sivasithamparan, K. (1981) Some effects of extracts from tree barks and sawdust on *Phytophthora cinnamomi* Rand. *Australian Plant Pathology* 10, 18–20.

Slattery, M., Rajbhandari, I., Wesson, K. (2001) Competition-mediated antibiotic induction in the marine bacterium *Streptomyces tenjimariensis*. *Microbial Ecology* 41, 90–96.

Spencer. S. and Benson, D.M. (1981) Root rot of *Aucuba japonica* caused by *Phytophthora cinnamomi* and *P. citricola* and suppressed with bark media. *Plant Disease* 65, 918–921.

Spencer, S. and Benson, D.M. (1982) Pine bark, hardwood bark compost, and peat amendment effects on development of *Phytophthora* spp. and lupine root rot. *Phytopathology* 72, 346–351.

Stephens, C.T., Herr, L.J., Hoitink, H.A.J., Schmitthenner, A.F. (1981) Suppression of *Rhizoctonia* damping-off by composted hardwood bark medium. *Plant Disease* 65, 796–797.

Szczech, M., Rodomanski, W., Brzeski, M.W., Smolinska, U., Kotowski, J.F. (1993) Suppressive effect of a commercial earthworm compost on some root infecting pathogens of cabbage and tomato. *Biology, Agriculture and Horticulture* 10, 47–52.

Ton, J. Mauch-Mani, B. (2004) β-amino-butyric acid-induced resistance against necrotrophic pathogens is based on ABA-dependent priming for callose. *The Plant Journal* 38, 119–130.

Trillas-Gay, M.I., Hoitink, H.A.J. and Madden, L.V. (1986) Nature of suppression of *Fusarium* wilt of radish in container medium amended with composted hardwood bark. *Plant Disease* 70, 1023–1027.

Tränkner, A. (1992) Use of agricultural and municipal organic wastes to develop suppressiveness to plant pathogens. In: Tjamos, E.C., Papavizas, G.C. and Cook, R.J. (eds) *Biological Control of Plant Diseases*. Plenum Press, New York, pp. 35–42.

Van Loon, L.C., Bakker, P.A.H.M. and Pieterse, C.M.J. (1998) Systemic resistance induced by rhizosphere bacteria. *Annual Review of Phytopathology* 36, 453–483.

Wakelin, S.A., McCarthy, T., Stewart, A., Walter, M. (1998) *In vitro* testing for biological control of *Aphanomyces euteiches* pea root rot. *Proceedings of 48th N.Z. Plant Protection Conference*, pp. 308–313.

Weller, D.M. and Cook, R.J. (1983) Suppression of take-all of wheat by seed treatments with fluorescent *Pseudomonads*. *Phytopathology* 73, 463–469.

Weller, D.M., Howie, W.J. and Cook, R.J. (1986) Relationship of *in vitro* inhibition of *Gaeumannomyces graminis* var. *tritici* and *in vivo* suppression of take-all by fluorescent pseudomonads. *Phytopathology* 75, abstract.

Weltzien, H.C. (1989) The effects of compost extracts on plant health. In: Global Perspective on Soil Agricultural Systems. *Proceedings of the sixth Intenational IFOAM Conference*, Santa Cruz, USA, pp. 551–553.

Weltzien, H.C. (1991) Biocontrol of foliar fungal diseases with compost extracts. In: Andrews, J.H. and Hirano, S.S. (eds.) *Microbial Ecology of Leaves*. Springer Verlag. New York, pp. 430–450.

Weltzien, H.C. and Ketterer, N. (1986) Control of Downy Mildew, *Plasmopara viticola* (de Bary) Berlese et de Toni, on grapevine leaves through water extracts from composted organic wastes. *Phytopathology* 76, 1104.

Whipps, J.M. (1997) Developments in the biological control of soil-borne plant pathogens. *Advances in Botanical Research* 26, 1–134.

Whipps, J.M. (2001) Microbial interactions and biocontrol in the rhizosphere. *Journal of Experimental Botany* 52, 487–511.

York, A. and Brinton, W. (1996). The basis for compost disease suppression in agriculture and horticulture. Conference on ecological agriculture, Pacific Grove, California.

Zhang, W., Dick, W.A. and Hoitink, H.A.J. (1996). Compost-induced systemic acquired resistance in cucumber to *Pythium* root rot and anthracnose. *Phytopathology* 86, 1066–1070.

Zhang, W., Han, D.Y., Dick, W.A., Davis, K.R. and Hoitink, H.A.J. (1998) Compost and compost water extract-induced systemic acquired resistance in cucumber and *Arabidopsis*. *Phytopathology* 88, 450–455.

13 Biotechnology: a Tool for Natural Product Synthesis

SANATH HETTIARACHI

Department of Botany, University of Ruhuna, Matara, Sri Lanka

Abstract

Molecular biology is arguably the fastest growing field in all biological sciences. New techniques are discovered and they soon find applications. The search for natural products has a long history, as illustrated by the dependence of traditional medicine on botanicals. With the finding that microorganisms can produce useful products such as penicillin, scientists took the challenge to explore the microbial world for new natural products. Although the search continues with increasing intensity, finding new and more useful products would not have matched with the effort without the support of biotechnology. Although the most powerful approach is genetic manipulation, other techniques such as mutagenesis, breeding and protoplast fusion and the relatively old biotechnology of plant tissue culture are very useful. These also include even more simple approaches such as optimizing culture conditions and design of fermenters. The combination of technologies together with innovative ideas has already increased the production level of already existing natural products and expanded the diversity of products obtainable from biological sources. In addition to harvesting products from living organisms either in the wild or in cultivation, the developments in metagenomics have also paved the way to harness the bioproduct-forming ability of unculturable microorganisms.

13.1 Introduction

By perusing the various definitions given by different authorities and in dictionaries, it becomes apparent that natural products are predominantly chemical compounds of biological origin and are extracted from plants and animals that produce them during secondary metabolism. None the less, naturally occurring mineral compounds may also be categorized as natural products. Among the myriad of natural compounds, many have pharmaceutical and other similar applications such as in pest and disease control in agriculture. Furthermore, natural antioxidants, bioflavours, biopreservatives, natural colourings, fragrances and microbial polysaccharides are becoming the choice to replace artificial chemicals in the food and cosmetic industries. Several contemporary approaches have widened the usefulness of natural compounds. For example, a new move employing rhizosphere metabolamics for bioremediation of polychlorinated biphenyls is suggested (Narasimhan *et al.*, 2003), where plants exuding high levels of phenylpropanoids have enhanced rhizosphere populations capable of degrading pollutants.

The study of the chemistry and extraction of natural products is so important that an entire discipline of chemistry is termed natural product chemistry. Many natural products have quite complex chemical structures, including stereochemistry, and hence their chemical synthesis is difficult. Yet some simpler compounds of natural origin are made by total synthesis and these are also treated as natural compounds. Scientists are also keen to elucidate the biochemical pathways of natural product formation. This information will lead to the development of technologies for the overproduction of already existing natural products as well as producing entirely new products through biological systems.

Plants and animals have coexisted with their pests, pathogens, grazers/predators and other competitors throughout evolution. In order for them to succeed in such a hostile atmosphere, coevolution has gifted the organisms with defence mechanisms of different natures, chemical defence being one of them. Therefore living beings are natural factories of chemical warfare. Scientific investigations guided by traditional knowledge have led to the identification of useful natural compounds for the benefit of humankind in medicine, veterinary medicine and in agriculture. Members of all classes of living beings from bacteria to higher plants and animals have been identified as sources of natural products. According to the accepted definitions of biotechnology, exploiting these organisms either from the wild or by cultivation for the production and extraction of products comes under biotechnology. This exploitation has already threatened some valuable plants and animals with local or even global extinction due to over-exploitation, for example for medicinal uses (Kala, 2005; Maundu *et al.*, 2006). The use of new biotechnological tools would certainly ease the pressure on natural populations.

Although biotechnology plays an important role in non-biological natural products, these are not included in this discussion.

13.2 Biotechnology in Genetic Diversity Revelation and Conservation

It is well known that different varieties / races of the same species have different chemical properties and hence their level of importance as sources of natural products may be different. Until recent times, the taxonomy and identification of plants, animals and microorganisms were based on morphological (vegetative and reproductive) and anatomical characters. However, these features usually cannot differentiate between two closely related individuals and hence cannot identify those individuals with the same morphology, but differences in their genetic makeup. Molecular biological tools, based on DNA, protein or secondary metabolites, provide additional taxonomic markers to supplement information gathered through classical taxonomical methods for further characterization. This enables the identification of variants not only at species or subspecies levels, but tracing down to variations between two individuals. Such biotechnological approaches thus enable the identification of more useful genotypes for exploitation in natural product synthesis. Identification of more useful genotypes is obviously very useful in the conservation of genetic diversity. Systematic studies can indicate which genomes to search, sample and study for useful products. Systematics also provides answers to questions relating to the evolution of chemical and physical structures and their synthesis or ontogeny.

The biotechnological tools available for systematics are protein profiles, polysaccharides, plasmids, DNA–DNA hybridization, various PCR-based techniques and DNA sequencing and alignment. With the assistance of bioinformatics, the molecular data are processed and phylogenetic relationships are generated. This facilitates the search for a better performer within producers of a known product or the search for new products, by narrowing down the scope of the search. Certain traits have evolved only once and therefore will only be present within a single clade or very closely related clades, rather than being present randomly across diverse organisms. For an example in bioproducts, taxol is known to be present in the yew family, Taxaceae. Therefore if one is interested in discovering other organisms producing taxol, the most logical approach would be to start the quest for taxol and other taxanes in plants closely related to Taxaceae family. The other sister family most closely related is Podocarpaceae and taxane has been found in *Podocarpus gracilior* for the first time outside Taxaceae (Stahlhut *et al.*, 1999). The story of mustard oil or mustard glucosinolates is a different one. With the exception of the genus *Drypetes* (a member of Malpighiales), all the other mustard oil producers are in families belonging to one ordinal clade, Brasicales. As order Brasicales and genus *Drypetes* are phylogentically distant, one can assume that mustard oil biosynthesis evolved twice (Rodman *et al.*, 1993). This is confirmed by the presence of two different biosynthetic pathways. Therefore, although the final product is the same, the evolution of the biosynthetic pathway is different. This information is extremely important in the search for alternative pathways for the synthesis of natural products (i.e. one has to search in distantly related clades).

Molecular tools are extremely sensitive in revealing minor differences in closely related genotypes within a species. As discussed elsewhere in this chapter, identifying genotypes of interest in bioproduct formation is particularly important for exploitation, but not restricted to that. This is equally or more important in conservation strategies. A genotype once lost shall not be able to be reconstituted. Exploring possibilities of making use of other biotechnological tools such as plant-cell and organ culture also help preserve the plants in the field. In addition, micropropagation using tissue culture techniques is useful in this context.

Recombinant gene technology in natural products

Recombinant gene technology is perhaps the most wonderful tool a biotechnologist ever had. Conventional breeding of plants and animals can bring the genes of two different but closely related individuals together through the fusion of two gametes in sexual reproduction. The nuclear DNA of eukaryotic organisms is organized into chromosomes and the incompatibility in chromosome numbers and chromosome morphology is the main barrier for bringing genes together from different parents to produce a hybrid. However, today's technology is such that the identification of DNA controlling a particular trait, isolation of that particular DNA, *in vitro* modification of that DNA, inserting the modified DNA into a desired organism to express the trait in the new organism, have become possible in an appropriately equipped laboratory. This is a straightforward approach when the final product is a peptide because engineering the base sequence in an open reading frame under proper control units is sufficient. It is a different matter when it comes to more complex molecules, such as secondary metabolites, which are products of a series of biochemical reactions each mediated by a specific enzyme. The maturation of the molecule may also involve chemical modifications within the environment of the producer cell. In order to make breakthroughs in the production of natural products, particularly those that are not peptides, using biotechnology, an understanding of the regulation of the secondary metabolite pathways involved at the levels of products, enzymes and genes, including aspects of transport and compartmentation in eukaryotic cells, are of paramount importance.

Antibiotics are natural products of microbial origin. However, these are naturally produced only in minute quantities that are not even detectable in nature. Overproducing individuals are present in nature and screening large populations can always yield better producers. Appropriate modifications to culture medium and culture conditions also result in higher production. Generating mutants artificially and screening increase the chances of finding such overproducers. A big leap in this direction was possible due to the new biotechnological tools such as genetic engineering. This possibility was already under consideration for more than two decades (Chater, 1990). There had been a slow progress as antibiotic synthesis occurs as a combination of the action of several genes.

While antibiotics are extremely useful in medicine and veterinary medicine, the development of resistance by the target pathogens is an increasing problem. Therefore heavy investments are made into research leading to the discovery of new antibiotics. However, the natural antibiotics belong to a few families and it may not be possible to isolate new and therapeutically important antibiotics belonging to new families. Once resistance is developed it is possible this resistance extends easily for that particular family of antibiotics. Therefore the modification of already known antibiotics to generate semisynthetic antibiotics, either chemically or enzymatically, may be more appealing.

As microorganisms can be cultivated in fermenters much more easily than plant cells and they are more amenable to genetic alterations via mutagenesis and genetic engineering, they have become factories of products originally produced by other organisms. This is a case of producing natural products in an 'unnatural' producer. The development of biotechnological approaches for enhanced production of such natural products is hindered by the complexity involving unknown regulatory genes and enzymes (Oksman-Caldentey and Inzé, 2004). In this context gene technology can help in overexpression, constitutive expression, breaking tissue specificity and so on, rather than expression in a different organism. This is possible when control mechanisms of expression are identified. It may be traced down to a particular molecule which either keeps the entire process switched off (suppressor) or required for switching on the system (inducer). Once the suppressor is knocked off or the inducer is constitutively expressed without tissue specificity, complex metabolites may be produced with a higher yield in the producer itself. One major limitation in plant cell suspension cultures is the tissue specificity of the metabolites. Therefore breaking tissue specificity should be given more attention in order to exploit the full potential of this technology.

Although animal cells can also be genetically manipulated, other problems associated with animal-cell and tissue culture are still limiting the formation of new products in animals. Nevertheless pharming, a technology still in its infancy, to produce pharmaceuticals in farm animals is a focal point in natural product formation with animals. What is basically done in pharming is to transform animals with genes coding for useful bioproducts, and allow expression and harvesting, for example by milking. Pharming has yielded drugs such as growth hormone, blood components such as haemoglobin, and large quantities of certain proteins needed for research. Although pharming is a novel production platform, it is yet to break through many hurdles such as regulatory barriers, consumer acceptance, environmental issues and ethical concerns. The term pharming is not restricted to production through animals, but also includes pharmaceutical production in genetically modified plants. Ramessar *et al.* (2008) reviewed the progress made in maize plants, the first plants used in pharming, as a platform for effective and safe molecular pharming.

Human insulin is the first mammalian protein expressed in bacteria and later became a huge contributor to the pharmaceutical industry. All human insulin preparations available in pharmacies are of microbial origin, based on two technologies using *Escherichia coli* as the expression system and different technology using yeasts to secrete the precursor of insulin (Petrides *et al.*, 1995). Another industrial success is the production of rennet or chymosin enzyme, which is important in making cheese. The original source was the stomach of calves. As the amount extracted from a single stomach is very low, production could not meet demand and hence the cost was very high. This was first cloned and expressed in *E. coli* (Nishimori *et al.*, 1982). Recombinant chymosin is now produced by two fungi, namely *Kluyveromyces lactis* and *Aspergillus niger*, in addition to *E. coli*, by different manufacturers (Neelakanthan *et al.*, 1999).

A comprehensive list of plant proteins that have been expressed in *E. coli* and in two yeasts, namely *Saccharomyces cerevisiae* and *Pichia pastoris*, is provided by Yesilirmak and Sayers (2009).

Whole plants can serve as factories of desired products – not only those that are present naturally, but also products that are engineered into them. Edible vaccine production is one such example. Antigenic components of pathogens are expressed in plants, and when eaten they can act as antigens to trigger antibody production, thereby developing immunity against the pathogen in question. Neither purification of the antigen nor formulation into vaccine is necessary. In engineering Hepatitis B edible vaccine into lupin (*Lupinus luteus*) and lettuce (*Lactuca sativa*), the gene coding for a viral envelope surface protein has been transferred from the viral genome to the plant genome with necessary modifications for expression in plants. When tissues of transgenic plants were fed to mice and human, they developed Hepatitis B virus specific antibodies to the plant-derived protein (Kapusta *et al.*, 1999). A fusion protein of components from two naturally occurring proteins, cholera toxin B subunit and insulin, has been expressed in potato tubers. This protein, when ingested with the potato tuber, can enter into the gut-associated lymphoid tissues, as it is linked to the carboxy-terminal sequence of the cholera toxin B subunit. When mice, having a diabetic mellitus condition due to insulin autoimmune disease, were fed with transgenic potato, a substantial reduction in pancreatic islet inflammation and a delay in the progression of clinical diabetes were observed (Arakawa *et al.*, 1998).

Nevertheless, the production of secondary metabolites depends mainly on pathway engineering / metabolic engineering, rather than engineering the gene responsible for the final product. Due to the complex nature and specificity of reactions at strain level rather than at species or broader taxonomic level, the progress has been slow. Transfer of the putrescine:SAM N-methyltransferase (PMT) gene from *Nicotiana tabacum* under the control of the CaM V 35S promoter into a *Duboisia* hybrid has increased the N-methylputrescine concentration in hairy root cultures. The increase has been reported as 2–4-fold in transgenic versus wild type. The enzyme catalyses the N-methylation of the diamine putrescine to N-methylputrescine, which is the first specific precursor of both tropane and pyridine-type alkaloids. Both of these types are present in *Duboisia* roots, however the concentrations of these alkaloids did not increase in the transgenic hairy root culture, despite higher levels of the precursor (Moyano *et al.*, 2002). Although the final product formation was not enhanced, this is a step forward. Working along similar lines it was also possible to engineer the enzyme hyoscyamine 6-hydroxylase from *Hyoscyamus niger*, under the regulation of the CaMV 35S promoter into a *Duboisia* hybrid and found one of their transgenic lines produced 74 mg/l scopolamine (Palazóna *et al.*, 2003). The enzyme was known to catalyse two steps in scopolamine biosynthesis.

In the above section the importance of modifications, either enzymatic or chemical, to natural products was mentioned. These 'unnatural' modifications to natural products can be done in a more 'natural' manner by combinatorial alterations by engineering genes relevant for the enzymes. An example of this approach is reported by McDaniel *et al*. (1999) who engineered erythromycin polyketide synthase to produce a library of over 50 macrolid antibiotics. Kantola *et al*. (2003) provided an analysis of available literature on this subject giving details of the gene clusters involved. Hopwood *et al*. (1985) was the first group to produce 'hybrid' antibiotics by transferring genes encoding enzymes making a set of antibiotics in one strain to another. This results in new antibiotics depending on the substrate specificity of the enzymes. Thus the possibility of finding more efficient antibiotics is made possible.

Metabolic engineering or combinatorial synthesis of traits from bacteria to plants has also been successful. For example, a bacterial gene encoding p-hydroxycinnamoyl-CoA hydratase/lyase (*HCHL*) has been expressed in *Beta vulgaris*. One line was able to accumulate the glucose ester of *p*-hydroxybenzoic acid (pHBA) at a rate of 14% of dry weight (Rahman *et al*., 2009).

Combinations of different approaches are of course possible and some trials have already given positive results. The effect of combining metabolic engineering with providing elicitors in the culture medium has been tested by Zhang *et al*. (2009) in flavonoid production by hairy root cultures of *Glycyrrhiza uralensis*. In metabolic engineering, overexpression of chalcone isomerase was achieved by *Agrobacterium*-mediated transformation with a 150% increase of total flavonoid over the wild type. When PEG8000 and yeast extract was added to the medium, the engineered line produced over 300% total flavonoids.

Plant tissue and cell culture in natural products

Many higher plants produce economically important organic compounds such as oils, resins, tannins, natural rubber, gums, waxes, dyes, flavours and fragrances, pharmaceuticals, and pesticides. Plants have been identified since ancient times as a source of remedies for ailments. Medicinal plants, even today, are given the highest importance in the pharmaceutical industry, despite the fact that some of these chemicals, which were first identified as botanicals, are synthesized chemically. It is a general rule in nature that if one has continued enthusiasm, patience and the right approach, one shall encounter a better performer of the function of interest. Natural product formation is no exception. Screening of natural plant populations has taken place throughout civilization and will continue. This is now happening in a more rigorous manner at the level of genetic diversity and identity with the aid of molecular biology and biotechnology tools. On the other hand, it is also possible to use the biotechnological tools for the generation of new varieties with more desirable traits, as discussed in the above section with regard to microorganisms. Once a better performer is identified, the next concern would be how to maintain the genetic constitution so that it would be used for a long period without genetic variation. Tissue culture is the first option available for the clonal propagation of plants within a short period of time in a small area. Similarly regenerated plants through micropropagation are extremely important to relieve the pressure on the wild plants exerted by overexploitation. Depending on the tissue specificity of the product, the harvesting may be destructive as vital parts, such as roots or bark, of the plant may have to be removed in harvesting, imposing additional heavy pressure on wild populations.

As mentioned in the review by Chaturavedi *et al.* (2007), the micropropagation and field establishment of *Dioscorea floribandu* in India has been a success story. They calculate the number of plantlets that could be obtained within a year starting from a single nodal cutting as 2,560,000 in comparison to a maximum of 10 plants from a single tuber which can be obtained after 3 years of growth in the field. Thus there cannot be any doubt about the power of micropropagation, especially for those plants that cannot be rapidly propagated otherwise. Even with those plants, the advantage of having propagation of true-to-type plants cannot be underestimated. The report criticizes the progress made in India in the micropropagation of medicinal plants, despite the fact that experiments in various institutes have come up with protocols ending up in the field of a long list of medicinal plants. This fact is mentioned here in order to highlight the inefficiency of technology transfer.

The extraction of natural products from wild or field-grown plants, however, suffers from many impediments such as low production, and no uniformity of production from place to place, from time to time and from one plant variety to another. Riker and Hildebrandt (1958) were probably the first scientists who had vision of the power of plant tissue culture in natural product synthesis. As tissue and cell cultures can be maintained under desirable culture conditions, once culture media and culture conditions for optimum growth and optimum production have been worked out, the above problems can be mitigated. When tissue is collected from cultivated or wild plants, there is a long waiting time from plantlet formation to product formation as extraction may be possible only after reaching maturity or some other status.

Among several other scientists working in the late 1970s and early 1980s, Berlin (1984) visualized plant tissue culture as a future potential process for natural product synthesis and stressed that more research was needed for the elucidation of regulatory controls in biosynthetic pathways. Alfermann and Petersen (1995) demonstrated the potential of plant tissue and cell cultures for natural product synthesis by referring to research by many scientists who were able to produce various metabolites at the laboratory scale.

It has been possible to manipulate biosynthesis through genetic manipulation, and overexpression and expression of new products has been possible in cultured cells and organs. Organogenesis may be required for generating certain tissue-specific substances (Rout *et al.*, 2000). The primary purpose of plant tissue culture was the exploitation of the totipotency of the plant cell for mass-scale propagation of useful varieties of plants. The plants obtained are expected to be genetically identical and hence form a clone. However, variations occur within the clone due to somaclonal variations as a result of replication errors in mitotic cell divisions. While it seems undesirable, the generation of somaclonal variations is intentionally done for the further improvement of the selected plant variety. The screening can be done in much the same way as for microorganisms – in Petri dishes rather than in a greenhouse or a field. Therefore overproducers and producers of new metabolites can be easily and rapidly selected. The generation of variants can also be achieved by somatic fusion of two protoplasts of cells from two different parents. The parents may be closely or distantly related. This can bring genetic information required, for example, for synthesis of one metabolite and an enzyme that can further transform it into a more useful product or products. This section of science is called combinatorial biosynthesis. The variants made by either method can usually be regenerated into whole plants or, if desired, can be maintained in cell or organ cultures.

Certain metabolites are tissue specific. Even for metabolites that are not tissue specific, the higher yield is given in compacted tissues rather than in loose-cell suspensions. For this, organogenesis is necessary. A very good solution is available as plant tissues transformed by *Agrobacterium rhizogenes* can grow indefinitely as roots in appropriate culture media. Once transformed, the cells can grow as hairy roots without the bacterium. When products are formed and secreted or made to secrete into the medium, product recovery is easy as it is not contaminated by cells. Further, continuous removal is also possible. While all these facts are fascinating and encouraging, the published literature shows only two plant metabolites that are produced on a commercial scale. Shikonin is the first botanical produced on a commercial scale using plant cell cultures of *Lithospermum erythrorhizon*. The specific productivity of shikonin has been increased 25-fold by simultaneous *in situ* extraction using n-hexadecane added before 15 days and immobilization of cells in algenate beads, in comparison with production in cell suspension culture (Kim and Chang, 2004).

Immobilized cells have shown to be better producers in general. When comparing yields of scopadulcic acid B, a diterpene that has antiviral and anti-tumour activity, production by *Scoparia dulcis* in suspension culture and cells immobilized in *Luffa* sponge, Mathew and Jayachandran (2009) noted 350.57 mg/g of cells by the 19th day by immobilized cells in contrast to 50.85 mg/g of cells after 30 days of incubation in suspension culture. This is a sevenfold increase with a shorter time of incubation. Continuous removal of products and inhibitory substances is an added advantage of using immobilized cells.

The only other compound produced using plant cell culture that has successfully entered the market is ginseng saponin originating from *Panax ginseng*. Further improvements to the yield have been shown in hairy root cultures fed with specific nitrogen and phosphorous sources (Jeong and Park, 2006).

Catharanthus roseus is one of the plants that showed success in the production of alkaloids in cell suspension culture. These are indole alkaloids, of which the bisindole alkaloids vinblastine and vincristine are antineoplastic medicines and the monoindole alkaloids (ajmalicine and serpentine) are anti-hypertension drugs. Since the first results published 30 years ago (Kurz *et al.*,

1980), continuous efforts have been made experimenting with various cell suspension culture techniques with modifications of culture conditions in order to maximize the production of desired compounds. The research focusing on improvements on both cell culture and bioreactor aspects is the subject of the review by Zhao and Verpoorte (2007). Nevertheless, further research is necessary to bring products of *C. roseus* to the commercial scale. Berberine, an isoquinolene alkaloid, has been produced in cell suspension culture of *Tinospora cordifolia* at a concentration of 5.5 mg g^{-1} dry wt in 24 days (Rao *et al.*, 2008). By screening eight cell lines, the authors were able to find one line that accumulated 13.9 mg g^{-1} dry weight of berberine. This is a 5–14-fold increase in product formation compared to that of the intact plant. This is an example demonstrating the potential of plant cell cultures and somaclonal variation for the natural product industry. A tenfold increase brings down the cost of the product in the market by at least the same magnitude.

Cyclotides are small cyclic peptides stabilized by disulfide bonds between six conserved cystine residues. Their biological activities include anti-HIV, antimicrobial and insecticidal actions, and hence they are important botanicals in biocontrol. In addition, due to their high stability, cyclotides are very good candidates for the development of drug delivery systems. Dörnenburg *et al.* (2008) developed techniques to produce cyclotides using callus, suspension culture and hydroponic cultures of *Oldenlandia affinis* and evaluated them for Kalata B1 accumulation. *In vitro* culture produced only up to 15% of Kalata B1 in comparison with plants grown in hydroponics. Further improvement is probable by manipulating culture conditions and the same group reported a higher rate of cell multiplication in a 25-l photobioreactor (Seydel *et al.*, 2009). They claim that this approach for harvesting Kalata B1 is more profitable than other methods, such as field cultivation and chemical synthesis.

Cell suspension culture is also useful in the synthesis of volatile oils. By manipulating culture conditions, Ishikura *et al.* (1984) obtained a yield of 0.005% to 0.01% of volatile oils of the fresh weight of the cells of *Cryptomeria japonica* cell suspension.

The progress of research in the development of hairy root technology for metabolite production has been reviewed by Guillon *et al.* in 2006. The delicate nature of the hairy roots is one of the major problems in maintaining hairy root cultures. Srivastava and Srivastava (2007) reviewed the problems and different technologies available to overcome them. The basic types of reactors are either liquid phase or gas phase or even a combination of the two. The nature, applications, perspectives and scale up are discussed therein.

The hairy root growth, production of the desired substance(s) and secretion may require different culture conditions. The production can be enhanced by providing the precursors at the right time and the right concentration. The addition of cadaverine to hairy root cultures of *Nicotiana rustica* shifted the alkaloid production in considerable favour of anabasine with a concomitant diminution of nicotine (Walton *et al.*, 1988), whereas the addition of menthol or geraniol at 25 mg l^{-1} to hairy root cultures of *Anethum graveolens* did not have an impact on the growth and resulted in the transformation of the compounds to volatile products (Faria *et al.*, 2009). In a recent experiment,

Hernandez-Vazquez *et al.* (2010) were able to increase the production of centellosides in cell cultures of *Centella asiatica* by feeding with α-amyrin together with DMSO to assist the penetration of α-amyrin in to the cells.

The secretion can be enhanced by manipulating the culture medium either by increasing the leakiness of tissues or by introducing a trapping system to trap the secreted metabolite. For example, Rudrappa *et al.* (2004) reported an increased recovery of betalaines (red natural pigment) production by hairy roots of *B. vulgaris* up to 97.2% by using alumina:silica (1:1) in the culture medium. The use of elicitors is a useful approach to enhance the productivity as some secondary metabolites are naturally produced in response to a signal. This has been shown practically by incorporating both live and autoclaved bacteria in the conversion of hyoscyamine to scopolamine by hairy root cultures of *Scopolia parviflora* (Jung *et al.*, 2003).

Podophyllotoxin is a useful botanical extracted from *Podophyllum hexandrum* for treating general warts. Chattopadhyay *et al.* (2002) used a 3-l bioreactor for the cell suspension culture of *P. hexandrum* and successfully recovered podophyllotoxin. It has been noted that providing coniferin as a precursor in the culture medium increases the productivity, but this is economically prohibitive. An innovative method to circumvent this problem is to co-cultivate *P. hexandrum* cell suspension with hairy roots of *Linum flavum*, which produces coniferin. This approach has increased the final product formation by 240% (Lin *et al.*, 2003).

A further step in bringing the substrate and enzymes from different sources has been made possible through combinatorial biosynthesis. Here, rather than cocultivating the two individual plants *in vitro*, the genetic traits controlling the production of one compound, which will be the precursor of the final product, and the enzyme required for the transformation are brought together into one cell. A somatic hybrid of two *Solanum* species producing solanidine and solanthrene (*S. tuberosum*) and tomatidine (*S. brevidens*) produced all three steroidal glycoalkaloid aglycones. In addition all hybrids also produced a totally new compound, demissidine (Laurila *et al.*, 1996). These authors propose a hypothesis to describe this which explains the production of demissidine by hydrogenating the double bond at position 5 of solanidine of *S. tuberosum* by a hydrogenase enzyme of *S. brevidens*.

13.3 Expanding Natural Product Diversity

In the quest for natural products for the benefit of humankind, it can often be thought that the existing natural product diversity is not sufficient. The generation of mutants is a tool available for increasing natural product diversity. Random mutagenesis is stimulated by treating with chemical mutagens or radiation. Although these are useful and have been employed for making variants and subsequent screening, with the better understanding of DNA and development of new biotechnological tools, specific, site-directed mutagenesis is possible. This has reduced the effort in screening a large number of mutants for the expected outcome.

In an attempt to restore the ability to produce taxol, an anticancer drug, to an endophytic fungus, *Tubercularia* sp., mutagenesis and genome shuffling in protoplasts resulted in no restoration, but the production of new metabolites (Wang *et al.*, 2010). One mutant produced three new and one already known sequiterpenoids, whereas another one made 18 novel compounds belonging to different classes and 10 already known compounds which were not present in the wild-type strain.

Altering the culture condition alone can induce the production of different natural products by an individual organism. If the mechanism of control of a metabolic pathway is understood, then breaking silence in expression should be simple. Sometimes the presence of another organism in the neighbourhood may influence the induction of a pathway leading to a previously unseen product. Therefore mixed fermentation can result in an increased concentration of already expressed or undetected or unexpressed metabolites in crude extract, and the production of new analogues of known metabolites due to combined or extended pathways (Pettit, 2009).

Each and every step of biochemical pathways are supposed be catalysed by a specific enzyme. Nevertheless, the size of the total genome, as unravelled by genome-sequencing projects, cannot account for the large number of metabolites present in organisms. Post-transcriptional modifications leading to the formation of isozymes can be partly responsible for the high diversity of metabolites, but this has now been mainly attributed to the low specificity of some enzymes, which means that one enzyme can convert several similar precursors or intermediates into several products causing ramifications in the next step, and finally ending up with several end products. The number of end products shall be much higher if there was no compartmentalization of substrates and enzymes within a cell and in different cells. Bringing a natural substrate in one cell in contact with an enzyme from another cell may result in entirely new 'natural' products. This is possible through breeding of sexually compatible individuals, or through protoplast fusion, which is possible between rather distant relatives. This technology is known as combinatorial biosynthesis or metabolic engineering or heterologous expression, and has been applied to microorganisms as well as to plants.

Polyketides are highly diverse natural products, including those important as antibiotics and other pharmaceuticals such as anticancer agents and immunosuppressors. The most important producers are actinomycetes, but other bacteria, fungi, plants and marine animals also contain polyketides. They are formed by the sequential polymerization of carboxylic acids, and as such carboxylic acid monomers become a source of variation in polyketides. The diversity has been possible due the modular nature of the polyketide synthases (PKSs). Once transcribed, the peptides assemble to make the functional enzyme consisting of different number of polypeptides, each having two modules. The large protein adds one monomer at each module to the growing chain. Mutagenesis in each module is possible and the outcome of each mutation and combinations of those can cause a great variety of polyketides (Kosla, 1997). The first polypeptide in the assembly of the megasynthase initiates polymerization and is the minimal PKS containing two enzymatic domains (ketosynthase (KS)_malonyl-CoA:ACP transferase (MAT) didomain) and an acyl-carrier protein (ACP) domain. Zhang *et al.* (2008) removed the minimal PKS from *Gibberella fujikuroi* in two formats and engineered it in *E. coli* – cyclized polyketide production was observed. In the fungus cyclization is C2–C7, whereas in *E. coli* cyclization of C9–C14 and C7–C12 were predominant. By expressing one of the two minimal PKSs (PKS_WJ) and downstream processing enzymes in *E. coli* they were able to synthesis nonaketide anthraquinone SEK26 in good amounts. This work also encourages scientists to investigate further the generation of variety through heterologous expression.

A very good example is the engineering of a new metabolic pathway in *E. coli* by a combination of mixing genes and modifying catalytic functions by *in vitro* evolution. In this experiment, phytoene desaturases were assembled by shuffling modules from different bacteria and expressing them in *E. coli*. The different strains generated were screened for new carotenoid pigments. One such chimeric enzyme had introduced six, in place of four, double bonds into phytoene. Another set of chimeric lycopene cyclases had also been made and engineered into the cells expressing the above novel phytoene, thus extending that pathway to produce a variety of coloured compounds. One such combination produced cyclic carotenoid torulene in *E. coli*. This experiment reported by Schmidt-Dannert (2000) suggests the unlimited possibilities of obtaining novel natural products by rational design.

Apart from the usual ribosomal polymerization of amino acids into peptides, a mechanism of non-ribosomal enzymatic polymerization of amino acids similar to that of polyketide synthesis is also present in living systems. For example, the well known hepatotoxin, microsystine produced by *Microsystis* spp., is formed by a microcystin synthetase complex consisting of peptide synthetases, polyketide synthases and hybrid enzymes (Dittmann *et al.*, 2001). Accordingly, heterologous expression in a similar way to PKS is possible in view of introducing new variants of microsystins in the quest for novel natural products.

A great proportion of microorganisms cannot be cultivated and hence no information on their potential for natural product formation is available. This portion reaches an astounding 90% or 99% of the total populations in soil. They belong to bacteria and archaea and also viruses. This also indicates the presence of an extremely large, yet untouched resource. Although the organisms are unculturable, their DNA can be isolated from environmental samples and amplified using PCR techniques. The isolation, amplification and study of DNA from a total community is referred to as metagenomics, or environmental or community genomics. Once DNA is amplified, purification and sequencing is straightforward. Sequencing helps in 'seeing' what is present, and at least how close they are to culturable, known members of the community. On the other hand, a functional approach, in which the DNA is expressed in surrogate hosts, is also very useful in finding out what they can do. When analysing clones of recombinant surrogate hosts for new products, there is a big chance of finding new bioproducts, perhaps some of which may be in entirely new families of compounds.

Fragments of DNA can be cloned using artificial bacterial chromosome (BAC) vectors. These are designed to carry larger fragments than those produced by PCR techniques. Rondon (2000) used this technique and revealed a great deal of information on phylogenetic, physical and functional properties of the metagenome. Some of the clones showed antibacterial and nuclease activity and the DNA sequences governing these characteristics were found to be different from those already known. This directly supports the idea that unculturable organisms provide a source of novel genes. Other functions observed were the production of amylase and lipase and hemolysis.

13.4 Conclusion

The search for natural products has been rekindled with realization of the vast biological diversity and genetic diversity that make biological systems an unlimited resource. Furthermore, the global society is demanding natural products and rejecting synthetic chemicals for all possible uses, including food and additives, drugs, cosmetics and so on. Although plants and animals had been a source of such products from ancient times, the horizons now have widened and continue to expand, particularly with the development of modern biotechnological tools. Man was unaware that he was exploiting microorganisms as they are invisible to the naked eye. Some products become redundant over time, as is the case for antibiotics for which resistance has developed in target (and non-target) organisms. Further new diseases emerge, new lifestyles demand new products, and biotechnology is needed to satisfy those demands. As new tools in biotechnology have been found, one can expect that the trend will continue. Exploring and exploiting new resources will be feasible in the future, as evidenced by metagenomics in new product formation which harness the biosynthetic capability of unculturable microorganisms. However, with all of these developments, the number of bioproducts that have already conquered the market is not encouraging. Therefore, there is a need for more intensive research on optimizing production of already identified bioproducts, with simultaneous research efforts on new product formation. It can also be noted that technology transfer is not occurring effectively, because the already available technologies remain in the laboratories where they were generated without reaching the industrial fermenters or industrial-scale farms.

References

Alfermann, A.W. and Petersen, M. (1995) Natural product formation by plant cell biotechnology: results and perspectives. *Plant Cell, Tissue and Organ Culture* 43, 199–205.

Arakawa T., Yu, J., Chong, D.K.X., Hough, J., Engen, P.C. and Langridg, W.H.R. (1998) A plant-based cholera toxin B subunit–insulin fusion protein protects against the development of autoimmune diabetes. *Nature Biotechnology* 16, 934–938.

Berlin, J. (1984) Plant cell cultures—a future source of natural products? *Endeavour* 8, 5–8.

Chater, K.F. (1990) The improving prospects for yield increase by genetic engineering in antibiotic-producing streptomycetes. *Biotechnology* 8, 115–121.

Chattopadhyay, S., Srivastava, A.K., Bhojwani, S.S., Virendra, S. and Bisaria, V.S. (2002) Production of podophyllotoxin by plant cell cultures of *Podophyllum hexandrum* in bioreactor, *Journal of Bioscience and Bioengineering* 93, 215–220.

Chaturavedi, H.C., Jain, M. and Kidwai, N.R. (2007) Cloning of medicinal plants through tissue culture: A review. *Indian Journal of Experimental Biology* 45, 937–948.

Dittman, E., Neilan, B.A. and Borner, T. (2001) Molecular biology of peptide and polyketide biosynthesis in cyanobacteria. *Applied Microbiology and Biotechnology* 57, 467–473.

Dörnenburg, H., Frickinger, P. and Seydel, P. (2008) Plant cell-based processes for cyclotides production, *Journal of Biotechnology* 135, 123–126.

Guillon, S., Tre´mouillaux-Guiller, J., Pati, P.K., Rideau, M. and Gantet, P. (2006) Hairy root research: recent scenario and exciting prospects. *Current Opinion in Plant Biology* 9, 341–346.

Faria, J.M.S., Nunes, I.S., A. Cristina Figueiredo, A.C., Pedro, L.G., Trindade, H. and Barroso, JG. (2009) Biotransformation of menthol and geraniol by hairy root cultures of *Anethum graveolens*: effect on growth and volatile components. *Biotechnology Letters* 31, 897–903.

Hernandez-Vazquez, L., Bonfill, M., Moyano, E., Cusido, R.M., Navarro-Ocaña, A. and Palazon, J. (2010) Conversion of α-amyrin into centellosides by plant cell cultures of *Centella asiatica*, *Biotechnology Letters* 32, 315–319.

Hopwood, D.A., Malpartida, F., Kieser, H.M., Ikeda, H., Duncan, J., Fujii, I., Rudd, B.A.M., Floss, H.G. and Omacrmura, S. (1985) Production of 'hybrid' antibiotics by genetic engineering. *Nature* 314, 642–644.

Ishikura, N., Kensuke, K. and Sugisawa, H. (1984) Volatile components in cell suspension cultures of *Cryptomeria japonica*, *Phytochemistry* 23, 2062–2063.

Jeong, G-T. and Park, D-H. (2006) Characteristics of transformed *Panax ginseng* C.A. Meyer hairy roots: Growth and nutrient profile. *Biotechnology and Bioprocess Engineering* 11, 43–47.

Jung, H-Y., Kang, S-M., Kang, Y-M., Kang, M-J., Yun D-J., Bahk, J-D., Yanga, J-K. and Choi, M-S. (2003) Enhanced production of scopolamine by bacterial elicitors in adventitious hairy root cultures of *Scopolia parviflora*. *Enzyme and Microbial Technology* 30, 987–990.

Kala, C.P. (2005) IUCN: Indigenous uses, population density, and conservation of threatened medicinal plants in protected areas of Indian Himalayas. *Conservation Biology* 19, 368–378.

Kantola, J., Kunnari, T., Mantsala, P. and Ylihonko, K. (2003) Expanding the scope of aromatic polyketides by combinatorial biosynthesis. *Combinatorial Chemistry & High Throughput Screening*, 6, 501–512.

Kapusta, J., Modelska, A., Figlerowicz, M., Pniewski, T., Letellier, M., Lisowa, O., Yusibove, V., Koprowski, H., Pluceinnicza and Legocki, A.B. (1999) A plant-derived edible vaccine against hepatitis B virus. *The FASEB Journal* 13, 1796–1799.

Kim, D.J. and Chang, H.M (2004) Enhanced shikonin production from *Lithospermum erythrorhizon* by *in situ* extraction and calcium alginate immobilization. *Biotechnology and Bioengineering* 36, 460–466.

Khosla, C. (1997) Harnessing the biosynthetic potential of modular polyketide synthases. *Chemical Reviews* 97, 2577–2590.

Kurz, W.G.W., Chatson, K.B., Constabel, F., Kutney, J.P., Choi, L.S.L., Kolodziejczyk, P., Sleigh, S.K., Stuart, K.L. and Worth, B.R. (1980) Alkaloid production in *Catharanthus roseus* cell cultures: initial studies on cell lines and their alkaloid content. *Phytochemistry* 19, 2583–2587.

Laurila, J., Laakso, I., Valkonena, J.P.T., Hiltunen, R. and E. Pehu, E. (1996) Formation of parental-type and novel glycoalkaloids in somatic hydrids between *Solanum brevidens* and *S. tuberosum*. *Plant Science* 118, 145–155.

Lin, H.W., Kwok, K.H. and Doran, P.M. (2003) Production of podophyllotoxin using cross species co-culture of *Linum flavum* hairy roots and *Podophyllum hexandrum* cell suspensions, *Biotechnology Progress* 19, 1417–1426.

McDaniel, R., Thamchaipenet, A., Gustafsson, C., Fu, H., Betlach, Me., Betlach, Ma. and Ashley, G. (1999) Multiple genetic modifications of the erythromycin polyketide synthase to produce a library of novel "unnatural" natural products. *Proceedings of the National Academy of Sciences USA* 96, 1846–1851.

Mathew, A.J. and Jayachandran, K. (2009) Production of scopadulcic acid B from *Scoparia dulcis* Linn. using a *Luffa* sponge bioreactor, Plant Cell, *Tissue and Organ Culture* 98, 197–203.

Maundu, P., Kariuki, P. and Eyog-Matig, O. (2006) Threats to Medicinal Plant Species – an African Perspective. In: Mittapala, S. (ed.) *Conserving Medicinal Species: Securing a Healthy Future.* IUCN: Ecosystems and Livelihoods Group, Asia, pp. 47–63.

Moyano, E., Fornalé, S., Palazón, J., Cusidó, R.M., Bagni, N. and Piñol, M.T. (2002) Alkaloid production in *Duboisia* hybrid hairy root cultures overexpressing the *pmt* gene. *Phytochemistry* 59, 697–702.

Narasimhan, K., Basheer, C., Bajic, V.B. and Swarup, S. (2003) Enhancement of plant-microbe interactions using a rhizosphere metabolomics-driven approach and its application in the removal of polychlorinated biphenyls. *Plant Physiology* 132, 146–153.

Neelakantan, S., Mohanty, A.K. and Kaushik, J.K. (1999) Production and use of microbial enzymes for dairy processing. *Current Science* 77, 143–148.

Nishimori, K., Kawaguchi, Y., Hidaka, M., Uozumi, T. and Beppu, T. (1982) Expression of cloned calf prochymosin gene sequence in *Escherichia coli. Gene* 19, 337–344.

Oksman-Caldentey, K-M. and Inzé, D. (2004) Plant cell factories in the post-genomic era: new ways to produce designer secondary metabolites, *Trends in Plant Science* 9, 433–440.

Palazóna, J., Moyano, E., Cusidó, R.M., Bonfill, M., Oksman-Caldenteyc, K-M. and Piñol, M.T. (2003) Alkaloid production in *Duboisia* hybrid hairy roots and plants overexpressing the *h6h* gene. *Plant Science* 165, 1289–1295.

Pettit, R.K. (2009) Mixed fermentation of natural product drug discovery. *Applied Microbiology and Biotechnology* 83, 19–25

Petrides, D., Sapidou, E. and Alandranis, J. (1995) Computer-aided process analysis and economic evaluation for biosynthetic human insulin production – a case study. *Biotechnology and Bioengineering* 48, 529–541.

Rahman, L., Kouno, H., Hashiguchi, Y., Yamamoto, H., Narbad, A., Parr, A., Walton, N., Ikenaga, T. and Kitamura, Y. (2009) *HCHL* expression in hairy roots of *Beta vulgaris* yields a high accumulation of p-hydroxybenzoic acid (pHBA) glucose ester, and linkage of pHBA into cell walls. *Bioresource Technology* 100, 4836–4842.

Ramessar, K., Sabalza, M., Capell, T. and Christou, P. (2008) Maize plants: An ideal production platform for effective and safe molecular pharming. *Plant Science* 174, 409–419.

Rao, B.R., D. Vijay Kumar, D., Amrutha R.N., Jalaja, N., Vaidyanath, K., Rao, A.M., Rao, S., Polavarapu, R. and Kishor, P.B.K. (2008) Effect of growth regulators, carbon source and cell aggregate size on berberine production from cell cultures of *Tinospora cordifolia* Miers. *Current Trends in Biotechnology and Pharmacy* 2, 269–276.

Riker, A.J. and Hildebrandt A.C. (1958) Plant tissue cultures open a botanical frontier. *Annual Review of Microbiology* 12, 469–490.

Rodman, J., Price, R.A., Karol, K., Conti, E., Sytsma, K.J. and Palmer, J.D. (1993) Nucleotide-sequences of the. rbcL gene indicate monophyly of mustard oil plants. *Annals of Missouri Botanical Gardens* 80, 686–699.

Rondon, M.R., August, P.R., Bettermann, A.D., Brady, S.F., Grossman, Liles, T.M., Loiacono, K.A., Lynch, B.A., MacNeil, I.A., Minor, C., Tiong, C.L., Gilman, M.

Osburne, M.S., Clardy, J., Handelsman, J. and Goodman, R.M. (2000) Cloning the soil metagenome: a strategy for accessing the genetic and functional diversity of uncultured microorganisms. *Applied and Environmental Microbiology* 66, 2541–2547.

Rout, G.R., Samantaray, S. and Das, P. (2000) *In vitro* manipulation and propagation of medicinal plants, *Biotechnology Advances* 18, 91–120.

Rudrappa, T., Neelwarne, B. and Aswathanarayana, R.G. (2004) *In situ* and *ex situ* adsorption and recovery of betalains from hairy root cultures of *Beta vulgaris*, *Biotechnology Progress* 20, 777–785.

Schmidt-Dannert, C. Umeno, D. and Arnold, F.H. (2000) Molecular breeding of carotenoid biosynthetic pathways. *Nature Biotechnology* 18, 750–753.

Seydel, P., Walter, C. and Dörnenburg, H. (2009) Scale-up of *Oldenlandia affinis* suspension cultures in photobioreactors for cyclotide production, *Engineering in Life Sciences* 9, 219–226.

Srivastava, S. and Srivastava, A.K. (2007) Hairy root culture for mass-production of high-value secondary metabolites, *Critical Reviews in Biotechnology* 27, 29–43.

Stahlhut, R., Park, G., Petersen, R., Ma, W. and Hylands, P. (1999) The occurrence of the anti-cancer diterpene taxol in *Podocarpus gracilior* Pilger (Podocarpaceae).

Biochemical Systematics and Ecology 27, 613–622.

Walton, N.J., Robins, R.J. and Rhodes, M.J.C. (1988) Perturbation of alkaloid production by cadaverine in hairy root cultures of *Nicotiana rustica*. *Plant Science* 54, 125–131.

Wang, M., Liu, S., Li, Y., Xu, R., Lu, C. and Shen, Y. (2010) Protoplast mutation and genome shuffling induce the endophytic fungus *Tubercularia* sp. TF5 to produce new compounds. *Current Microbiology* DOI: 10.1007/s00284-010-9604-7.

Yesilirmak, F. and Sayers, Z. (2009) Heterologous expression of plant genes. *International Journal of Plant Genomics* DOI:10.1155/2009/296482.

Zhao, J. and Verpoorte, R. (2007) Manipulating indole alkaloid production by *Catharanthus roseus* cell cultures in bioreactors: from biochemical processing to metabolic engineering. *Phytochemistry Reviews* 6, 435–457.

Zhang, H-C., Liu, J-M., Lu, H-Y. and Gao, S-L. (2009) Enhanced flavonoid production in hairy root cultures of *Glycyrrhiza uralensis* Fisch by combining the over-expression of *chalcone isomerase* gene with the elicitation treatment. *Plant Cell Reports* 28, 1205–1213.

Zhang, W., Li, Y. and Tang Y. (2008) Engineered biosynthesis of bacterial aromatic polyketides in *Escherichia coli*. *Proceedings of the National Academy of Sciences (USA)* 105, 20683–20688.

Index

Page numbers in *italic* indicate figures, those in **bold** indicate tables.